# Amateur Radio Essentials

*Edited by*

George Brown, PhD, CEng, FIEE, M5ACN/G1VCY

Radio Society of Great Britain

Published by the
Radio Society of Great Britain,
Lambda House, Cranborne Road, Potters Bar, Herts. EN6 3JE, UK

First Published 2006



ISBN 1-905086-12-1
EAN 9781-9050-8612-2

Typography: Chris Danby, G0DWV.
Production: Mark Allgar, M1MPA
Cover Design: Dorotea Vizer

Printed in Great Britain by Nuffield Press of Abingdon, Oxon.



This book has a suporting website http://www.rsgb.org/books/extra/essentials.htm

Any corrections and points of clarification that have not been incorporated in this printing of the book can be found here along with any supporting material that may have become available

# Contents

# Preface

This book is intended to fulfil a gap in the literature for a small reference book. Not a 'data' reference book, but one in which both beginners and seasoned amateur radio enthusiasts can find information on a variety of subjects that the standard tomes may gloss over or omit altogether.

It uses carefully-selected *RadCom* articles as its source material, and brings together basic information on subjects as diverse as filters and Fourier Transforms, speech processing and screening.

There is also a section on designs for 13 test equipment projects, ranging from a time-domain reflectometer to a simple magnetometer.

Only four aerial designs are discussed, it having been decided that the author's previous two popular books (*The International Antenna Collection* and *The International Antenna Collection 2*), together with the host of other books on the subject, had effectively exhausted the field! One of the types covered is the Fractal Antenna, which deserves more coverage and research than it has achieved so far.

Articles are also included about operating; not how to arrange your equipment and what to say on your first contact, but things such as what to expect from Sporadic-E on VHF, how to detect meteors, and how to improve your language skills.

Some general articles are presented, including one on the fascinating subject of long-delay echoes.

Finally, two articles are included on the subject of computers in the shack. How to network them, and how to interface them with your transceiver.

I must admit to the enormous pleasure I have had in compiling this book. The range of subjects covered obliged me to read each article carefully, not only from the editor's normal stance of weeding out grammatical errors and the odd typo, but from the viewpoint of an active radio amateur. You, too, will find things in here that you didn't know, or that you had forgotten. Perhaps even things that you had misunderstood for many years, and now have the opportunity to put right.

On more than one occasion, while collecting articles, I was lured away to try things I was reading about. By the end, I had added some more aspects to my long list of amateur radio interests! I hope this happens to you.

My thanks must go to the authors of these original articles for having taken the time and trouble to try to interest other like-minded people in what they have done, or in subjects that particularly interest them. Where would amateur radio be without such people? We are, after all, engaged in communication, and this must not be restricted to radio; the written word can be even more powerful in getting over new ideas – your ideas!

*George Brown,*
*M5ACN/G1VCY*
*Potters Bar, 2006*

# Chapter 1 AN INTRODUCTION TO ATV

**by Ian Waters, G3KKD**

One of the specialist modes of communication within amateur radio is amateur television. It takes two forms; Slow-Scan and Fast-Scan.

In Slow-Scan, colour *still* images, producing modulating frequencies within the audio band, are transmitted using either FM at VHF/UHF or SSB on HF. In the latter case, worldwide contacts are possible. Pictures are generated by a computer from stored digitised images from a camera or from a flat-bed-type scanner. Received images are processed and displayed by the computer.

In Fast-Scan, colour *moving* images with standards similar to broadcasting in the operator's country are used. As the video modulating frequencies extend to 5MHz, transmission must be in the UHF/SHF bands.

On 70cm, amplitude-modulation is employed where, due to bandwidth constraints, only monochrome pictures may be used. On 23cm and on microwaves, frequency-modulation with colour and a sound sub-carrier is employed. Ranges of typically 50km are possible, with greater distances during 'lifts'. In many parts of the country, repeaters operating in the 23cm and 3cm bands, or sometimes with input on one and output on the other band, have been installed at elevated locations. Unlike voice repeaters intended mainly for use by mobiles, ATV repeaters provide better coverage and increased range for fixed stations. With all stations in a group beaming at the repeater, a picture transmitted by any one station may be seen by all the others.
Other TV amateurs are not on the air but specialise in developing video techniques, preserving historic cameras, or using the old Baird 30-line and EMI 405-line systems.

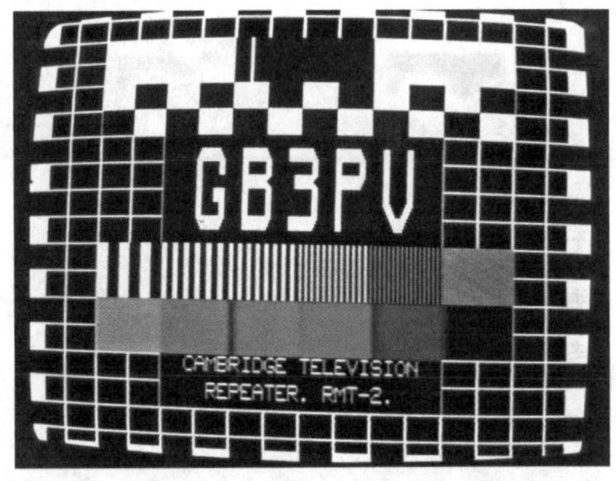

**Test card radiated by Cambridge ATV repeater GB3PV, when in beacon mode, ie not being accessed. The original photograph was in colour.**

The special interests of TV amateurs are catered for by The British Amateur Television Club (BATC) [1], which is affiliated to the RSGB. This article will concentrate on fast-scan TV.

### THE ATV QSO

Those not familiar with ATV may wonder what a QSO is like and what ATV operators get up to. It is usual to use one or more video cameras to take pictures, usually in colour, of the shack in which the operator may be seen speaking to his contact. Anything being discussed, perhaps a component, a piece of equipment or circuit diagram may be brought into close-up.

The only limits on content are those imposed by the licence. Amateur-made video tapes, perhaps taken by the operator during a visit or on holiday, are often transmitted, but commercially-produced tapes and the re-transmission of broadcast programmes are not allowed. Some operators have a camera installed on their aerial which can be panned round by a rotator to give a panoramic view of the district. It is usual for the sound side of the contact to be duplex, with the outgoing sound carried on a sub-carrier on the vision channel while return sound is on 2m using the ATV talk-back frequency of 144.750MHz (although other frequencies are used in some areas).

Occasionally, full duplex vision and sound is employed, with the reverse circuit being on another band. As mentioned, in many areas, operation is both direct station-to-station and via a repeater. Repeater groups often nominate specific times each week when stations try to be active and join in a net. A particularly good use of TV is when two or more operators are designing some piece of equipment or tracing a fault.

One may instruct the other, via the talk-back channel, to connect an oscilloscope or other test gear to a particular point while observing the result shown over the air, so analysing the problem.

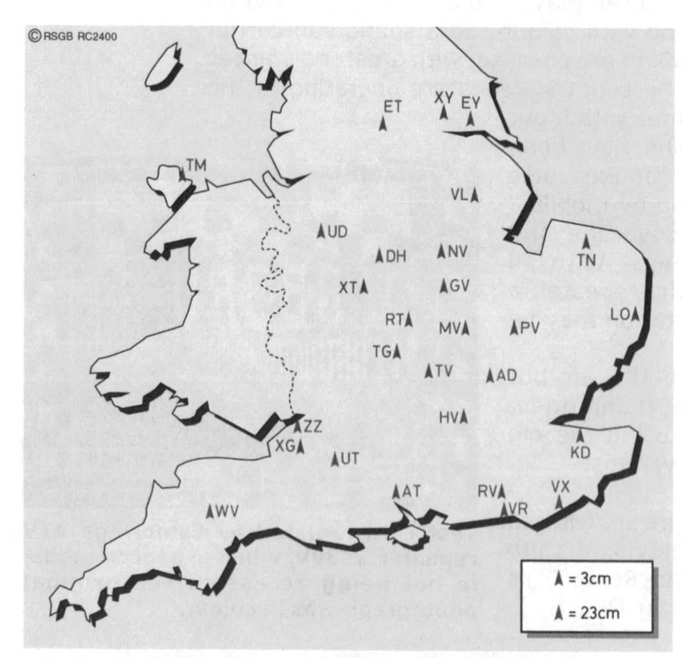


**Fig 1: ATV repeaters in the UK (October 1999).**

## GETTING STARTED

So you think this sounds interesting and would like to have a go?

How to do it depends rather a lot on where you live. If you are fortunate to live within the coverage area of one of the ATV repeaters in the UK (see **Fig 1**), the first thing is to listen on the talk-back frequency, get some idea of local activity and make contact with some ATV operators.

You will find them friendly, helpful, and very willing to assist a newcomer to join their ranks.

Although not strictly line of sight, propagation on these bands is very dependent on the lie of the land. Most repeater groups have coverage maps which will enable you to see if a signal is likely to be available at your QTH. However, as these are usually generated by path-prediction software backed up by some verification measurements, they give a good general indication, but should not be taken as absolute.

If the map shows you to be in a coverage area, and you have an unobstructed view in the direction of the repeater, you will probably be OK. Equally, there are many cases in which the software suggests an inadequate signal but satisfactory results can be obtained. Repeaters on 23cm generally have a better coverage than those on 3cm. In case of doubt, many repeater groups have portable equipment and will be willing to visit you and carry out on-site tests. It is far better to carry out preliminary investigations before making or buying equipment and installing it, which could lead to disappointment if you happen to be living in a black hole!

If you do not live in a repeater area, all is not lost. There may well be activity on a direct station-to-station basis within range of you. Again, listen on the talk-back frequency and ask questions on the air to see what may be happening. Are there any stations, and on what band(s) are they active? If you find that there are possible contacts and you can get access to a computer with path-prediction software, try entering your national grid reference and then theirs to see what the path looks like, whether it is obstructed or not and what path loss is calculated for the band in question. Local knowledge and observations will enable you to get a feel for whether there is likely to be significant extra attenuation due to obstructions. Knowing the estimated path loss you can calculate the probable picture quality you may receive as follows – on the 23cm band, take the transmitter power (in dBm), add the effective transmitting aerial gain (gain dBd minus feeder loss) deduct the path loss, then add the effective receiving aerial gain (gain dBd minus feeder loss). This gives the power at the receiver input in dBm. ATV pictures are reported on a scale of P0 to P5. P1 is a picture just resolvable in the noise, while P5 is a clear noise free colour picture. A typical ATV receiver with a front end noise figure of 1dB and a bandwidth of 14MHz, will give a P1 picture with an input of about -99dBm. It will require an increase of some 34dB, ie an input of about -76dBm, to give a P5 picture.

## WHAT EQUIPMENT?

A block diagram of a simple ATV receiving station is given in **Fig 2**, while a photograph of this equipment is shown below.

**Look-through picture of G3KKD, transmitted via GB3PV at a distance of 12.7 km and received back in the shack. The original photograph was in colour.**

### Aerial

The aerial you will require varies enormously with your circumstances. If you are fortunate to be, say, 10km from a 23cm repeater with an unobstructed path, you may be able to receive a P5 picture with a dipole literally only 22cm long, a metre or so off the ground. If you are further away and/or have an obstructed path, you may require an aerial with 22dBd or more of gain. Designs for building various types have been published, while some Yagi arrays may be purchased.

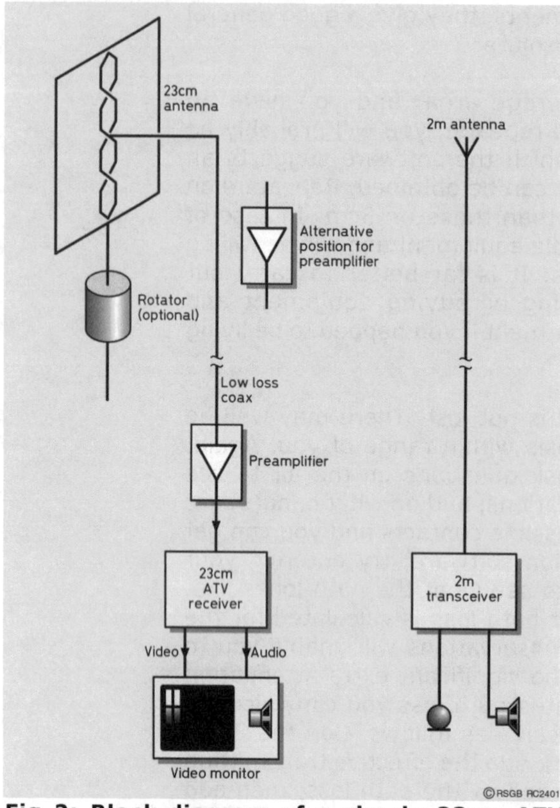


**Fig 2: Block diagram of a simple 23cm ATV receiving station.**

Polarisation is horizontal. Generally, it is best to erect an aerial with as much gain as possible as high and in the clear as possible. If you intend to operate only via a repeater, the aerial may be aligned and locked but, if a range of contacts is possible, a rotator will be necessary. Remember, high gain aerials at these frequencies are *very* directive. Also, there must be some compromise over height, as cable attenuation is high. For shorter runs, UR-67-type cable may be used but, for longer runs, it is preferable to employ better quality LDF4-50 with a flexible tail of UR-67 if required for a rotator. If you have a weak signal and/or are forced to use a long cable run, you may consider a masthead pre-amplifier, which will improve matters considerably.

However, mast-head pre-amplifiers complicate things when it comes to transmitting.

## Receiver

Here again, you have a make-or-buy option. You may wish to make a receiver following one of the designs that have been published. Alternatively, you may assemble a receiver from boards that may be bought. Although it is not the ideal solution, perhaps the quickest and easiest route is to adapt a broadcast satellite receiver indoor unit, such as the Amstrad SRX200, which provided the IF amplification and demodulation in the satellite system. The IF tuning covers the 23cm band. Units of this sort are available at rallies for a few pounds.

If using a satellite unit, several points need to be remembered:

1. These units output DC on their RF input sockets, to feed the LNB. If you do not wish to use it to drive a preamplifier, it *must* be disconnected, otherwise the aerial could short it out and cause damage.

2. An indoor unit alone does not have enough gain for ATV reception, except with an extremely strong signal. You will need a high gain, low noise pre-amplifier (typically a noise figure around 1dB and gain of about 40dB). Again, these can be made or bought.

3. While it is possible to tune a satellite receiver to the ATV channel using its tuning push buttons, it is better to disconnect the original

tuning arrangement and substitute a simple potentiometer tuning control. It is also necessary to select a 6.0MHz sound sub-carrier.

4. As the deviation used for satellite transmissions is about twice that standardised for ATV, the unit's IF bandwidth will be too wide. This leads to some reduction in signal/ noise and less immunity to adjacent channel interference, ie radar or other amateur stations. The level of the demodulated video will also be low, but this can usually be accommodated by the monitor contrast control. Many stations use satellite receivers – they are certainly a very good way to get started.

**Monitor**

You may either use a video monitor, such as those supplied for use with older computers which output standard 625-line video, or a television set equipped with a SCART connector to which composite video and sound from your receiver can be connected. Have a dedicated monitor, don't be tempted to use the domestic TV set, it will lead to strife!

**READY TO GO**

With the above items plus a 2m transceiver, you are now equipped to join the fun with ATV, at least as a receiving station. After gaining some experience, you may wish to progress to transmitting.

Many designs for aerials and other equipment mentioned here have been featured at some time in *CQ-TV*, the BATC's journal. If you're fired-up by what you have read here, do consider joining.

I would like to thank my friends in the Cambridge ATV net for their help in producing this article.

**REFERENCE**

[1] BATC Membership Secretary, Dave Lawton, G0ANO, Greenhurst, Pinewood Road, High Wycombe, Bucks HP12 4DD. Tel: 01494 528899. E-mail: memsec@batc.org.uk

# AN INTRODUCTION TO AGC

**by Peter Finbow, G0DEH**

I'm sure we have all seen the initials AGC on the front of a receiver. If you have looked closely at the control you are likely to have seen that it has at least two positions. These are 'slow' and 'fast'. Some receivers also have an 'off' position. A few of the more modern pieces of equipment also have an 'auto' position.

This article will explain what AGC is, why we use it, and a few tips on how to use it.

## WHAT IS AGC?

AGC or Automatic Gain Control is a method of controlling the gain of the amplifiers in the IF/RF amplifier chain of a receiver. In some cases, AGC is also applied to the audio stages.

So why do we use AGC? If you think about the signal being input to the RF stages of a receiver, it varies with time. This is due to changes in the propagation conditions to which the signal is subjected, giving rise to the phenomenon we call fading (QSB). These variations may make the signal weaker or stronger.

We are going to take a slight detour here and talk about amplifiers for a few moments. When designing them we consider the maximum output signal achievable within acceptable distortion limits. The parameters change with different transistors, power supply voltages and biasing.

Once these parameters (and others) have been decided upon, the amplifier will have a good distortion-versus-input-signal-amplitude characteristic.

If the situation arises (and it will) that the input signal rises above the level for which the amplifier was designed, the distortion level will rise. Eventually the signal will be so distorted that the output will be unusable. To prevent this, we apply feedback to the amplifier to reduce its gain. You could look at AGC as a 'system' feedback.

## HOW AGC WORKS

If you look at **Fig 1** you will see a block diagram of an AGC feedback loop as often implemented in amateur equipment. It is important to remember that all of the stages are RF decoupled, to prevent interaction (the decoupling has not been shown for the sake of clarity). The AGC amplifier provides enough loop gain (overall gain in the AGC feedback loop) to ensure the AGC characteristic (**Fig 2**) is achieved.

The loop begins to work once a certain signal level (called the AGC threshold) is reached at the input to the AGC amplifier. After the threshold

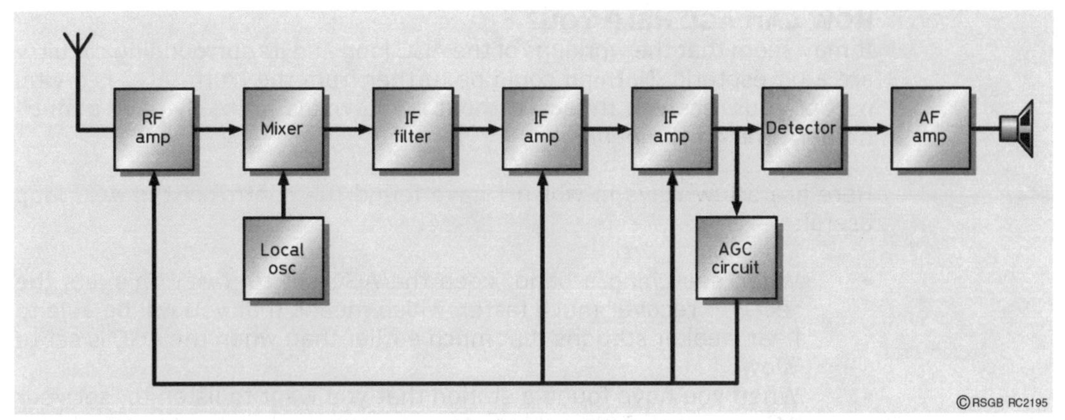


**Fig 1: Block diagram of an AGC loop, as often implemented in amateur equipment.**

is exceeded, the audio output slowly rises. Normally this rise is quite small. In most cases, operators find that too little or too much gain is undesirable.

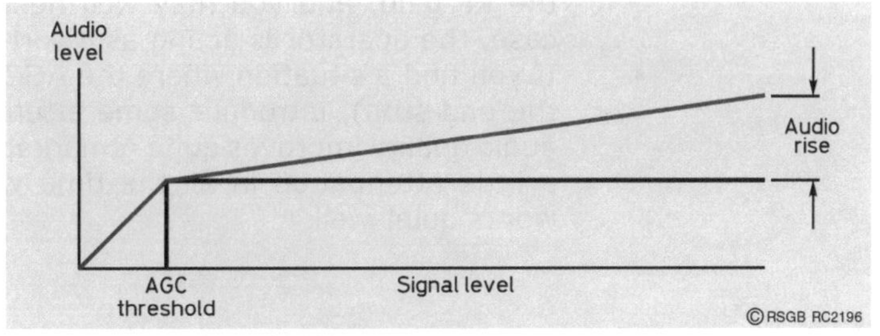


**Fig 2: AGC characteristic curve.**

Once the loop begins to work, it reduces the gain of the RF/IF amplifiers to prevent distortion. There is a point, however, when the system can't handle any more signal. The result is distortion. Some designs of AGC loop delay the action on the RF stages of the receiver. This is so that the gain of this stage will not start to reduce before a larger signal appears.

This also improves the noise performance of the RF/IF amplifiers. The normal way to do this is to use a diode, which means that there has to be an AGC signal greater than the 0.6V forward voltage drop across the diode before the RF amplifier is acted on by the loop.

The 'time constants' (how fast the AGC loop works) are controlled by RC circuits. These can be set so that the 'attack' time is not too fast. The 'hang' (or 'decay') time of the loop is normally controlled in the same way. If the RC circuits have switchable values, the time constants of the whole loop can be varied. This allows the user to adjust AGC attack or decay times to suit conditions.

## PROBLEMS

AGC loops always work better with slowly-changing signal levels. We do not have the ideal situation in amateur communication, because CW and SSB are fast-changing signals. This is not the problem it might seem to be. Most modern systems operate well.

**HOW CAN AGC HELP YOU?**

It may seem that the workings of the AGC loop and its surrounding circuitry are a bit esoteric. Nothing could be further from the truth. AGC is a vital part of your receiver. In fact, without it you would probably have a much harder time on the bands.

Here are a few ways in which I have found the control of the AGC loop useful:

- When searching a band, keep the AGC set to 'fast'. This lets the receiver recover much faster, which means that you will be able to hear weaker stations that much earlier than when the AGC is set to 'slow'.
- When you have found a station that you want to listen to, set your AGC to 'slow'. This will tend to even-out small fades and avoid AGC 'pumping'.
- Sometimes (especially in contests), there can be a problem with close-in stations breaking through into the pass-band of the receiver. This drives the AGC into operation, masking the weaker (wanted) station. In this case, try switching the AGC 'off' (if you can). Reduce the RF gain, and you may well hear the weaker station. In this case, the operator is acting as a sort of human AGC.
- If you find a situation where the AGC is swamped (the needle hits the end-stop), introduce some attenuation. You will find that the audio quality improves quite remarkably; indeed, some people keep a little attenuation in all the time when using 40m or 80m – it works quiet well.

# AN INTRODUCTION TO BALUNS

**by Anthony B Plant, G3NXC**

The prime purpose of a balun (a contraction of BALanced-to-UNbalanced) is to allow an unbalanced source to drive a balanced load, or vice versa. Some types of balun will also yield an impedance transformation, but this should be regarded as a secondary function.

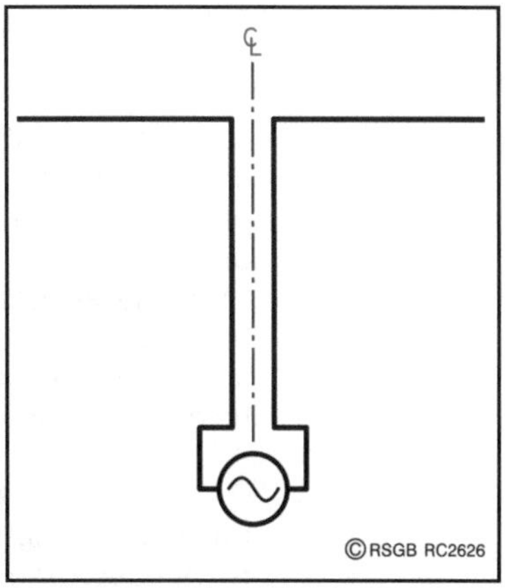

## BALANCED SYSTEM

Before getting into the details of baluns, it is necessary to understand just what is meant by a balanced load, and why feeding such a load from an unbalanced source can create problems. **Fig 1** shows a typical balanced load. The arrangement is symmetrical about the centre line. Each point on the left hand side is mirrored by an equivalent point on the right hand side, where the currents and voltages are equal in amplitude but opposite in phase.

©RSGB RC2626

**Fig 1: The standard dipole is electrically symmetrical about the centre line.**

In the dipole itself, the currents in the two legs create fields which add together to generate the usual 'figure of eight' radiation pattern. The fields generated by each half of the feeder, though, cancel out each other so that there is no radiation from it.

## FEEDING VIA COAX

Now consider the same dipole fed through a length of coaxial cable from a typical transmitter, as shown in **Fig 2(a)**. The current flowing in the inner conductor of the cable has only one destination, the left hand leg of the dipole. That flowing in the outer of the cable, though, has two destinations – the right-hand leg of the dipole and back down the outside of the cable to ground. **Fig 2(b)** shows a somewhat simplified representation of the various current paths, with I3 being that flowing back down the outside of the coaxial cable. The result of having a path for I3 is that a top-fed vertical aerial is, in effect, put in parallel with the right hand leg of the dipole. This vertical aerial will, of course, radiate.

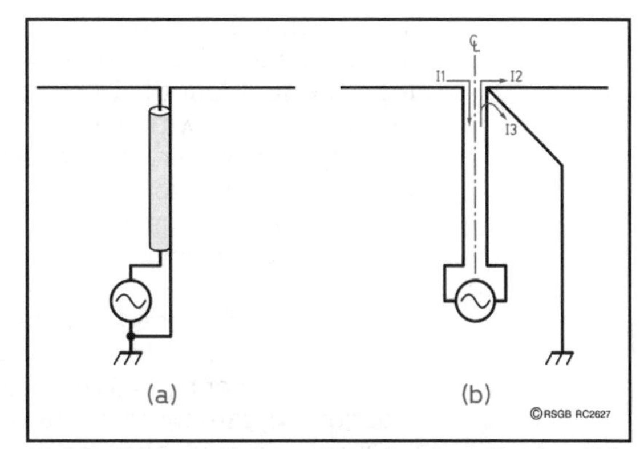


**Fig 2: If coax is used to feed a dipole, this generates a current path to earth, effectively resulting in a top-fed vertical aerial.**

The amplitude of I3 is dependent upon the length of cable being used. If it is an odd multiple of $\lambda/4$, the feed impedance of the effective vertical aerial is very high, so I3 is low. Under these conditions the unbalancing effect on the dipole is insignificant. If, though, the cable is a multiple of $\lambda/2$, the feed impedance is low and I3 is, therefore, high. As would be expected, the unbalancing effect is also high.

## POTENTIAL PROBLEMS

Having explained the situation, the obvious question to be answered is: does it really matter? In order to answer this question, we need to consider two separate aspects: firstly the effect on the aerial's ability to radiate and, secondly, what might be called the system effects - such matters as EMC, VSWR measurement and so on.

Feeding a normal wire dipole via coax causes some change to the radiation pattern, but this may not be particularly significant. There can, in fact, be some benefit in that radiation from the vertical aerial represented by the outer of the coax fills in, to some extent, the nulls off the ends of the dipole. Should the dipole be part of an array, a Yagi beam for instance, then the effects of feeding directly via coax can be quite significant. The forward gain will be reduced, as will the front-to-back ratio, and some side lobes will also appear. Similar problems will also occur with other forms of array based on balanced radiators, such as the cubical quad.

Of the system effects, EMC is probably the one of greatest concern. In normal installations the coax will enter the shack and be in close proximity to housewiring. Radiofrequency energy radiated from the outer of the coax is highly likely to find its way into all the surrounding wiring and cause breakthrough problems.

Under normal circumstances, the VSWR on a transmission line is dependent only upon the load impedance and the characteristic impedance of the line. Changing the line length should not cause the VSWR, as indicated by the usual VSWR meter, to change.

Sometimes, though, it is found that the VSWR reading *does* change quite significantly with line length. This variation of reading is indicating that there is something odd about the setup. Feeding a balanced load via coax represents one of the possible oddities.

Fig 2(b) shows that the actual load is made up of the dipole plus a top-fed vertical aerial. Also, as mentioned previously, the level of I3 is dependent upon the line length. In effect, then, the load impedance at the aerial terminals varies with line length. With VSWR being dependent upon this load impedance, varying the line length will cause the VSWR meter reading to change. It is also possible that I3 flowing back down the outer of the coax will affect the operation of the VSWR meter, thus causing an additional source of confusion.

So, having established that feeding a balanced load via coax causes the radiation pattern of the aerial to change, the EMC threat to increase and the measurement of VSWR to become somewhat unpredictable, the next question is: can anything be done to improve the situation? Fortunately the answer is yes, and one of the solutions is to use a balun.

## NARROWBAND BALUNS

There are many types of balun, each having its advantages and disadvantages. Let's look first at some narrow band arrangements – these being suitable for single-band aerials.

**Fig 3(a)** shows a very simple arrangement using an additional half-wavelength of coax. One feature of a half-wave line is that the voltage at the output is equal in amplitude to that at the input, but with the phase reversed. With the arrangement shown, the voltages at A and B are equal in amplitude but of opposite phase. The voltage between A and B is twice that of the input. As a result, the load impedance for a 50Ω input impedance must be 200Ω. It should be noted that the aerial is not connected to the outer of the coax. If the inner of the coax does not have a DC path to earth then there can be problems with static build-up, particularly when there are electrical storms in the locality.

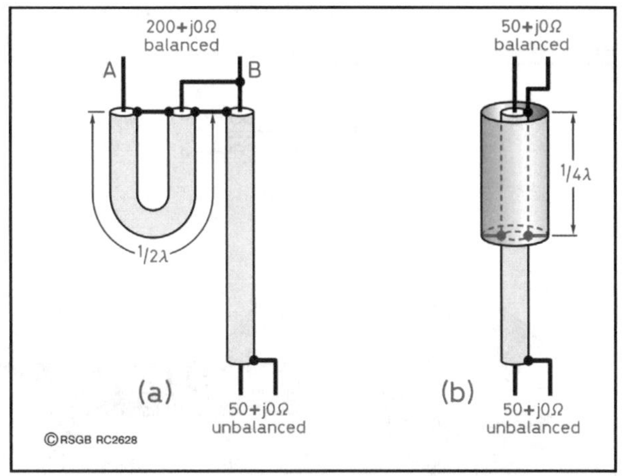


**Fig 3: Narrow-band baluns can be made quite simply. A half-wavelength of coax, connected as (a), generates a 180° phase shift. A quarter-wavelength sleeve (b), effectively decouples the last part of the coax.**

If a 1:1 impedance transformation is required, the arrangement shown in **Fig 3(b)** will suit the need. This uses a λ/4 sleeve effectively to decouple the last section of the coax. Note that the top end of the sleeve needs to be well-insulated and the bottom end connected to the outer of the coax.

Although these two baluns have the virtue of simplicity, they are only really suitable for the higher frequencies owing to the lengths of line needed to make them.

## BROADBAND BALUNS

These can be one of two basic types: those which force the currents in the two halves of the aerial to be equal in amplitude but of opposite phase; and those which force the voltages to have this relationship. If the aerial is truly balanced, both achieve the same effect. One problem with wire aerials at the lower frequencies is that it is difficult, for many reasons, to achieve a fully-balanced arrangement. The current balun ensures that, in such cases, the currents in both conductors of the feeder are equal in amplitude.

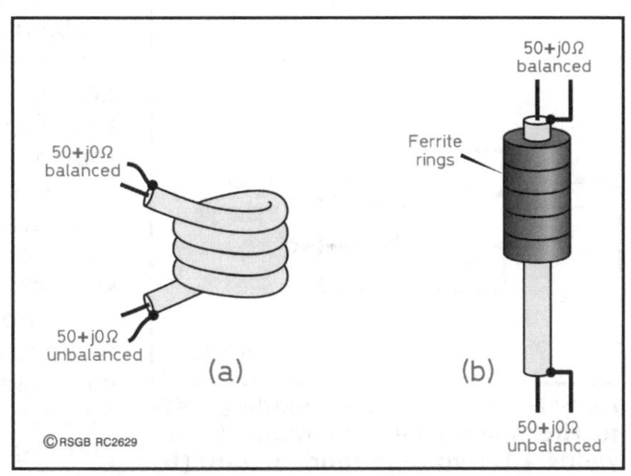


**Fig 4: Two simple current-mode baluns. Both use inductance to reduce the current on the coax outer.**

**Fig 4** shows two simple types of current-mode balun, both of which provide a 1:1

impedance ratio. These work on the principle of providing an impedance to restrict the flow of an out-of-balance current. In the case of **Fig 4(a)** this is achieved by coiling up the coax near to the feed-point of the aerial. The coil has no effect on the normal signal flowing up the coax but looks like an inductance to any current trying to return via the outer. The same effect is achieved in **Fig 4(b)** by threading ferrite rings over the coax.

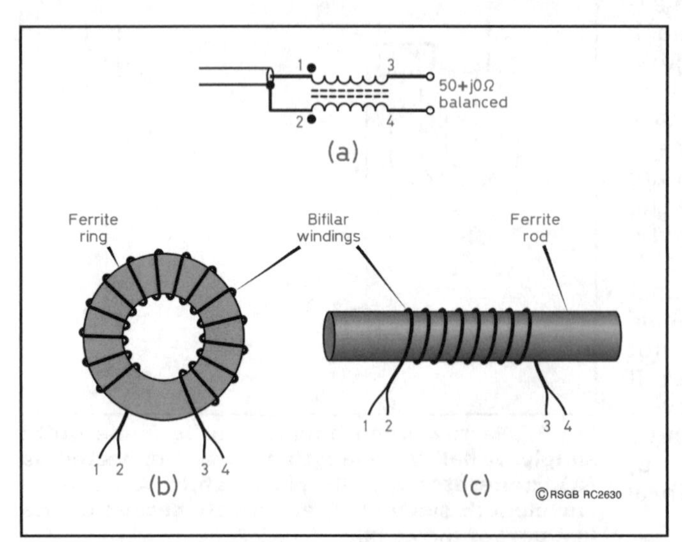


**Fig 5: A bifilar winding on a ferrite core, either a toroid or a rod, yields a simple current-mode balun. Note that the dots indicate the beginnings of the windings.**

One difficulty with the arrangement shown in Fig 4(a) is that it can be difficult to make it work effectively over a wide frequency range - say 3.5 to 30MHz. Providing sufficient turns to cope with 3.5MHz is likely to result in the interwinding capacitance being too high for effective operation at 30MHz. The situation can be improved by winding the turns on a ferrite rod ring.

A current-mode balun can also be constructed by using a bifilar winding on a toroid or ferrite rod, as shown in **Fig 5**. It must be understood that these bifilar windings act like transmission lines, which can limit the performance of the arrangements in some circumstances and yield rather erratic results. Ruthroff [1] advocated the addition of a third winding to the simple bifilar winding of **Fig 5(a)** to yield the arrangement shown in **Fig 6(a)**.

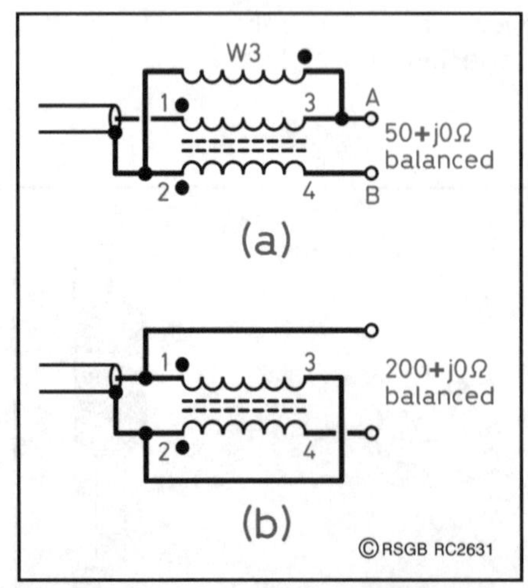


**Fig 6: (a) Adding a third winding, W3, to the simple current-mode balun yields a 1:1 voltage-mode balun. (b) Connecting the two bifilar windings in a different way gives a 4:1 impedance step-up design.**

Although this third winding overcomes some of the problems of the two-winding arrangement, it has the effect of turning the balun into a voltage-mode device. Windings 1-3 and W3 act like an auto-transformer, so that the voltage at point A is half that of the input. The voltage at point B will also be half that of the input, but with a phase reversal. This arrangement tends to be regarded as the 'standard' for 1:1 impedance ratio baluns. The simplest form of voltage-mode balun, albeit with a 4:1 impedance step-up, uses the same basic arrangement as Fig 5(a), but with the windings connected in a different way, as shown in **Fig 6(b)**. Construction is as for the examples in **Fig 5(b)** and **Fig 5(c)**.

There are two, often conflicting, criteria associated with the design of voltage-mode baluns. Firstly, the inductive reactance of the windings should be high; and secondly, the leakage reactance should be low - compared with the load impedance in each case. The first usually determines the low-frequency limit of operation whilst the second determines the high-frequency limit.

**CONCLUSIONS**

This has been a fairly brief introduction to the subject and has (quite deliberately!) begged the question of which design is 'best'. The reason for this omission is that the use of baluns tends to result in compromises having to be made: what works well in one application may be a total failure in others. The sensible approach is to try a few different ideas and select the one which gives the best performance in your particular setup. Fortunately, the components used are reasonably inexpensive and can easily be re-cycled for the different arrangements.

For those wanting to have a go, [2 - 7] list articles and books containing more details of the different arrangements. You might be bemused by the fact that some authors will be enthusiastic about a particular arrangement whilst others regard it with horror. Take due note of any objections to the various designs, but do not let these put you off trying them.

**REFERENCES**

[1] C L Ruthroff, 'Some Broad-Band Transformers', *Proc IRE*, Vol 47, August 1959.
[2] Ian White, G3SEK, 'Balance-to-Unbalance Transformers', *Radio Communication*, December 1989 (highly recommended reading).
[3] *Radio Communications Handbook*, RSGB.
[4] *ARRL Handbook*,(ARRL).
[5] *HF Antennas for All Locations*, RSGB.
[6] *Backyard Antennas*, RSGB.
[7] *Transmission Line Transformers*, ARRL.
[8] *Reflections Transmission Lines and Antennas*, ARRL.

# CRYSTAL SETS - RECRUITING OR EXPERIMENTAL AID?

**by Pat Hawker, G3VA, from 'Technical Topics'**

An item 'Headphone Adapter' in the December 1998 'TT' noted that the absence of high-impedance headphones has made it more difficult for the novice radio constructor to take the first steps of building his own crystal set. The venerable crystal set has continued to take a useful educational role.

This item attracted a surprising amount of correspondence, including several long letters and numerous enclosures from Vic Pierson, G3MXV, dealing not only with the question of headphones, but also extolling generally the value of the humble crystal set, not only as an educational tool, but also as still offering experimental opportunities.

I was reminded of this material by a letter in *QST* (September 1999, p24) from Kerry Michael, KC0EYD, published under the heading 'Attracting New Voices'. This pointed out that children as young as five years old are extremely curious as to how things work, but have only a very short attention span: "Unless they are incredibly interested, four minutes is the longest they will listen to Mum or Dad explaining how antennas work. There's a better chance to teach older kids, but it is possible with the younger ones also. These kids are eager to learn anything. It's the perfect time to introduce them to amateur radio!

"But how? In my opinion, there is no better way to introduce kids to electronics than a crystal radio set. They're inexpensive, the components are easy to find and relatively simple to assemble. All the parts can usually be found at a local Radio Shack [Tandy in the UK – *Ed*.] or science store and kits are also available. Curious youngsters will be thrilled when they hear radio stations through a set they built themselves, or even with a little help!

"Controversial as this may seem, Morse code is another powerful attraction for many kids. My eight-year-old brother ran away with learning the code. He loves the idea of knowing a 'secret' alphabet (well, at least semi-secret!). Take it from a 17-year-old, video games are not the be-all and end-all of the 90s kid [or hopefully in the New Millennium - *G3VA*.]. Getting kids interested in radio isn't hard at all! All you have to do is take a little time and show them the way."

I must admit that some 30-odd years ago, I succeeded in getting my son interested in some aspects of radio and electrical construction, but primarily his interest was in model aircraft and I failed lamentably when it came to Morse and lost him to the then already-expanding computer industry, linked to airlines. Perhaps I did not start him on radio early enough! To my shame I never even tried with my daughter!

G3MXV confesses that there have been three periods in his life when building crystal sets have taken up most of his spare time; first as a youngster of 6 or 7 years; as a father with two young boys (neither interested); and as a grandfather with one grandson, similarly uninterested. But, despite the failures, crystal sets still intrigue him. He tells the following story:

"As a boy, in Manchester, I was repeatedly told by my elders that in their day at the birth of radio broadcasting, it was possible for all the family to enjoy a wireless programme simply by placing a single earpiece of the crystal-set inside a pudding dish to distribute the sound. This spurred me on in an attempt to achieve greater output from the circuits I built.

"Despite making large basket-weave coils, trying an assortment of crystal detectors (readily available as late as 1936) and pleading at every birthday for more sensitive headphones, I had no success. It was some consolation later to discover my parents had lived only a few miles from the first Manchester transmitter, call-sign 2ZY, and had used fully the permitted outdoor aerial length of 100ft." [I had reasonable success with the pudding basin experiment, being fortunate to live only a few miles from the high-power BBC twin station at Washford Cross, Somerset - but also found that the old horn-style loudspeaker with in-effect a large earphone-type transducer, was much more sensitive and more efficient than a moving-iron or moving-coil loudspeaker of the 1930s. But my greatest success was picking up Radio Roma and Zeesen on a crystal set, by fitting a shortwave coil and using a good length antenna – *G3VA*.]

G3MXV continues: "I did not give up on the crystal-set, however, and eventually had two sons to be introduced to the miracle of 'simple radio'. There were now germanium signal diodes, a valued stock of genuine litzendraht wire (not the loosely-bunched imitation), ferrites, pot cores, and a nice long-wire amateur radio antenna available. In accordance with the KISS principle, our variable capacitors were made from aluminium kitchen foil wrapped around two thick cards, one having an extra cover of cling-film as dielectric. The usual 1 – 3nF capacitor shunting the headphones had been rejected in my early days as unaffordable and having no benefit whatsoever - it makes no perceivable difference. All this was good fun for my boys, but with transistor sets readily available and the output from a crystal set so puny, how could one expect it to hold their interest for long?

"When the time came to try again with my grandson, a major short-coming seemed to be the absence of my old pair of 4000Ω S G Brown headphones (long gone), or even a near-equivalent. The junk box offered only low-Z ex-government headphones, left-overs from the many Morse practice oscillators I had been called upon to make... Step-down transformers would at best lose 50% of the output. Against all conventional theory, I connected a single low-Z earpiece to the crystal set. The result was astonishing, I called my wife into the shack to listen to music from Radio Leicester on medium-wave, with the earpiece sitting on the table. At long last I had achieved the ability to listen to a crystal set without wearing headphones, albeit from a local transmitter. Being already aware of the sensitivity of these particular headphones – a single earpiece could fill a hall when driven by my Morse oscillator at its resonant frequency and acoustically loaded, I had to wonder where the secret really lay.

"Is it purely a question of sensitivity? What about the loading effect on the high-*Q* tuning coils that we have struggled to produce? But, of course, the loading is only during the half-cycle when the diode is conducting [Because of the 0.7V junction barrier, conduction will be significantly less than a half-cycle so that the diode will not load the tuning coil to anything like the extent generally assumed - *G3VA*]. And one recalls the 1917 book *The Elementary Principles of Wireless Telegraphy* (2nd edition) by R D Bangay, page 136. Under the heading 'High Resistance Telephones', para 502: 'Unfortunately, however, as we reduce the size of the wire (*wound on the magnet to increase the number of turns, space being at a premium*) so do we increase the resistance per turn of that wire, and therefore decrease the amount of current which would pass through it for a given voltage. Therefore, unless the current at our disposal is already limited by some external resistances, we shall not gain anything by increasing the number of turns if, at the same time, we increase the resistance of the (*headphone)* coil in proportion.' The words in the italics are mine [*G3MXV's*] otherwise quoted directly from the book. Most significant is the final reference to the resistance of the headphone.

"When I try to interest other radio amateurs in their 70s, I get a sideways look, so I always insist: 'Please try it for yourself, it takes but minutes to clip the parts together.' Sadly, so far, no takers. Yet nothing special is required to prove my point, no high-*Q* coil, just a bunched winding of plastic-covered wire around a cardboard former. A very high L to C ratio in the resonant circuit is suggested in the Bangay book, something I have not yet tried. Any appropriate diode will do, eg OA5/ OA95, an ancient 500pF air-spaced variable capacitor and *specifically* an ex-Army earpiece: Type No DLR No 5S (external marking), other markings ITBA5 (external), YA 5275 (internal). Measured DC resistance: 25Ω (varies slightly with sample). Measured impedance: approx 450Ω at 1200Hz. Best perceived performance and impedance is when acoustically loaded, eg if earpiece is placed face down on a flat surface with a small air gap.

"These particular headphones were a glut on the surplus market for many years, instantly identified by their wire head-frame and khaki-coloured headband. There must still be a lot around. Surely somebody will spare a few minutes to prove this intriguing anomaly and perhaps offer a reasoned explanation."

In a later letter, G3MXV reported making two further discoveries, again going counter to perceived theory and begging a theoretical explanation. Both are concerned with 'loading' the resonant tuned circuit and its effect on sensitivity/selectivity, though he has long ignored the need for tapping the detector down the tuning coil or connecting a bypass capacitor across the output. He has also found that he can connect all his four DLR5 low-impedance earpieces in parallel and still have acceptable reception. He draws attention to 'The Xtal Set Society' founded 1991, based in the USA and providing News-sheets on the Internet (**www.midnightscience.com/**) with a number of publications on crystal set projects, construction and history. He also reports that GWM Radio

Ltd (*RadCom* advertisers) have (or had) some 20 of the ex-Army DLR5 headsets in stock. G3MXV writes: "The outcome of this fascinating business will probably be an anticlimax. My growing belief is that the manufacturer of these headphones hit upon a unique design involving the dynamics of the diaphragm, sometimes referred to as 'motional resistance' and perhaps taking into account audio harmonics due to the magnet core. The end paragraph on page 11 of David Owen's 'Alternating Current Measurements' (Methuens's Monographs on Physical Subjects, 1937), throws some light."

He also draws attention to 'The Crystal Radio' - subtitled 'What's Old is New Again!' by Dave Evison, W7DE (*QST*, December 1997, pp56 – 57), although this presents the conventional circuit diagram (**Fig 1(a)**) with the diode tapped down the coil and a 1000pF capacitor connected across the headphones. W7DE emphasises that building a crystal set is one of the few ways of "reaching today's kids" with the true magic of radio buried under layers and layers of complicated hardware that needs to be stripped away to reveal radio in its purest form.

But G3MXV has surely shown that even the humble crystal set can throw up some interesting experiments that can confound accepted practice as well as (possibly) providing a means of interesting youngsters in radio in their pre-teen years.

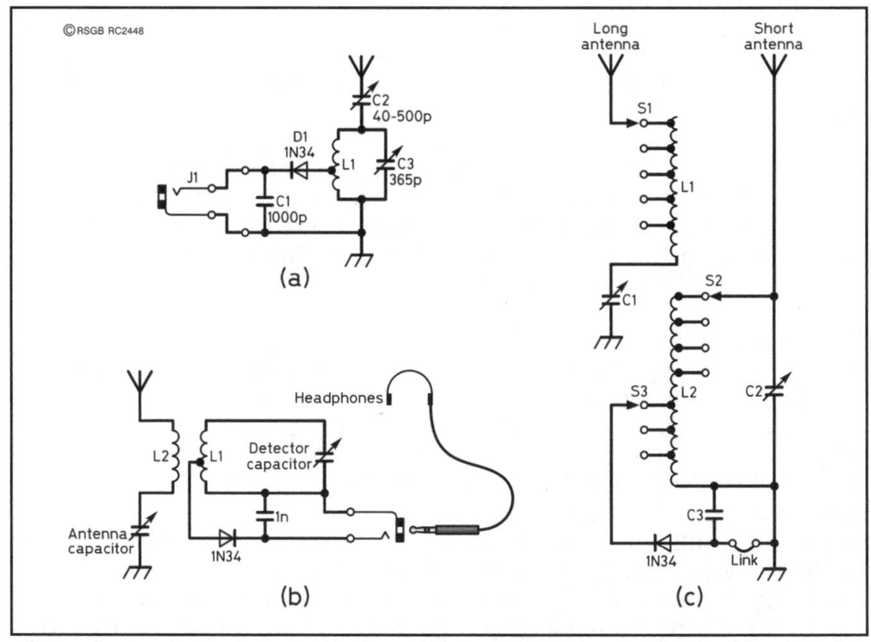


**Fig 1: Some typical recently-described crystal-set receivers, all of which tap the detector across only part of the coil and connect a fixed capacitor across the headphones - practices deplored by G3MXV. (a) from *QST*, December 1997. (b) From *73 Amateur Radio Today*, September 1998. (c) From *QCWA Journal*, Spring 1999. All intended for use with 2000Ω headphones, but see text for G3MXV's notes on the surprising results he achieves using ex-Army low-Z headphones.**

# AN INTRODUCTION TO DECIBELS

**by Jeff Black, G0UKA**

When faced with calculations involving dB (decibel) functions, the majority will suddenly be possessed of a desire to mow the lawn, take up flower arranging or iron a dozen or so shirts! In fact, decibel notation is the ideal way to understand the relationship between two different values - regardless of what these values relate to. The purpose of this article is to explain how dB are calculated and used (and sometimes misused!).

**WHAT ARE DECIBELS?**

The first - and probably the most important - thing to establish is that dB on its own is just a ratio. Decibel calculations require the dB to relate to something. This is why RF output is often shown as dBW - the output ratio *relative* to 1W.

However, dB could, in theory, refer to absolutely anything. For example, let's start with dB 'tea-bags' as a reference. Since an increase of three dB equates to doubling the value (I'll explain why later), increasing 'tea-bags' by 3dB doubles the number of tea-bags. I agree that this example is strange, unless you are into home-brew equipment [Ugh! - *Ed*], but the same principle applies to power output. Increasing 10W by 3dB gives you 20W.

Equally, decreasing by 3dB will halve your power to 5W. In the real world, once the subject matter has been established, the reference value (whether 'watts', 'tea bags', or anything else) is often dropped. So the specification for a linear amplifier may refer to a 20dB gain; this would mean that the amplifier's output is 100 times its input.

To summarise, for dB to possess an absolute value, it must relate to a known quantity (eg dBW). On its own, it is simply a ratio of two quantities, whatever they may be.

This idea of a ratio is important, because fixed value calculations can potentially misrepresent what is actually happening. For example, increasing the power output of a transceiver from 2W to 4W has doubled its output (+3dB). However, increasing the output of a 100W transceiver by the same amount (ie 2 watts) only increases the power by a paltry 2% (+0.086dB). To achieve the same 'relative' improvement, the transceiver's output would need to increase to 200W.

Now for the mathematical bit. No, don't go out to mow the lawn, the maths is not as bad as it looks, and there are some short cuts to dB calculations.

Firstly, the formulae:

1. Direct conversion from dBW to Power:
   Power (W) = antilog (dBW/10)

2. Direct conversion from power to dBW:
   dBW = 10 (log W)

3. Difference (ie ratio) between two power levels (W1 and W2) in dB:
   Ratio (dB) = 10 (log W1 / W2)

| dB | Mathematical relationship | |
|---|---|---|
| +30 | One thousand times | x 1000 |
| +20 | One hundred times | x 100 |
| +10 | Ten times | x 10 |
| +3 | Two times | x 2 |
| +1 | One and a quarter times | x $^{5}/_{4}$ |
| 0 | No change | x 1 |
| -1 | Four fifths | x $^{4}/_{5}$ |
| -3 | One half | x $^{1}/_{2}$ |
| -10 | One tenth | x $^{1}/_{10}$ |
| -20 | One hundredth | x $^{1}/_{100}$ |
| -30 | One thousandth | x $^{1}/_{1000}$ |

**Table 1: Short cuts for converting between power and decibels.**

## SHORT CUTS

OK, that was the nasty bit. Fortunately, there are some short cuts, which can make life much easier. Since dBW is, by definition, dB relative to 1W, a value of 0dBW is equal to 1W (ie zero difference to your starting point – see **Table 1**).

From formula (1), +1dBW = 1.2589W; +2dBW = 1.5849W; +3dBW = 1.9953W.

The first short cut is: multiply the power by 10 for every 10dB added. This is an easy one to remember.

From this you will see that adding 20dB will multiply your power by 100; if you start with 4W and increase by 20dB, you end up with 400W.

The second short cut is: double the power for every 3dB added. This is not 100% accurate, but close enough for all but the purists.

The third short cut is: multiply the power by 1.25 for every 1dB added (see Table 1). Once again, this is not 100% accurate, but close enough.

**APPLICATION**

The maximum legal power limit on most bands is +26dBW. If 0dBW is 1W, then +20dBW is 100W. Double this to 200W for +23dBW, and again for +26dBW, giving the 400W maximum which we all know, love, and wouldn't dream of exceeding. Table 1 and **Table 2** provide more detail, and **Fig 1** a graphical method of conversion.

| dBW | Short cut | Equating to |
|---|---|---|
| +10 | +10 | x 10 |
| +9 | +10 -1 | x 10 x 4/5 |
| +8 | +10 -1 -1 | x 10 x 4/5 x 4/5 |
| +7 | +10 -3 | x 10 x 1/2 |
| +6 | +3 +3 | x 2 x 2 |
| +5 | +3 +3 -1 | x 2 x 2 x 4/5 |
| +4 | +3 +1 | x 2 x 11/4 |
| +3 | +3 | x 2 |
| +2 | +3 -1 | x 2 x 4/5 |
| +1 | +1 | x 11/4 |
| 0 | -— | x 1 |
| -1 | -1 | x 4/5 |
| -2 | -3 +1 | x 1/4 x 11/2 |
| -3 | -3 | x 1/2 |
| -4 | -3 -1 | x 1/2 x 4/5 |
| -5 | -3 -3 +1 | x 1/2 x 1/2 x 11/4 |
| -6 | -3 -3 | x 1/2 x 1/2 |
| -7 | -3 -3 -1 | x 1/2 x 1/2 x 4/5 |
| -8 | -10 +1 +1 | x 1/10 x 11/4 x 11/4 |
| -9 | -10 +1 | x 1/10 x 11/4 |
| -10 | -10 | x 1/10 |

**Table 2: How you can calculate power for every whole dB via the short cut method.**

For those of you with sadistic tendencies and a decent calculator, the legal maximum of 26dBW actually equates to 398.107W, and not the 400W often quoted. However, I don't think the DTI will be too severe on a station which is transmitting only 0.02dBW over the permitted maximum! (Should I now wait for a visit from someone large and official, wanting to cart me off for encouraging amateurs to operate illegally?)

Unless you have at least a general understanding of dB calculations, it is very easy to be misled by some manufacturers' claims about the dB gain of aerials. Thankfully, deliberate misrepresentation is rare within the amateur world, but the occasional undefined dB gain figure is not unknown. The 'gain' of different types of aerial will often determine their sales potential, since we would all like minute aerials with huge gain characteristics.

The gain figures for aerials are usually specified in dBi or dBd. dBi relates to gain over a theoretical 'isotropic' aerial, which is one which radiates equally in all directions; the second relates to gain over a 'dipole', which

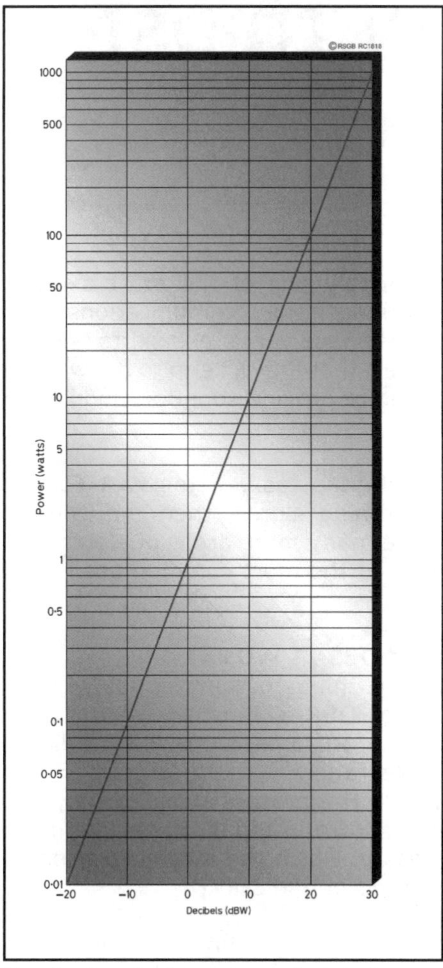


**Fig 1: Graph for converting between power and decibels.**

doesn't. Since a dipole's gain is some 2.1dB better than an isotropic aerial, specifications given in dBi could give the impression of greater gain. For example, a beam aerial with a gain of 3dBi is in fact only 0.9dB better than a dipole. If its gain were 3dBd, it would have twice the gain of a dipole. When purchasing aerials, if the manufacturer indicates gain figures in dB only, ask if it is dBd or dBi - you may save yourself an expensive disappointment.

# AN INTRODUCTION TO DSP

**by Peter Finbow, G0DEH**

Digital Signal Processing (DSP) is probably one of the most exciting technologies to be adopted by those working in the communications field. It allows for the construction of filters that have very steep skirts (sides), which reduce the harmful effects of signals that are close-in to the wanted signal. DSP techniques can even be used to provide superior modulation of an analogue transmitter.

In fact, DSP is not restricted to the filtering task. If the Fast Fourier Transform (FFT) is applied to the incoming signal, then the familiar 'waterfall' display is produced. There are many more areas of electronics that use DSP. In this article we will look at the most common use within amateur radio – the digital filter.

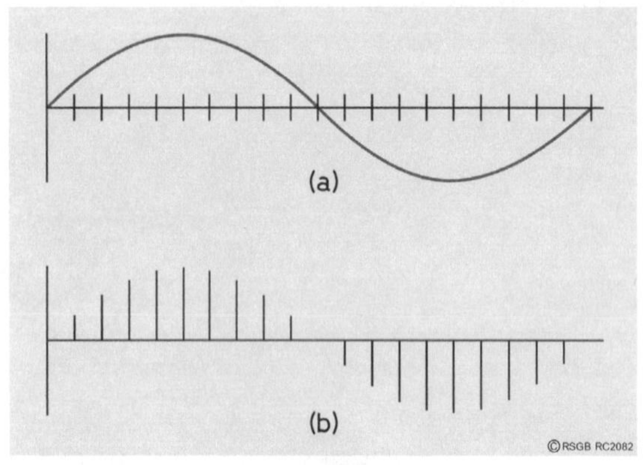


**Fig 1: (a) Input signal plus sample time marks. (b) Output from the sampler. Each sample has the same amplitude that the signal had at sample times.**

## HOW DOES IT WORK?

The first thing to think about with all types of DSP is how do we turn the incoming analogue signal into a digital representation that the processor can work on? This process is called 'sampling' and is shown in **Fig 1**. Sampling is the process in which the incoming signal is measured periodically.

The digital number that represents each 'sample' is then fed on to the rest of the processor, as shown in **Fig 2**.

It is important to remember that the interval between samples is critical. For example, if the sampling circuit can only sample every 2μs and a signal with a period of 1μs is applied, the sampled values will be unrepresentative. This will lead to a very poor reproduction when the time comes to output the signal from the processor. However, high-speed sampling devices are expensive. For this reason, most amateur radio DSPs work at audio frequencies, but this is rapidly changing as the price of

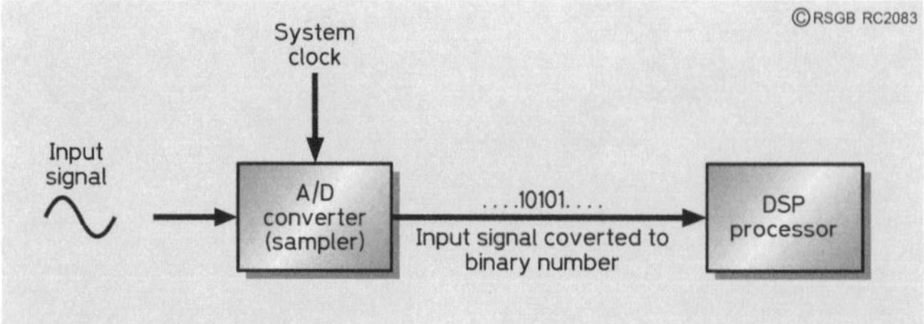

**Fig 2: Sampled input is converted into a stream of digital data for processing.**

devices starts to fall. A simplified block diagram of a digital signal processor is shown in **Fig 3**.

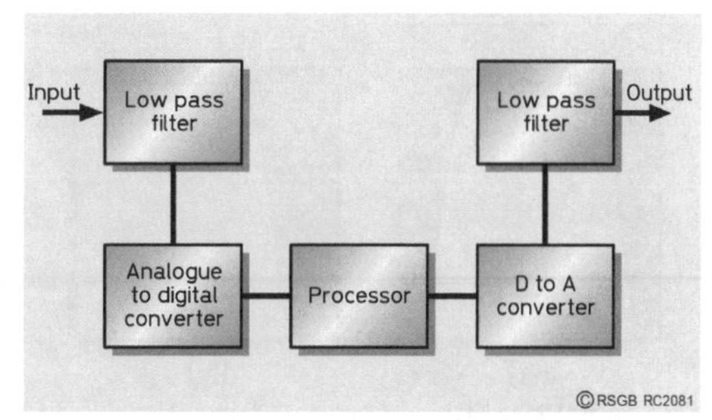


**Fig 3: Block diagram of a DSP filter.**

## NUMBER-CRUNCHING

Having converted the incoming analogue signal to an accurate digital number, the processor must now do its job. This is where the user has an opportunity to take control of the processor and set up the parameters that he / she wants. So what exactly can you, as a user, control?

Probably the most important adjustment is the frequency pass-band. This is done by setting the operating frequencies of a low-pass and a high-pass filter. These filters (high-pass at the low-frequency end and low-pass at the high-frequency end) define the pass-band. The minimum pass-band depends on the mode in use and will generally be narrower on the CW settings than the voice modes. The actual bandwidth is for the user to decide and can be as low as desired, provided the signal can still be understood.

## COMMERCIAL MODELS

There are several useful little switches on commercial processors. First is the auto-notch filter. Most of us are familiar with the standard notch filter which is used to remove an interfering heterodyne (whistle). This is a very handy function, especially on the 40m and 80m bands which are plagued with 'tuner-uppers'. One need only hear the dreaded heterodyne, switch on the DSP auto-notch filter and, Hey Presto, no more heterodyne.

Theoretically, the notch filter could be used to remove an infinite number of heterodynes, but this would use up the whole pass-band and leave no signal from the wanted station. Another point worth remembering is that the notch doesn't work on CW, or rather it does - too well. If engaged during a CW contact, the station you are listening to will vanish and all you will hear will be a few key clicks.

There now follows an assortment of functions, and it really does depend on which model you buy as to which ones you get.

- Random noise suppression (may cause slight distortion).
- Mains noise reduction.
- Noise reduction filters are included on some transceivers and a few DSP separates.

The point of the NR filter is to reduce interference on the operating frequency. These filters should include band-pass filter for interference rejection, low-cut (high-pass) filters for high-frequency emphasis, mid-cut (band-stop) filters for high- and low-frequency emphasis, and high-cut (low-pass) filters for low-frequency emphasis. See **Fig 4** for details. The point of all of these filters is to improve the intelligibility of the incoming signal in the presence of interference. Like all things, filters are not 100% effective and they can make the incoming signal sound a little strange.

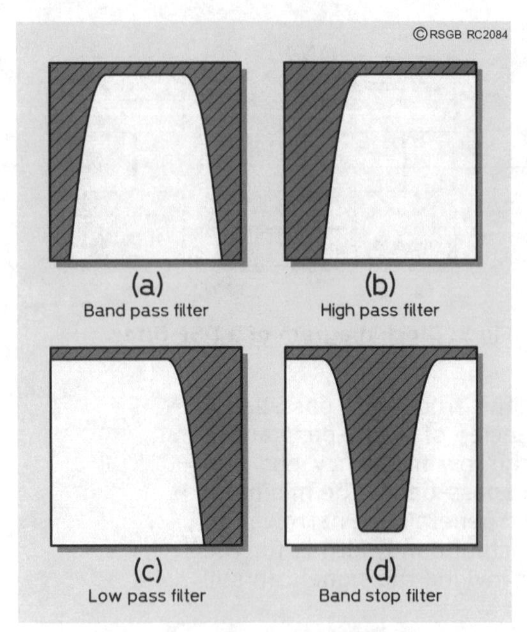


**Fig 4: The effect of various DSP filters. The cut-off frequencies can usually be set (programmed) by the user.**

With the parameters set by the user, the DSP microprocessor processes the incoming digital numbers that come from the sampler. This is possible only because the data stream is a binary one (1s and 0s) and because each filter is in fact a set of instructions to the processor on what to do with the data stream. In reality, the filters are mathematical operations executed on the data stream. Once all of the relevant processing is done, the data is sent to a Digital-to-Analogue converter, where it is output as an analogue signal for delivery to a speaker or headphones.

## ALTERNATIVE DSP

You will probably have realised by now that the DSP processor is in fact a computer. It has a rather exclusive purpose, but it uses and handles data just like any other computer. This opens a rather interesting possibility. Why don't we use a home computer as the basis of a DSP system?

In fact it is possible to use a PC as a DSP if your sound card has a microphone input and it is fitted with a DSP chip. You will need software, some of which is available as shareware and freeware. By using a computer, you are not restricted to simple audio work. You will be able to investigate the incoming signal and produce various displays of the spectral properties of the signal under investigation. It should be mentioned that the guys on 136kHz use DSP.

## POSSIBLE PROBLEMS

There are some possible problems with DSP. On certain types of input, the processor outputs garbled audio. In my experience, this happens in about 4% of contacts. Random noise filters can take some of the 'body' out of the recovered audio. In each case the cure is to switch off the filter.

## NOT THE FINAL SOLUTION

In the final analysis, the current generation of DSP equipment is not the final solution, although it does represent a vast improvement in filter technology in that parameters can be set by the user who can tailor responses to his / her needs. The future holds the promise of faster processors working at IF.

The current generation of HF transceivers may be the last to use conventional crystal analogue filters exclusively. In fact, the time is coming when DSP filters will replace them. DSP will, in the end, even find its way into handheld equipment for VHF and UHF. Another stroke for the digital world and a loss of the old ways.

# AN INTRODUCTION TO FETs

**by Peter Buchan, G3INR**

A charge-controlled device similar to the Field Effect Transistor (FET) was explored in 1928, but it was not until 1958 that a practical FET was developed. In the 1960s they became generally available and nowadays are to be found in most electronic equipment, especially so in such things as wrist watches, pocket calculators, mobile phones, computers, etc, not to mention our world of communication.

The FET is a semiconductor device which depends upon an electric field to control current flow. There are two common types of FET, the *Junction Field Effect Transistor,* abbreviated JFET, and the *Metal Oxide Semiconductor Field Effect Transistor,* abbreviated MOSFET (also known as an IGFET, *Insulated Gate Field Effect Transistor)*. Both of these FETs can be made using p-type or n-type semiconductor material, though the majority use n-type.

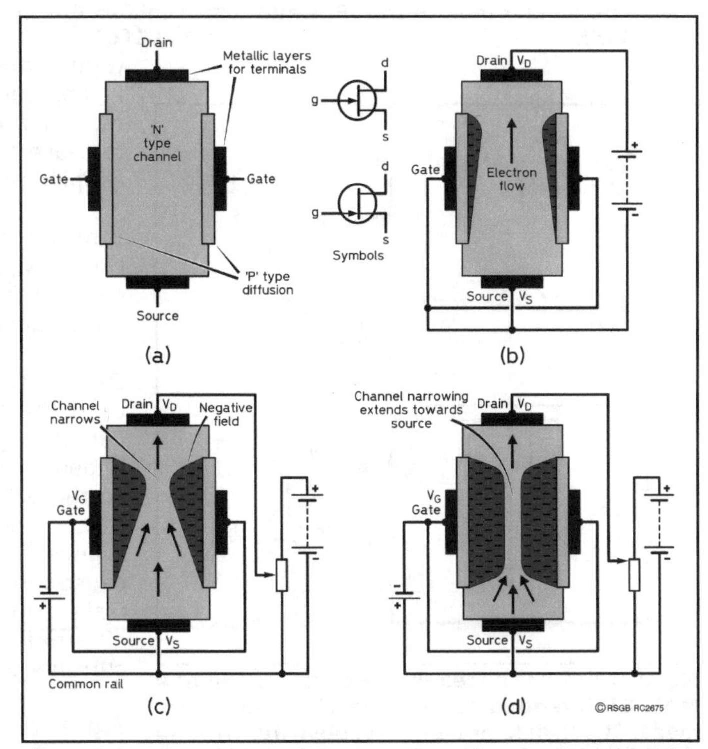


**Fig 1: (a) General idea of the construction of a FET. (b) With the Gate tied to the Source and a potential applied across Drain and Source, current (electrons) flows. Negative charge moves into the channel from the reverse-biased p-n junction. (c) Negative potential on the Gate causes constriction of the channel, being greatest at the Drain end. (d) Increasing Drain-to-Source potential causes the constriction to move toward the Source, although the channel width remains constant. Therefore the channel resistance increases linearly.**

## HOW IT WORKS

The JFET, the most straightforward to understand, is constructed from a tiny 'bar' of n-type semiconductor with a connection at each end. On opposite sides of the bar, p-type semiconductor is diffused into the n-type, these also having connections. One end of the bar is called the *Source* and the other the *Drain*. The p-type sections are called the *Gate*. The section between the source and the drain is known as a *channel*, along which the current (electrons) will flow, the rate of flow

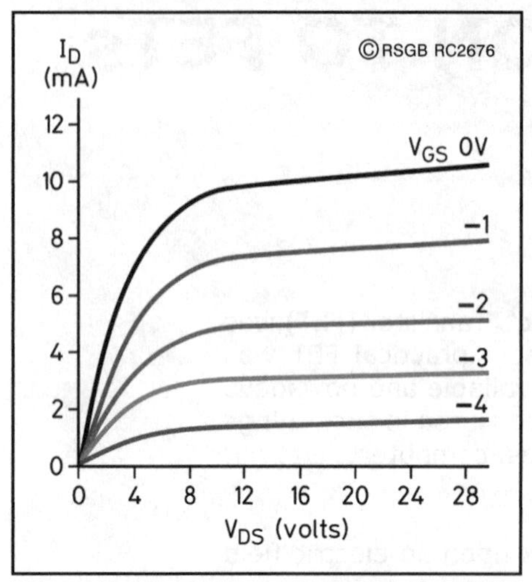


**Fig 2: Typical characteristic of a FET, which are similar in both the JFET and the MOSFET.**

being governed by the Gate (see **Fig 1(a)**). It is possible for current to flow either way, but since the JFET is constructed to ensure the capacitance between the Gate and the Drain is least, the Drain should always be positive with respect to the source. This particular JFET would be called an n-channel FET.

If both Gate terminals are connected together and to the Source, and a positive voltage is applied across the Drain and Source ($V_{DS}$), current ($I_{DS}$) will flow in the channel.

The Gate potential ($V_{GS}$) will be 0V, as shown in **Fig 1(b)**. Should $V_{GS}$ be made negative, the channel current will be reduced. With a negative $V_{GS}$ between Gate and Source, a negative electric field surrounds the p-type diffusion and encroaches into the channel, reducing its width, as **Fig 1(c)** shows. By increasing $V_{GS}$, the channel width may be reduced to the point where $I_{DS}$ approaches zero (**Fig 1(d)**), a further reduction of channel width causing $I_{DS}$ to cease altogether. This point is known as 'pinch-off' ($V_p$).

It may not be immediately obvious that the p-type diffusion into the n-type channel forms a semiconductor diode, a p-n junction.

Due to the polarity of the potentials used, the p-n junction is reverse-biased. For this reason, the input resistance / impedance is very high, $10^8\Omega$ or greater, so the Gate draws virtually no current (except for leakage). Normally, the Gate is never forward-biased, since this would cause high current to flow through the Gate and into the channel. For a p-channel JFET, all the voltages and currents would be reversed.

**Fig 2** shows that at a fixed Gate bias voltage, when $V_{DS}$ is increased beyond a certain point, the curve levels off and remains almost constant. This indicates that the output resistance / impedance is high.

**Fig 3: (a) General idea of the construction of a MOSFET intended to work in the 'depletion' mode. It consists of a p-type substrate with an n-type channel diffused into it, an insulating layer laid on the top with a metal surface for the Gate connection, and the n-type connections for the Source and Drain protruding through. (b) Showing potentials connected across Drain and Source, and a negative bias potential on the Gate. The field surrounding the Gate is shown, and the arrows indicate positive current carriers (holes) attracted up into the n-type channel, causing 'depletion'.**

### MOSFETS & IGFETS

The characteristics of these are very similar to the JFET, the outstanding difference being the input resistance / impedance which is considerably greater, with figures equal or greater than $10^{12}\Omega$. This is due to the 'insulated gate' of this device, which ensures the input leakage current approaches zero. The construction of these FETs is quite different.

For an n-channel IGFET, the device is built-up on a p-type substrate into which an n-type channel is diffused. As with the JFET, terminals are connected to each end of the channel. On top of the channel, a thin layer of silicon dioxide, $SiO_2$ (basically glass), is formed, with an additional thin layer of metal to provide a Gate connection: **Fig 3(a)**. This truly insulates the Gate from the channel, so much so that the input impedance/ resistance is now $10^{12}\Omega$ or even more. This FET has four terminals, one of which is connected to the P-type substrate, plus the other three; Source, Gate, and Drain. In operation, the substrate is connected to the Source or a point of lower potential (in some FETs the substrate is connected internally to the Source).

In a similar manner to the JFET, a positive voltage is applied across the Drain and Source. If the Gate is tied to the Source, ie $V_{GS}$ = 0V, a current, $I_{DS}$, will flow. However, if the Gate is taken negative with respect to the Source, though the field produced in the channel is negative (as before), the effect is to draw positive current carriers (holes) up into the channel from the p-type substrate, reducing the net current flow in the channel. This FET, like the JFET, is operating in the 'depletion' mode, ie the channel current is depleted. See **Fig 3(b)**.

Another form of IGFET is constructed, still using a p-type substrate, but instead of diffusing an n-type channel, two 'wells' of n-type material are diffused into the p-type substrate (see **Fig 4(a)**). As before, a silicon-dioxide insulating layer with a thin layer of metal for the Gate connection is laid down on the surface of the p-type substrate between the two n-type wells. Connections are made to the two n-type wells, one being the Source and the other the Drain. Connecting the Gate and Source together and applying a positive voltage across Drain and Source does not cause current to flow, despite $V_{GS}$ being 0V! If, however, the Gate is taken positive with respect to the Source, current will start to flow between Source and Drain. The positive voltage on the Gate creates a positive electric field in the p-type substrate, repelling the positive holes and creating a channel, allowing a current (electrons)

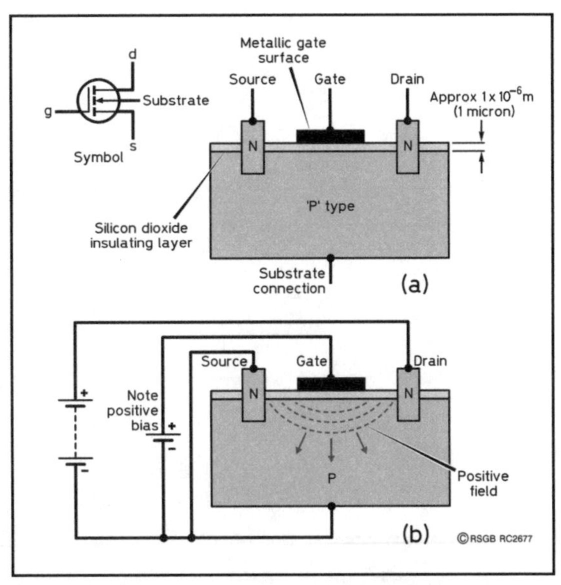


**Fig 4: (a) Again the general idea of the construction of a MOSFET, but this time intended to work in the 'enhancement' mode. Two n-type 'wells' are diffused into the p-type substrate. As before, an insulating layer is provided with a metal Gate connection. (b) With the positive bias potential, the field surrounding the Gate is a positive one, which tends to drive away positive current carriers (holes) down into the substrate (arrows), thereby creating a channel for the negative current carriers (electrons) to form a channel between the two n-type wells. In other words, the conductivity is 'enhanced'.**

to flow from Source to Drain. The substrate and Source are connected as before. This FET operates in the 'enhancement' mode (see **Fig 4(b)**).

**CHARACTERISTICS**

As far as characteristic curves are concerned, the shape of the JFET and MOSFET curves are very similar, it is the input impedance / resistance that distinguishes one type from the other. One important precaution that should be observed when dealing with MOSFETs is to remember that, with the Gate impedance / resistance being so high, a static charge can destroy the FET in a microsecond, so it is wise to take precautions as per the manufacturers' instructions.

There is a great deal more that can be learned about FETs. Also, there are many other types, such as the GaAsFET (Gallium Arsenide), used at microwave frequencies.

# AN INTRODUCTION TO FILTERING

by Jeff Black, G0UKA

It is a good idea to think of filtering the output of your transmitter, even if you feel that your signal is 'clean' anyway. In fact, the majority of modern rigs have low-pass (HF rigs) or band-pass (VHF/UHF rigs) filters built into them; but don't rely 100% on these, because not all rigs have them and those that do are not necessarily very good or efficient.

## BASIC TYPES

There are four basic types of filter.

### Low-pass

It is formed by an *inductor* in series with the signal path, and a *capacitor* on either side of it going to ground (or to the earth braid of the co-axial cable), **Fig 1(a)**. As the name implies, this type of filter will *attenuate* (the technical term for 'reduce') all frequencies *above* a certain point, **Fig 1(b)**, depending on the values of the L and C used.

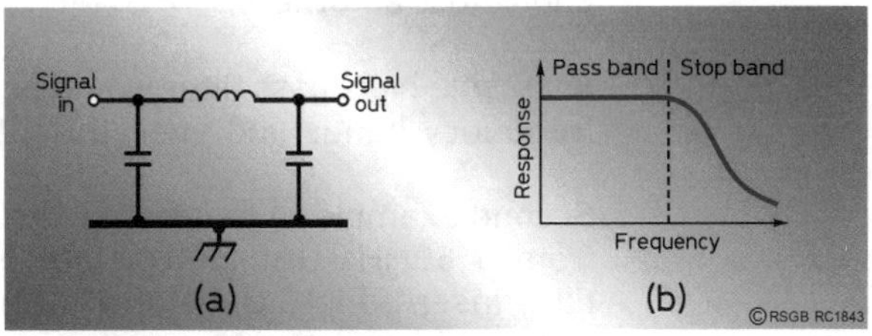

**Fig 1: Basic low-pass filter, typically employed on an HF transmitter.**

### High-pass

It is formed by a *capacitor* in series with the signal path, and an *inductor* on either side of it going to ground (or to the earth braid of the co-axial cable), **Fig 2(a)**. As the name implies, this type of filter will *attenuate* all frequencies *below* a certain point, **Fig 2(b)**, depending on the values of the L and C used.

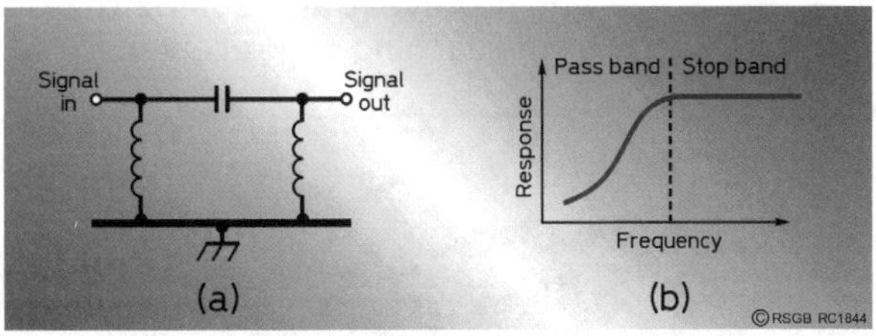

**Fig 2: Basic high-pass filter, typically employed on a TV set.**

### Band-pass

It is formed by an *acceptor* (series-tuned) circuit in series with the signal path, with two *rejectors* (parallel tuned circuits) on either side of it, going to ground (**Fig 3(a)**). This filter will only allow a narrow band of frequencies to pass through **(Fig 3(b))**, while shunting to ground all the others.

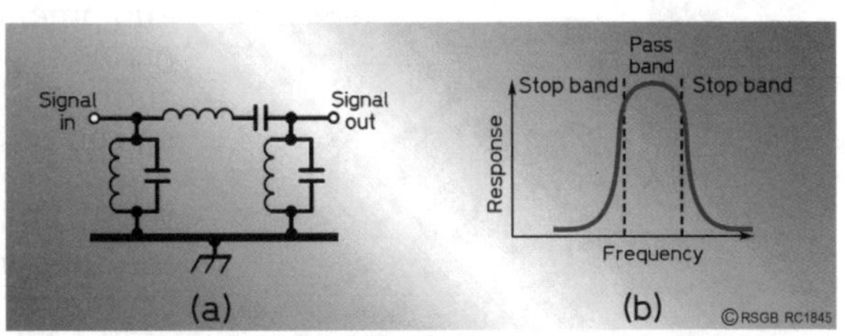


**Fig 3: Basic band-pass filter, often used to filter the output of a VHF transmitter.**

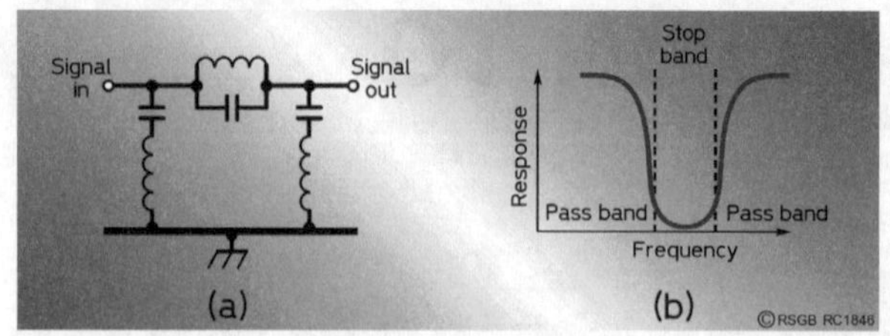


**Fig 4: Basic band-stop filter, often used on a broadcast receiver to 'kill' interference from transmissions on one amateur band.**

### Band-stop (notch)

It is formed by a *rejector* (parallel-tuned) circuit in series with the signal path, with two *acceptor* (series-tuned) circuits on either side of it, going to ground **(Fig 4(a))**. This filter will *attenuate* a narrow band of frequencies **(Fig 4(b))**, while allowing all other frequencies to pass through it.

## USES

For the first example, let us say that you are using HF and causing problems on the television set next door. The first thing to do is try a *Low Pass* filter on the output of yourtransmitter.

If this makes no difference to the breakthrough, try a band-pass or high-pass filter in the aerial lead of the TV set. A suitable band-pass filter would be tuned to accept only the television broadcast frequency bands and reject all others.

A suitable high-pass filter would be tuned to accept the television broadcast frequency bands and everything above them.

Second example: A band-stop filter might be used to prevent breakthrough from a 50MHz transmitter into a VHF broadcast receiver. A band-stop filter inserted into the transmitter's aerial lead, tuned to 100MHz, would prevent any second harmonic (2 x 50MHz = 100MHz, which is right in the middle of the VHF broadcast band!) from being radiated.

A band-stop filter inserted into the receiver's aerial lead, tuned to 50MHz, would prevent any 50MHz that is picked up on the VHF broadcast aerial from reaching the receiver.

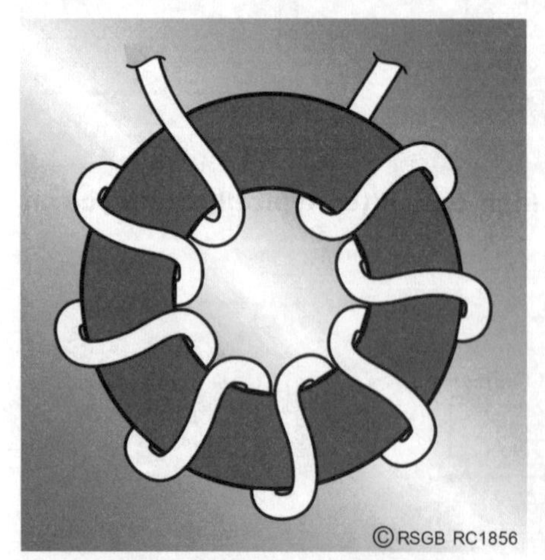


**Fig 5: A toroidal choke filter, which can be employed on a mains lead, speaker lead, or coaxial cable.**

## OTHER TYPES

There are several other 'special' types of filter.

### Toroidal choke

Shown in **Fig 5**, this is one of the most common filters used by radio amateurs. When used in the radio station installation, a toroidal choke is often placed on the mains leads of transmitting equipment, to prevent RF from getting into the mains supply. When used on domestic equipment suffering breakthrough, a toroidal choke can be placed on a mains lead, or TV / radio aerial lead.

When RF interference is picked up on cables, all the conductors (ie the wires) are affected equally. On mains cable, the interfering signal will appear on the live and neutral; on coax it will appear on both the centre conductor and the braid. The easiest way to understand how a toroidal choke works is to think in terms of it '*blocking*' the interference, thus preventing it from proceeding.

Imagine that the signal to a television receiver is flowing in one direction down the inner of the cable, but flowing in the opposite direction on the braid. These wanted signals are said to be in *antiphase* (because they are going in opposite directions). As I said previously, interfering signals affect *all* the conductors equally.

Unwanted signals therefore flow in the same direction, so are *in phase*. The effect of a toroidal choke - usually wound on a ferrite ring - is to cancel the in-phase signals, while leaving the antiphase signals unaffected. On a mains cable it works the same way, because whichever way current is flowing in the live wire, it will always be flowing in the opposite direction in the neutral wire (ie *antiphase*), whereas the interfering signal travels the same way on both. This argument also holds true for loudspeaker leads.

**Braid-breaker**

The most common type of braid-breaker uses a modified high-pass filter. The circuit, shown in **Fig 6**, is similar to the high pass filter in Fig 2, but with an extra capacitor in the braid. Note the presence of the resistor, R. It has a high value and is there to prevent static electricity building up on the aerial and giving someone a nasty shock.

Any piece of wire up in the air will start to build up static (it happens in dry weather, thunderstorms, etc), so if you actually break the braid of an aerial there is no way of grounding the build-up. The high value resistor supplies a discharge path, without significantly affecting the efficiency of the filter.

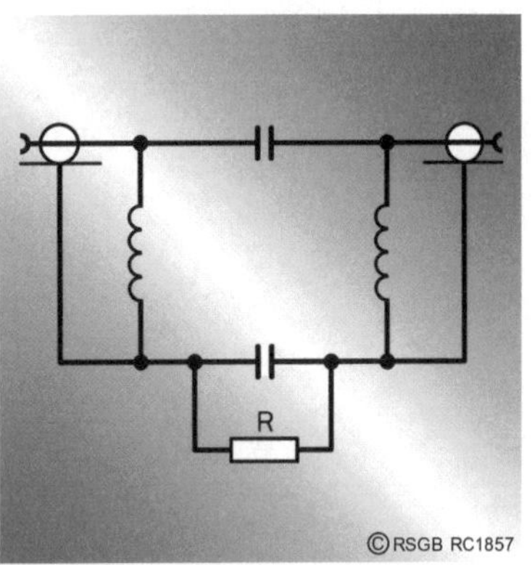

**Fig 6: Braid-breaker filter, very useful on a TV set!**

**Quartz crystal**

Beyond the scope of 'An Introduction to. . .', quartz crystals are often used – usually in combination – to form very narrow band-pass or notch filters at RF.

**GENERAL COMMENTS**

Your transmitter is probably expecting to see a characteristic impedance of 50Ω from its aerial system. If you insert a filter in line with the aerial, the filter becomes part of the aerial system. Therefore, for the frequencies that the filter is allowing to pass through, it *must* also have a characteristic impedance of 50Ω, to maintain a good aerial 'match' and prevent a high SWR.

All filters have some level of *insertion loss* - the signal loss you get from passing a signal through anything. A well-made transmitting filter will have an impedance of 50Ω and would typically show an insertion loss of 1dB or less. A braid-breaker filter for a TV set – as shown in Fig 6 – might well have a higher loss, but this is unlikely to affect reception (except in fringe areas).

Remember that TVs use 75Ω (not 50Ω) impedance for their aerials. In theory, you cannot use the same filters on both your transmitter and a TV. In reality though, impedance matching for receive is far less critical than it is for transmit, so you may well be able to get away with a bit of mismatching.

**FINALE**

Don't forget – always try to remove any problems on your own equipment before attacking someone else's!

# AN INTRODUCTION TO GAMMA-MATCHING

**by Peter Buchan, G3INR**

Many long- and medium-wave broadcast stations use vertical, ground-mounted aerials, the base of the aerial being well-bonded to a large earth mat extending outward for many metres. Energy is 'shunt-fed' to the vertical portion of the aerial through a transmission line, one conductor of which is connected to the base of the aerial and the other tapped up the aerial until a good match to the transmission line is obtained (**Fig 1(a)**). This method of feeding energy to the aerial is satisfactory, but an improvement is seen if the transmission line is dropped vertically from the tap at a constant distance from the aerial (**Fig 1(b)**). The distance at which the vertical portion of the transmission line is held from the aerial is dependent on the diameters of the aerial and of the transmission line conductor. In practice, a rigid tubular conductor would be used to form the vertical portion of the transmission line and held in place with stand off insulators (**Fig 1(c)**).

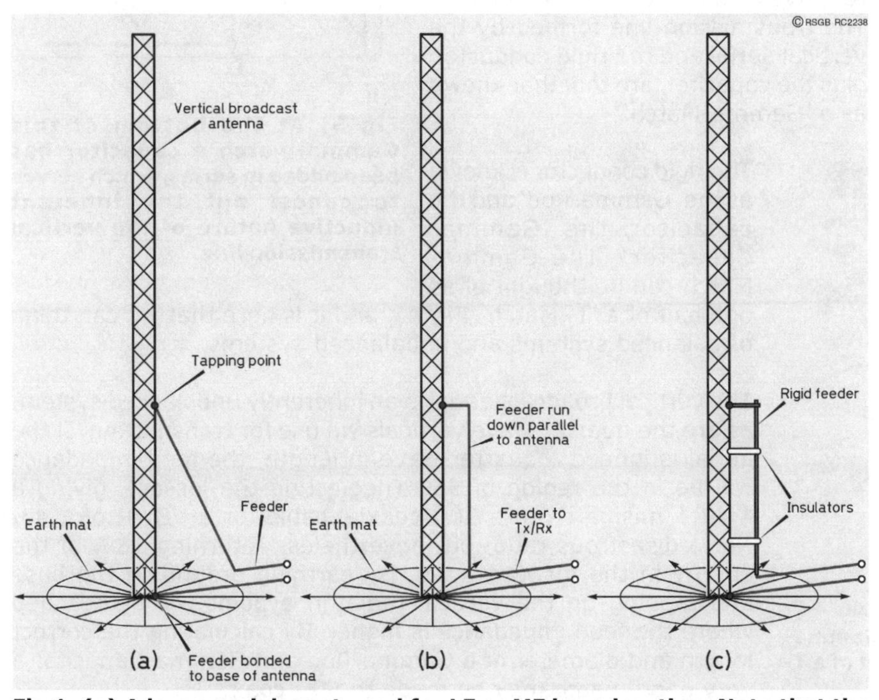


**Fig 1: (a) A large aerial mast used for LF or MF broadcasting. Note that the transmission line has been taken away from the tapping at an angle. This method mirrors the balanced 'Delta' type of match. (b) The feeder has now been taken vertically down and parallel to the mast. (c) A rigid length of transmission line has been installed and held parallel to the mast by stand-off insulators.**

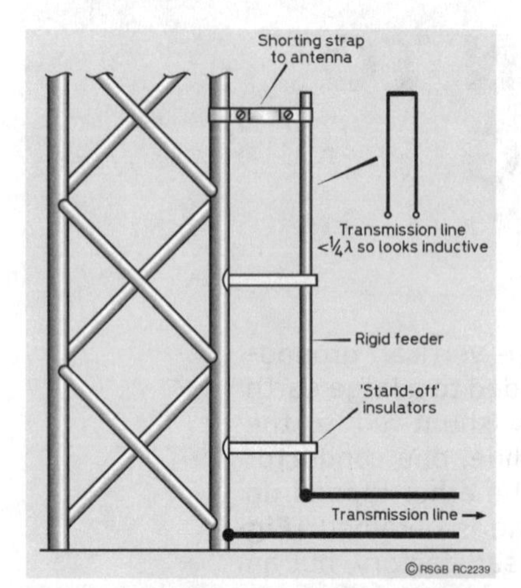


Fig 2: A section of mast, including the Gamma-matching section in more detail. The shorting strap is shown at the top of the rigid vertical transmission line.

With the rigid conductor well-shorted to the aerial at the tapping point, and moving vertically downward, this looks like a section of transmission line itself (**Fig 2**). Since this will be less than a quarter-wave in length, the open end at the bottom will look inductive [1]. Provided the rigid vertical conductor has been chosen with due consideration to the diameter of the vertical aerial, a satisfactory match is produced. However, a further refinement is made by inserting a series capacitor between the end of the transmission line and the bottom of the rigid conductor (**Fig 3**). This capacitor serves to 'tune out' the inductance along the rigid conductor and the vertical aerial, making the system resonant and hence offering a resistive load to the transmission line. The transmission line formed by the vertical aerial and the rigid conductor, plus the capacitor, are together known as a 'Gamma-Match'.

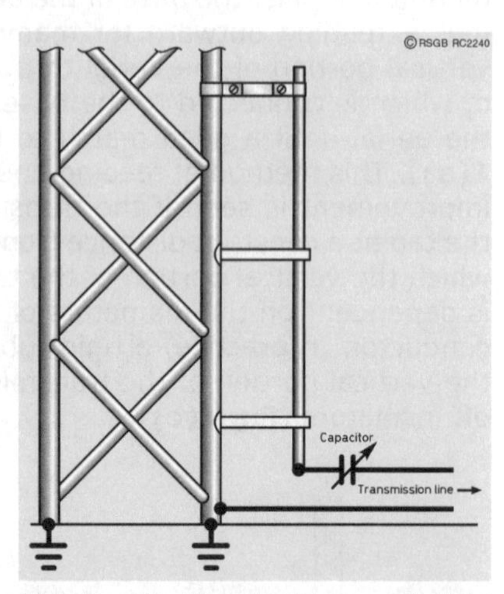


Fig 3: At the bottom of this Gamma-match a capacitor has been added in series, which serves to cancel out the inherent inductive nature of the vertical transmission line.

The rigid conductor is known as the 'Gamma-Rod' and the capacitor, the 'Gamma-Capacitor'. The Gamma-Match can be thought of as one half of a 'T'-Match (**Fig 4**), and it is here that we can think of balanced systems and unbalanced systems.

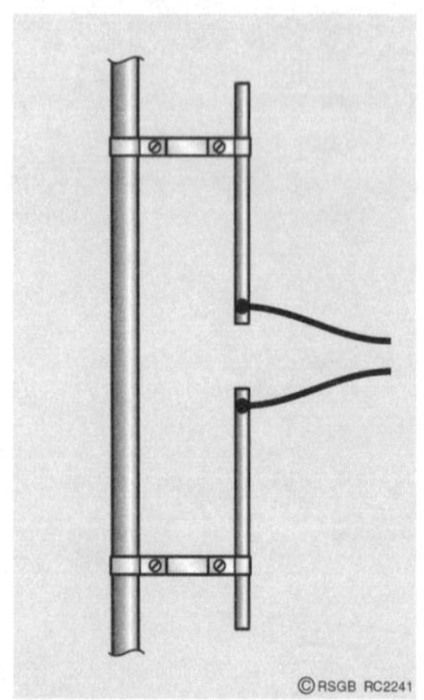


Fig 4: How the Gamma-match is one half of a T-match.

The vertical broadcast aerial is an inherently unbalanced system, as are the quarter-wave verticals we use for transmitting. If the aerial is indeed a quarter-wave in length, the feed impedance will be in the region of 35Ω (neglecting the losses), giving a 1.43:1 mismatch to a 50Ω coaxial cable, or a VSWR of 1.43. Not a disastrous ratio, but nevertheless returning 18% of the energy in the forward wave. By earthing/grounding the base and tapping up the vertical radiator, a point will be reached where the feed impedance is higher. By calculating the correct length and diameter of a Gamma-Rod and Gamma-Capacitor a very good match may be made to 50Ω coax.

## BEAMS

Considering the T-Match and balanced systems, we can turn to rotary beams, especially those made of tubing or 'plumber's-delight' construction (**Fig 5**). A rotary dipole will have a feed impedance between 15 and 75Ω

[2], depending on the height above ground in terms of wavelength. When extra elements are added, the feed impedance of the driven element or dipole falls [3].

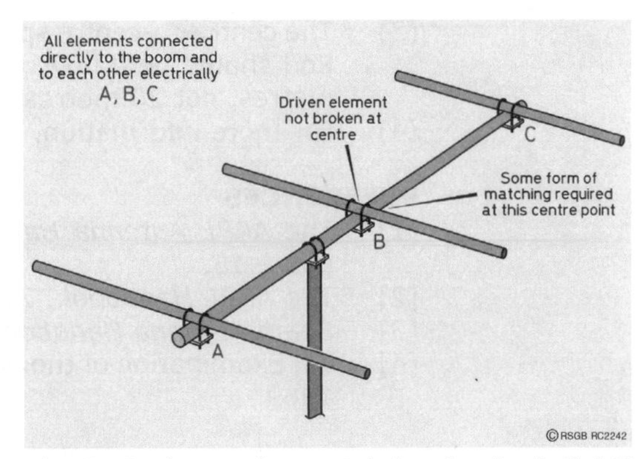


**Fig 5: A three-element 'plumber's-delight' rotary beam.**

Some constructors of beams fold the driven element, taking note of the diameter of the upper and lower conductors. Hence the feed impedance of the driven element - on its own - rises. On adding the other elements, the feed impedance will fall again. Provided the correct diameter conductors have been chosen, a good match may be obtained to a popular transmission line.

However, for a plumbers-delight-constructed beam, where the driven element is bolted directly to the metal boom, it is much more convenient to consider a T-match or Gamma-match. It should be noted that although the Gamma-Match provides a satisfactory match of an aerial to a transmission line, it does not eliminate aerial currents forming on the transmission line.

## T-MATCHING

A T-match (**Fig 6**) is used for a balanced transmission line. Nowadays, coax is the preferred choice for a transmission line, therefore the Gamma-Match is the one to use (**Fig 7**). As already pointed out, the driven element is bolted directly to the metal boom. To all intents and purposes, the feed-point is grounded/ earthed/ shorted to the boom. Hence this can be our starting point. Constructing a Gamma-Rod of the correct dimensions and choosing a suitable capacitor, the coax braiding is secured to the centre point of the driven element and the inner connected to end of the Gamma-Rod. The aerial may now be matched to the transmission line.

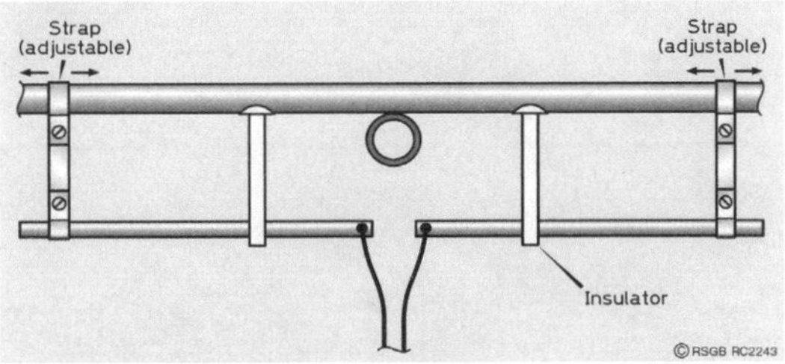

**Fig 6: Showing how a 'plumber's-delight' beam may make use of a T-match.**

## RULES OF THUMB

The dimensions for a Gamma-Match system may be obtained in the following way (for the 14MHz band):

(1) Assume the driven element is resonant. The length of the Gamma-Rod should be 0.04 to 0.05L (where L is the wavelength in use).

(2) The diameter of the Gamma-Rod should be 0.33 to 0.5 that of the driven element.

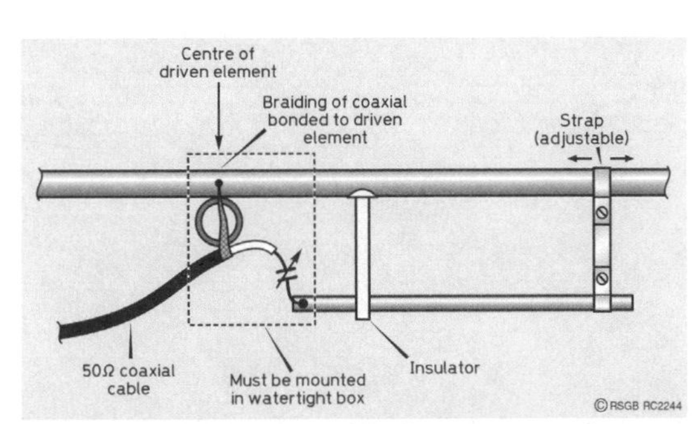


**Fig 7: The Gamma-match connected to an unbroken driven element . The shorting strap may be adjusted during matching.**

(3) The centre-to-centre spacing of the driven element and the Gamma-Rod should be $0.007\lambda$. (Note, for example, that 14.15MHz is 21.2 metres, not 20 metres!)

(4) For more information, see [4].

**REFERENCES**

[1] The *ARRL Antenna Handbook*, 14th edition, 3rd printing (1984), pp3 - 15.

[2] The *ARRL Handbook*, 75th edition (1997), pp20 - 22.

[3] *Beam Antenna Handbook*, William Orr, W6SAI, 1st edition, p51.

[4] 'An Examination of the Gamma-Match', D Healey, W3PG, *QST*, April 1969.

# AN INTRODUCTION TO INDUCTANCE

**by Steve Ortmayer, G4RAW**

Mathematical calculations are something many radio amateurs shy away from, but they really don't have to be that difficult.

Let's take a look at the fomula used to calculate the inductance of the air-cored coil shown in **Fig 1**, and use it to determine the upper and lower limits of resonance when coupled to a variable capacitor.

In our example, we will use an inductor which is 1.5in in diameter, and consists of 170 turns of 26SWG (0.018in diameter) wire, close spaced.

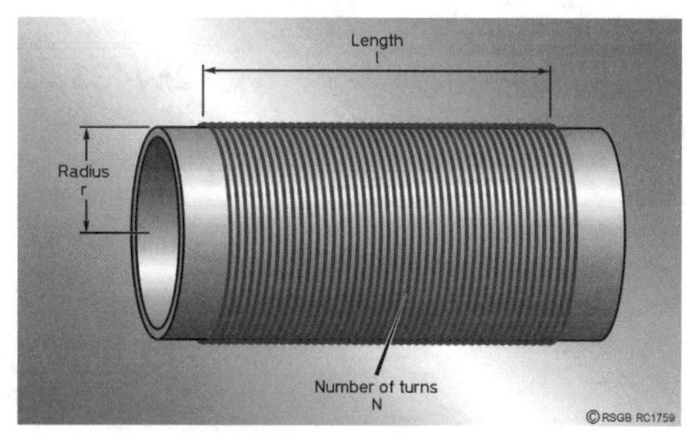


**Fig 1: The dimensions of the coil.**

## CALCULATING INDUCTANCE

The formula for calculating inductance is:

$$L(\mu H) = \frac{N^2 r^2}{9r + 10l}$$

where: $r$ = radius of coil (in inches);
$l$ = length of coil (in inches);
$N$ = number of turns.

First of all, the length of the coil = 170 x 0.018 = 3.06in

Therefore:

$$L = \frac{170^2 \times 0.75^2}{9 \times 0.75 + 10 \times 3.06} = \frac{16256}{37.35} = 435\mu H.$$

## CALCULATING RESONANCE

First, we need to know the maximum and minimum values of the variable capacitor, and something which is not universally understood is that variable capacitors have residual capacitance when the vanes are fully unmeshed. In our example, let's take a capacitor which has a maximum value of 360pF. When it is set to minimum, the capacitance is 20pF.

The formula for resonance is:

$$f(\text{Hz}) = \frac{1}{2\pi\sqrt{LC}}.$$

At maximum capacitance (360pF), the resonant frequency is:

$$f = \frac{1}{2\pi\sqrt{435 \times 10^{-6} \times 360 \times 10^{-12}}} = 402183.5\text{Hz},$$

or about 402kHz.

At minimum capacitance (20pF), the resonant frequency is:

$$f = \frac{1}{2\pi\sqrt{435 \times 10^{-6} \times 20 \times 10^{-12}}} = 1706320\text{Hz},$$

or about 1.7MHz.

This tuned circuit would be suitable for tuning the entire medium-wave broadcast band, which extends from about 540kHz to 1.6MHz.

# AN INTRODUCTION TO INSTABILITY

**by Jeff Black, G0UKA**

If your transmitter or receiver uses a VFO for generating the frequency on which you wish to transmit or listen, it should be constructed from good quality components and be of 'sound mechanical construction' (a phrase often used in the examinations). In simple terms, if you use cheap and nasty components, their values may alter as a result of heat (particularly capacitors) or vibration (especially inductors).

## DRIFT

Drift is the word used to describe a signal that gradually changes frequency. If drift occurs during transmission, it is a real pain in the neck for the receiving station, because it has to alter frequency continually to hear the transmitting station. On CW, the Morse 'tone' will alter and, if you are using a very narrow filter to pick one station out of a whole load of them on a busy band, you could easily lose the station entirely. On SSB, the result of drift would be that a station which at first sounds fine when tuned in, will gradually become 'Donald Duckish' as it goes off frequency.

At best, drift is inconvenient and a thorough nuisance, using unnecessary amounts of bandwidth and upsetting other operators whose frequencies may be drifted onto or across. At worst, drift may cover several kilohertz and a transmission could wind up outside the amateur band (punishable by a visit from the Radio Interference Service and a fine of several hundred pounds!).

Drift is usually caused by heating effects on the components in the VFO. **Fig 1** shows an example of this. It can also be caused by inadequate stabilisation in the part of the power supply running the VFO. In other words, a stable, unchanging, frequency is dependent not only on good components in the oscillator itself, but also on the supply voltage. If this voltage alters, so will the oscillator's frequency.

Long term drift, over the course of years, is often caused by the ageing of components in a VFO.

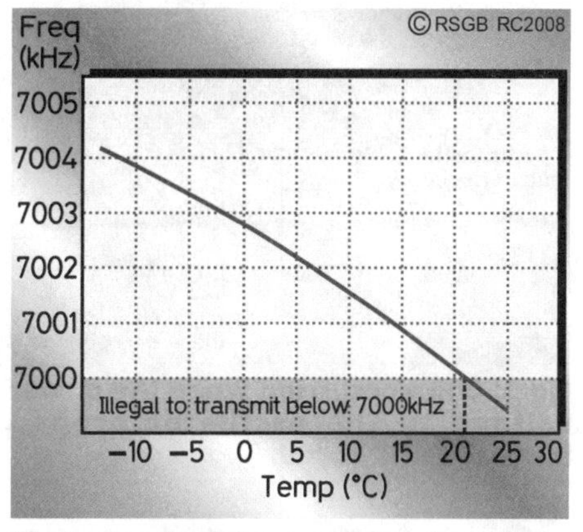


**Fig 1: Graph of the possible output from a VFO which does not possess temperature compensation, showing that it would be illegal to transmit if the temperature rose above 21°C.**

## MICROPHONY

Components which are not mechanically sound can be affected by vibration transferred to them through physical contact or through the air. This is most prevalent in inductors or connecting wires inside oscillators.

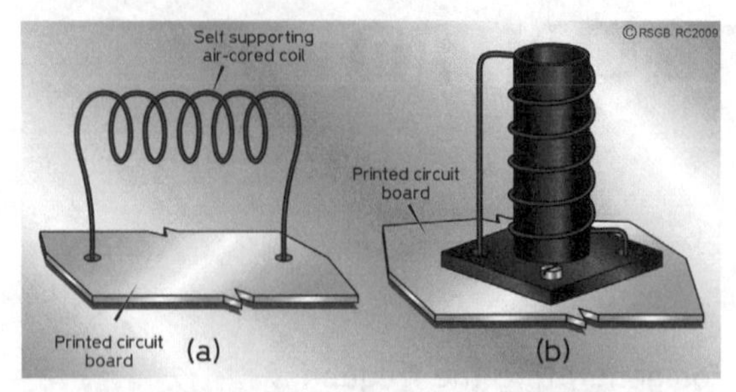


**Fig 2: (a) A self-supporting, air-cored coil will vibrate readily, which in turn will affect the output frequency of an oscillator. (b) A coil made from thick wire, wound tightly on a former which is fixed down, will hardly vibrate at all.**

To prevent this, good mechanical construction is called for when building an oscillator. Inductors should be made from wire which is *tightly* wound on a former, and then usually covered in a sealant. The former must then be fixed down, to prevent it from moving.

These points are illustrated in **Fig 2**. Any signal-carrying wires in an oscillator should be thick, short and straight. These are usually good enough reasons to build your VFO on a PCB, which will largely (or completely) eliminate signal-carrying wires.

## CHIRP

This occurs in A1A (CW) mode. It describes the very rapid change in frequency (and as a result the received tone) at the moment of 'key down'. There are three main causes of this effect.

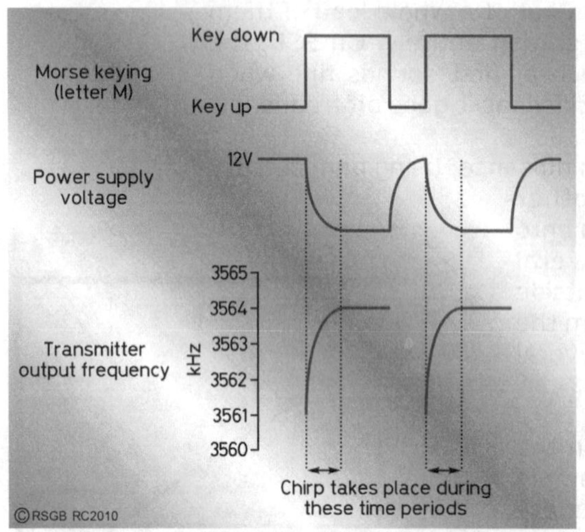


**Fig 3: Chirp occurring when the power supply voltage to the oscillator drops as the transmitter is keyed.**

1. At the instant you press the Morse key, the transmitter is switched on. If your power supply cannot deliver adequate current at a stable voltage, the sudden power surge will cause the voltage to dip, resulting in a frequency change in the VFO (the same as for drift, but it happens faster). This is illustrated in **Fig 3**.

2. Keying the output of the oscillator itself, instead of having a buffer to protect the oscillator from unnecessary loading. An oscillator is normally a very sensitive bit of circuitry, and placing the Morse key (which is, after all, only a very simple type of switch) immediately after it will cause a varying load to be placed on it. If the oscillator circuit is followed by a buffer - often just a single transistor - it is the buffer which will carry the varying load and not the sensitive oscillator circuit.

3. RF Feedback. This occurs when the transmitted frequency from the PA gets picked up by the oscillator and causes a frequency change. The PA may be generating hundreds of watts output, while the oscillator can only produce milliwatts. Therefore, unless there is good screening round the oscillator and care is taken in filtering and screening its power leads, other wires, etc, this sort of unwanted pick-up can occur.

## SPURIOUS OSCILLATION

Amplifier stages have a rather nasty habit of oscillating, if enough of their output manages to be picked up by their own input. This is called '*feedback*' and if it gets bad enough, can destroy a very expensive bit of equipment. The frequency of this oscillation is not predictable and could be as powerful as the frequency you actually want to transmit! As usual, good screening and decoupling (using capacitors to 'ground' the unwanted signals) is very important.

If both the input and output of the amplifier are tuned to the same frequency, this sort of 'self-oscillation' can occur at or near the frequency you actually want. Sometimes it will show itself as an output signal even when you are not actually transmitting; sometimes it will occur only when you are transmitting and results in rather 'burbly' speech (like someone trying to talk under water), or as a very rough CW note when using Morse.

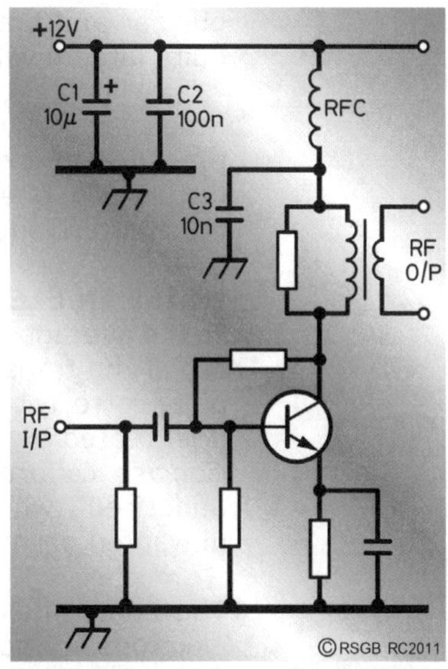


**Fig 4: The driver circuit of a multi-band transmitter. The combination of the RFC and C1, C2 and C3 helps to decouple the amplifier at all the relevant frequencies, but if not properly designed can form tuned circuits and oscillate at other frequencies.**

Oscillations can happen at quite low frequencies (say only a couple of kilohertz) and could therefore affect your transmitter's audio stages, producing audio modulation and sidebands where you don't want them.

The odd thing is that your attempts to decouple circuits to avoid parasitic oscillations on one frequency could actually produce parasitic oscillations on other frequencies.

The RFC (a type of inductor) and the multiple decoupling capacitors shown in **Fig 4** are intended to prevent the amplifier from becoming unstable and going into parasitic oscillation. However, the RFC and decoupling capacitors can produce their own tuned circuit and oscillate!

Even the wires linking one bit of a circuit with another could cause some problems, in which case low-*Q* inductors and/or ferrite beads can also help cure or prevent parasitic oscillations.

# AN INTRODUCTION TO NOISE

**by George Brown, M5ACN / G1VCY**

Noise is a part of all our lives; we live with it largely without recognising its effects. How many times do we say "Pardon?" because we haven't heard something above a noise in the background? It is the cleverness of our ear and brain combination that allows us to hear most things over quite large levels of noise.

No electronic system can yet equal man's capability to see or hear in the presence of extraneous inputs, so it comes as no surprise that the disturbance of electronic systems by electronic noise is an area requiring much attention in their design.

## NOISE IN ELECTRONICS

To the average radio amateur, the word 'noise' simply means what comes out of the transceiver speaker when no station is being received. This is quite correct, but the way in which noise arises in the first place and how it is treated by amplifiers takes us to the root cause of the problem. The *Concise Oxford Dictionary* defines noise as "any sound... especially undesired", while science dictionaries focus on electronic noise as "any unwanted or interfering current or voltage".

The commonest source of electronic noise is the random motion of electrons in a conductor. This motion is caused by the temperature of the conductor, and increases when the temperature increases. So, any current flowing in the conductor carries fluctuations caused by this thermally-induced motion. The fluctuations are obviously very small, but are sufficient to produce noise voltages across conductors. This type of noise is called *Johnson Noise* or *Thermal Noise*.

The random motion can be visualised simplistically as a vibration about a mean position; the vibration is random in terms of direction, amplitude and frequency, and changes in temperature cause changes in these parameters. The RMS noise voltage, $V$, generated in a conductor is given by the surprisingly-simple formula:

$$V = \sqrt{4kTR\Delta f}\ , \qquad (1)$$

where:

- $k$ is Boltzmann's constant; it has the value $1.38 \times 10^{-23}$ J $K^{-1}$ (read as Joules per Kelvin);
- $T$ is the absolute temperature;
- $R$ is the conductor's resistance in ohms;
- $\Delta f$ is the bandwidth (Hz) of the system which is measuring the Johnson Noise.

We have already seen that the noise will increase with temperature and resistance, but just as important is $\Delta f$, the system bandwidth. Suppose we have a noiseless amplifier (this is impossible, but is a useful hypothesis!) of bandwidth 1kHz, and at its input we have a resistor generating Johnson Noise. We can measure the noise voltage at the output of the amplifier to be *V*out. If we now vary the amplifier bandwidth *only* - until it is now 4kHz - we will find that the output noise voltage has been *doubled* to 2*V*out. This is because the bandwidth is under the square-root sign. This shows that a system should not have a bandwidth greater than is needed by the signal, because the noise content will be increased and the signal will not.

## EQUIVALENT NOISE TEMPERATURE

This is a useful concept, describing the amount of noise contributed by an amplifier, for example. Consider the circuit of **Fig 1**. Here, a resistor is connected across the input of an amplifier. The resistor produces Johnson Noise, which is fed into the amplifier. At the output of the amplifier is more noise, but we have no way of knowing how much of the noise has come from the resistor and how much from the amplifier itself. However, from equation (1) we can calculate the temperature to which the resistor should be heated to produce the same amount of noise as is coming from the output of the amplifier. This temperature is called the *equivalent noise temperature*.

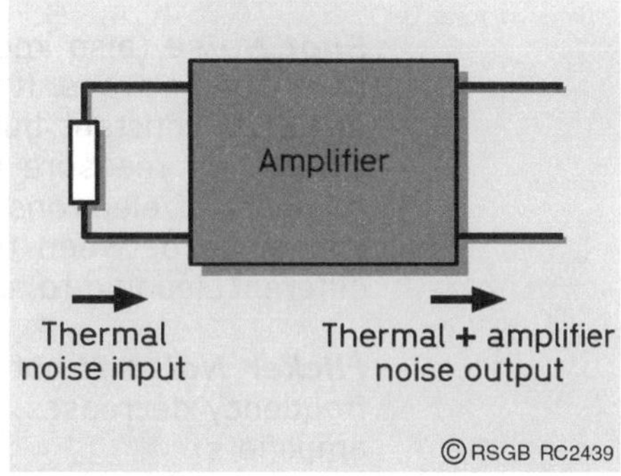


**Fig 1: Illustrating the concept of thermal noise.**

## NOISE FACTOR

All electronic devices use charge carriers (such as electrons) in order to work. All will generate noise because of this, but not necessarily in the same way or in the same amounts. An amplifier with no input will produce noise at its output; each component, each wire in the circuit will contribute something towards it. **Fig 2** shows noise as displayed on an oscilloscope, and explains the name given to this type of display – 'grass'. Its basic characteristics do not change even when the timebase speed is changed. Johnson Noise is also called 'white noise' because it contains all frequencies in equal amounts. Any wanted signal having an amplitude smaller than the 'grass' will be lost.

All electronic processes generate noise, and the *noise factor* is used to quantify the amount of noise added to a signal by a system.

$$\text{NoiseFactor} = \frac{\text{Input S/N ratio}}{\text{Output S/N ratio}}.$$

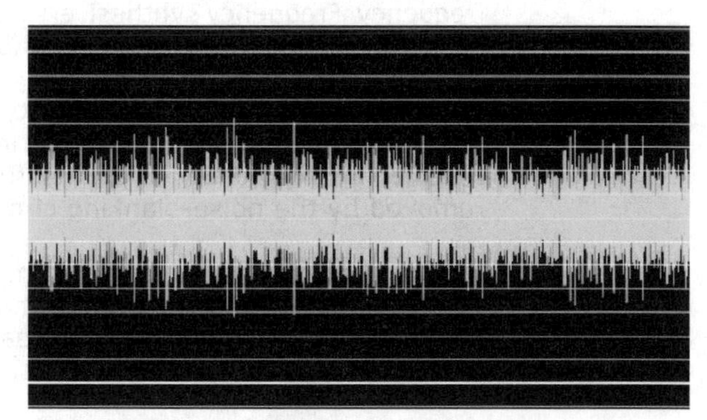

**Fig 2: Noise, as displayed on an oscilloscope.**

### SIGNAL-TO-NOISE RATIO

This ratio, (S/N), simply measures the ratio by which the signal is greater (or less) than the noise at a particular point in a system. Because all systems *add* noise to the incoming signal, the output S/N is always *less than* the input S/N

### OTHER NOISES

***Johnson Noise*** is due to the random motion of electrons due to their thermal energies. Other types of noise are generated because of the discrete nature of the electron (ie assuming that the electron is a very small particle). These noise types were noticed originally in thermionic valves, where the current in the valve was a beam of individual electrons being fired at the anode and sometimes intercepted by other electrodes.

***Shot Noise*** (also known as Schottky Noise) is caused by the discrete nature of electrons. It is normally assumed that the current in a transistor or FET is constant but, because the current is a flow of electrons, the current we measure is only the statistical time-average of the *actual* numbers of electrons flowing per second. For very small currents, the difference between the average and the actual values can be quite different, leading to significant current perturbations.

***Flicker Noise*** is not fully understood, except that it increases as the frequency decreases, and thus is important principally in audio and DC amplifiers.

***Partition Noise*** is generated in valves where the beam of electrons from the cathode is divided between, for example, the anode and the screen grid. The discrete nature of the beam again results in the division of current being a statistical process, producing different noise currents in the cathode, screen grid and anode leads.

***Induced Noise*** occurs mainly in valves operating in the VHF / UHF / microwave region. The existence of an electron beam in the *vicinity* of another electrode can induce noise into the circuit of that electrode. In microwave valves, this is the dominant process by which beam noise is coupled into the output circuit.

***Phase Noise*** emanates from oscillator circuits as sidebands on the carrier frequency. Frequency synthesisers are particularly prone to this, and cause design problems in transmitters and receivers.

***Ignition Noise***, unlike the other types of noise discussed so far, comes under the commoner heading of interference, and is suffered by any receiver in the vicinity of an insufficiently-suppressed car engine. It is removed by the noise-blanking circuits [1] in most receivers.

***Atmospheric Noise*** is a persistent problem at frequencies below about 10MHz, and is some 40dB greater than thermal noise at these frequencies. Above 10MHz, its magnitude decreases at about 20dB for each doubling of frequency (20dB per *octave*) [2].

***Galactic Noise*** from outer space is about 20dB above thermal noise at frequencies around 20MHz. Above this frequency it diminishes at 20dB per *decade* [2]. It is the means whereby radio astronomy is possible.

***Transmitter Noise*** is present in all transmitted signals. Any phase noise in the carrier is mixed with the phase noise of local oscillators and mixers to produce a complex set of coherent sidebands together with any wideband noise which may be present. Similar problems exist in the receiver.

## FINALE

In a concise introduction such as this, the concept of noise can be only superficially addressed. Those readers wishing to delve further into this arcane world are recommended further reading [3].

## REFERENCES

[1] P Hawker, *Amateur Radio Techniques*, RSGB 1991, p111.
[2] *Reference Data for Radio Engineers*, H W Sams & Co, 1977, p 29-2.
[3] *The VHF / UHF DX Book*, DIR Publishing, 1992, pp 4-2 – 4-14.

# AN INTRODUCTION TO *Q*

**by Peter Martinez, G3PLX**

The components list for a transmitter or receiver circuit, or almost anything that has tuned circuits in it, will often specify that the inductors or capacitors used should have 'a high *Q*', 'a *Q* greater than 50', or something similar. Component catalogues will often quote the *Q* factor of these components, but the higher-*Q* ones are often large and expensive. What is *Q* all about?

**QUALITY**

*Q* stands for quality, and the *Q*-factor of a tuned circuit is a measure of its quality. A perfect inductor would exhibit only inductance - the property whereby the AC current in a circuit lags the voltage by precisely 90° - and the current in a perfect capacitor would lead by precisely 90°. If we combine both in a tuned circuit, we get a complete 180° cancellation at one frequency. The *Q*-factor tells us something about how good this cancellation is.

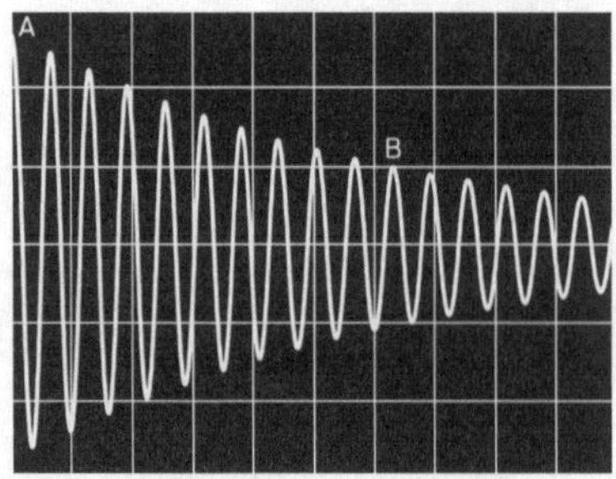


**Fig 1: Decaying oscillation in a tuned circuit, excited by an oscilloscope timebase pulse. This shows an initial amplitude at point A of 2.7V, decaying after 10 cycles to 1V at point B, so the circuit has a *Q* of 10.**

A tuned circuit made from a perfect inductor and a perfect capacitor would, if we started it off by momentarily connecting a battery across it, oscillate for ever at its resonant frequency. Practical tuned circuits don't oscillate for very long, the amplitude decaying quite quickly. The lower the *Q*, the quicker it decays. The *Q*-factor is the amplitude of oscillation divided by the amount of decay in one cycle. For example, if one peak of the oscillation measured 10V and the next peak was 9V, the *Q*-factor would be 10.

This isn't easy to measure, but the mathematics of decaying oscillations shows that if a signal decays by a factor of $1/Q$ in one cycle, it decays by a factor of 2.718 after *Q* cycles. This does give a way to measure the *Q*-factor of a tuned circuit on the bench if we connect an oscilloscope across it and pulse it, for example with the timebase of the oscilloscope itself. Set the oscilloscope controls so that the oscillation is 2.7cm in amplitude at the left-hand edge of the screen, as in **Fig 1**, and count the number of cycles from there to the point where the oscillation is 1cm in amplitude. The result is the magnitude of the *Q*-factor.

**SIGNIFICANCE**

There are two reasons why we need to keep an eye on the *Q*-factor of components in radio circuits.

The first is the main reason we have tuned circuits at all: to pass one frequency and suppress those either side of it. The higher the *Q* of a tuned circuit, the better it will reject signals either side of its resonant frequency, that is, the narrower will be its pass band. Mathematics shows that the *Q*-factor is equal to the resonant frequency divided by the width of the pass-band, measured at the points either side where the signal has dropped to 71% of the centre value (also known as the 3dB bandwidth). This is illustrated in **Fig 2**.

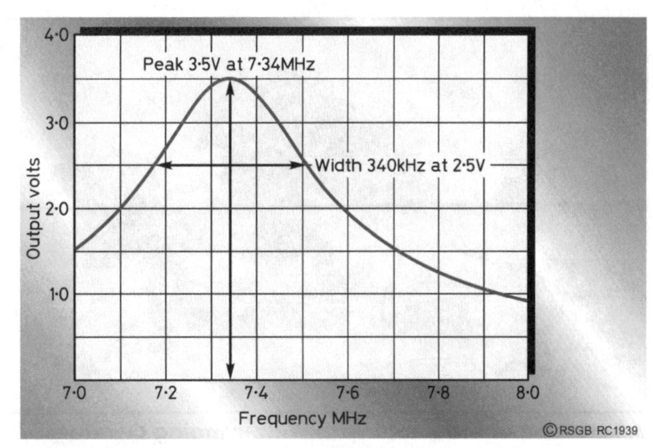


**Fig 2: Output voltage of tuned circuit over the range 7 – 8MHz, showing a bandwidth of about 340kHz centred on 7.34MHz, indicating a *Q* of 21.6.**

The second reason that we want to watch our Qs is because we are also watching our Ps, by which I mean powers. Our perfect tuned circuit continued to oscillate for ever because there is no power lost in it.

The oscillation in a real tuned circuit decays because the stored energy in the tuned circuit 'leaks away', either as heat lost in the resistance of the wire and the leakage in the capacitor, or indeed because we have our own resistive load across it. Ideally, we want to maximise the power in the load and minimise the lost power. If the tuned circuits in our low-noise pre-amps had too low a *Q*-factor, we would not be able to hear weak signals, and if the tuned circuits in our power amplifiers or antenna tuning units had too low a *Q*, the power dissipated in them could cause them to overheat.

## MEASURING Q

Since perfect inductors and capacitors don't dissipate any heat and resistors do, another way to find the *Q*-factor of a tuned circuit is to measure its 'equivalent resistance'. This is the resistance which we would have to add to a perfect circuit to make it as bad as the one we have. We can do this for a series tuned circuit, for example, by connecting a variable resistor in series with it (see **Fig 3**) and placing the combined circuit across a signal generator. With an oscilloscope or diode probe, we adjust the variable resistor until the voltage across it is the same as that across the tuned circuit at resonance. The resistor must now have the same resistance as the tuned circuit and we can measure it separately on a DC test meter.

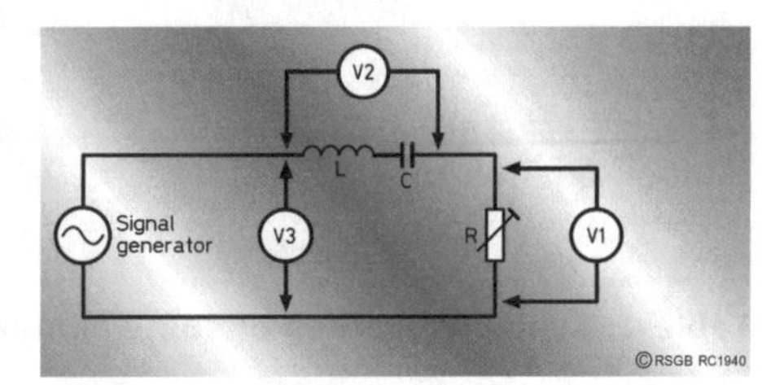


**Fig 3: Measuring *Q* by loss resistance. Adjust R until V1 = V2 or, more easily, until V1 is half of V3, then measure R and calculate $Q = 2\pi fL/R$.**

Another way to measure *Q* if we have a signal generator that we can couple loosely to the tuned circuit. We can sweep the frequency through resonance and measure the frequencies at which the voltage across the

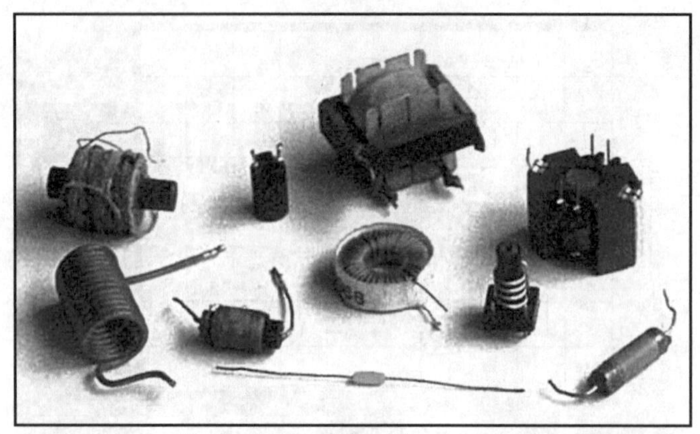

**A selection of inductors with wide-ranging *Q*-values.**

tuned circuit is a maximum and 71% of maximum either side. We can also turn this relationship round the other way to calculate the bandwidth, if we know the *Q*-factor and the frequency.

Yet more mathematics shows that the *Q*-factor of the series tuned circuit is equal to the reactance of the inductor at resonance (or the reactance of the capacitor at resonance, since they are equal) divided by the resistance. We can either use this method to measure the *Q*-factor of a given tuned circuit, or by turning the equation round we can calculate how much resistance will be introduced into our signal path as a result of adding the tuned circuit.

We have now answered the question in the first paragraph. A perfect tuned circuit has a bandwidth of zero, no resistance, and infinite *Q*. Practical tuned circuits possess resistance, which lowers the *Q*. We have seen three ways to measure *Q* and seen how to use it to calculate bandwidth and loss. Although we normally only talk about the *Q* of tuned circuits, we can actually measure the *Q*-factor of almost any kind of resonant system. A few surprises are included in **Table 1**.

| Item | Frequency | *Q*-factor |
|---|---|---|
| Small wire-ended 47μH fixed inductor | 1MHz | 50 |
| Toko coil can with iron dust core | 1MHz | 80 |
| Ferrite bead with one turn | 1MHz | 5 |
| Ferrite bead with one turn | 50MHz | 0.1 |
| Big air-wound PA coil, 14SWG wire | 14MHz | 100 |
| Low-K ceramic capacitors | 10MHz | 60 |
| High-K ceramic capacitors | 10MHz | 20 |
| Polyester capacitors | 100kHz | 150 |
| Silver mica capacitors | 10MHz | 200 |
| Quartz crystal | 10MHz | 40,000 |
| Clock pendulum | 1Hz | 100 |
| Tenor saxophone | 440Hz | 30 |

**Table 1: Typical *Q*-values of some common items.**

# AN INTRODUCTION TO RADIATION RESISTANCE

**by Peter Buchan, G3INR**

Radiation resistance is a most interesting and very important phenomenon associated with aerials, but it is one that is not always fully understood. Assume, for example, that we are testing a transmitter and the indications are that it is supplying 100W to a resonant λ/2 aerial at a height of λ/2, fed at the centre with 72Ω flat twin cable. Appropriate steps have been taken to ensure that the balanced feeder and the unbalanced nature of the transmitter output have been catered for, also the VSWR is very low. If the aerial were replaced with a non-reactive 72Ω resistor sufficiently large to dissipate 100W, provided the frequency to which the transmitter was tuned was left unchanged, the change would be undetectable at the transmitter end of the feeder.

This is the way that the Radiation Resistance of an aerial is described, perhaps elaborated a little, but essentially the same. For example: "*The total amount of energy radiated from a transmitting aerial can be measured in terms of a Radiation Resistance which is the resistance that, when replacing the aerial at the end of the feeder, will consume the same amount of power that is actually radiated*" is typical of the many text books dealing with aerials. While conveniently describing the phenomenon, the impression given might lead you to assume that the Radiation Resistance is in fact a resistor.

**ALTERNATIVE VIEW**

Looking at an aerial in a somewhat different way from that which we are accustomed, it might described as '*a region of transition between a wave*

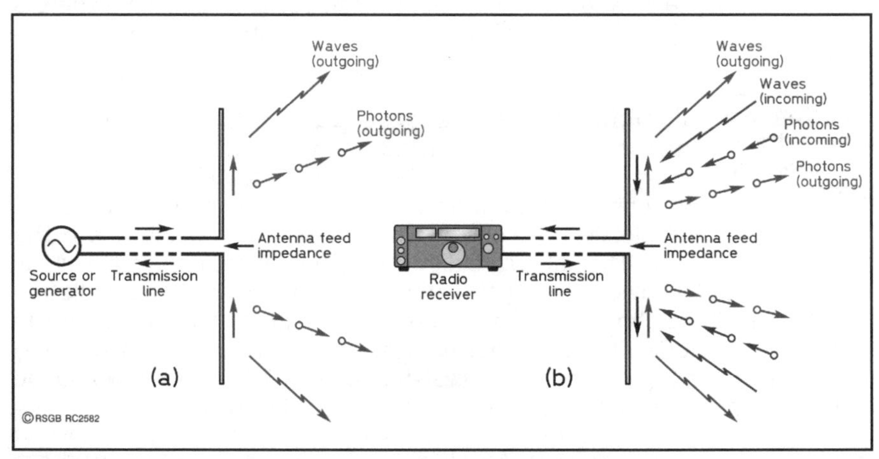


**Fig 1: Energy (a) leaving, and (b) leaving and being received on a dipole aerial. A photon is the quantum unit of 'electromagnetic energy'.**

*guided by a transmission line and a free-space wave'*, or perhaps *'an aerial interfaces between electrons on conductors and photons in space'* (**Fig 1**). Both these quotations make you realise that an aerial is something rather special and that the point of departure of our electromagnetic energy cannot be just a resistor, for if it were, the energy would be dissipated at the site of the aerial and there would be none left to journey onwards.

The complex nature of an aerial when radiating energy, and indeed when receiving it, has been investigated mathematically. It is to the mathematician that we owe the calculation of the radiation resistance of many kinds of aerial, though the dipole is the fundamental building block. For example, the radiation resistance of 'short dipoles' has been carried out and produced the following results:

| Length | Radiation Resistance |
|---|---|
| λ/10 | 7.9Ω |
| λ/100 | 0.08Ω |

The radiation resistance of a short dipole is therefore small.

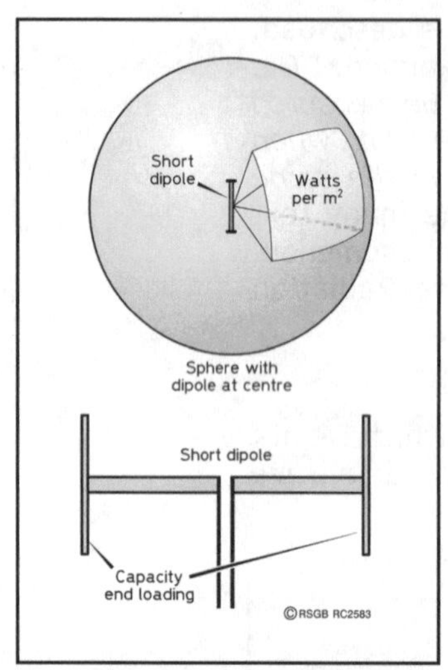


**Fig 2: Radiation resistance is calculated by placing a dipole at the centre of a sphere that is large with respect to one wavelength.**

## CALCULATIONS

To calculate radiation resistance, the dipole is considered to be at the centre of a sphere (**Fig 2**) that is large with respect to wavelength, and here the 'Poynting Vector' of the 'far field' (ie many wavelengths from the dipole) is used to obtain the total power radiated. Assuming no losses, this power is equal to the power fed to the dipole. From Ohm's Law it can be calculated that the power, *P*, must be equal to the square of the RMS current flowing in the dipole, times a resistance, *R*, called the Radiation Resistance. This total power is the rate at which energy is streaming out of the sphere surrounding the dipole. For these calculations the length of the dipole is considered to be *very much less* than the wavelength used in the calculation. All very theoretical, perhaps, but nevertheless shown to be sufficiently near the truth when investigated in a practical situation.

Virtually the same procedure (although different mathematically) is carried out for a half-wave dipole, but this time the Radiation Resistance is found to be 72Ω. This is considered to be at the current maximum point on the dipole, ie the centre, which is also the point where the feeder is connected. The half-wave dipole may be utilised on *odd harmonic frequencies*, the lowest being the third harmonic where the radiation resistance will be about 90Ω, on the fifth about 120Ω - all of these at the centre of the aerial. For even harmonics, the centre point of the aerial will have a high (or very high) resistance, the 72Ω feeder would be unusable and open wire line would have to be utilised.

This, of course, describes the multi-band aerial. The popular G5RV is non-resonant, except perhaps nearly so on 20m, (about three half-waves), but on 80m the centre impedance will be something like 30Ω resistance and -500Ω reactance (ie 30 - j500Ω).

**PRACTICALITIES**

When looking at a practical aerial, there will be a number of other points which must be taken into consideration, not the least of which is the environment in which the aerial is erected. The working value of the radiation resistance depends on:

- the relation to ground of the aerial,
- the ratio of conductor diameter to length,
- the proximity of other conducting objects such as masts, buildings, house wiring, telephone wires, etc.

For example, the feed point impedance of a practical half wave dipole is about 73 plus 42.5Ω of inductive reactance (end effect), ie 73 + j42.5Ω.

In addition to the radiated energy, energy is also lost in the resistance of the aerial wire, the resistance of the ground, along with dielectric losses in trees, insulators, and losses in other objects with an imperfect dielectric (**Fig 3**). The losses are brought together and included with the natural radiation resistance when describing the 'feed impedance', which is often used in practice. It is perhaps evident that the true radiation resistance of a particular aerial is not normally measurable, due to the factors mentioned above, hence the feed impedance is equal to the radiation resistance plus the loss resistance, ie

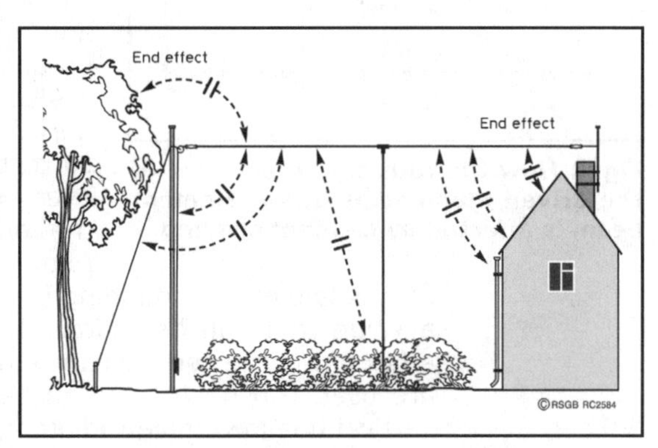


**Fig 3: How radiated energy is wasted due to dielectric losses.**

$$R_{aerial} = R_{radiation} + R_{losses},$$

where $R_{aerial}$ represents the feed impedance. The efficiency, $\eta$, of the system is given by:

$$\eta = \frac{R_{radiation}}{R_{radiation} + R_{losses}},$$

and is usually very high where dipoles and other centre-fed systems are concerned (in the region of 90%).

Another type of half-wave aerial is the folded dipole. This aerial has a somewhat different feed impedance characteristic. A two-wire folded dipole has a feed impedance of about 280Ω, though it is generally taken as 300Ω. The feed impedance of a folded dipole varies in a nonlinear fashion.

For a three-wire half-wave folded dipole, the feed impedance is 630Ω. In general, for a half-wave folded dipole with *n* wires, the feed impedance is $70n^2\Omega$.

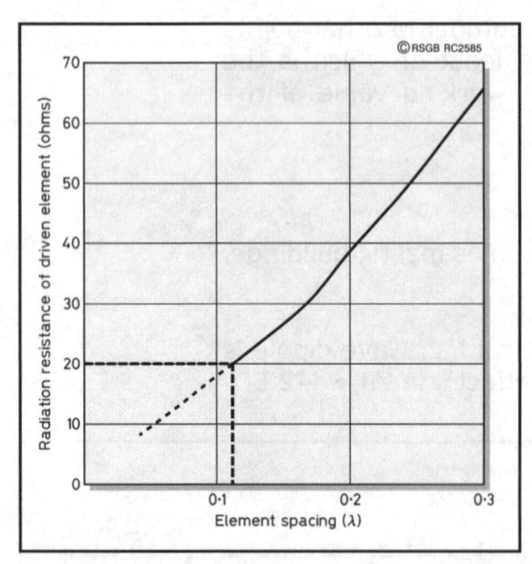


**Fig 4: How the radiation resistance of the driven element of a two-element beam is affected by element spacing.**

Beam aerials are a case where the proximity of reflecting and directing elements change the feed impedance substantially . For a two element beam with 0.3λ spacing, the dipole or driven element feed impedance falls from about 72Ω to about 65Ω, due to mutual coupling. As the spacing is further decreased to the optimum of 0.11λ, the feed impedance falls to 20Ω (**Fig 4**). This indicates that the presence of other aerials nearby any aerial in use will most certainly have an effect on its feed impedance.

Turning from the horizontal aerial to the vertical and, in particular, the λ/4 vertical fed against ground or radials; in theory, this has a radiation resistance of about 35Ω (half that of the horizontal λ/2 dipole). While this may well be correct, the actual feed impedance is rarely – if ever – near to 35Ω. The problem lies mainly with the conductivity of the ground beneath the aerial. Experiments have shown that a number of radials laid out surrounding the vertical will bring the feed impedance to a value that can be matched, and the aerial will accept energy. The efficiency of the vertical can be rather poor but, provided sufficient radials are used (up to 40), it can be brought up to about 70%. Whilst the vertical has low-angle radiation, good for long-distance communication, the nature of the vertical aggravates the mutual coupling, hence the presence of trees, foliage, etc, makes the losses greater, which of course affects the feed impedance.

**SUMMARY**

There is a considerable variety of different types of aerial and as many different radiation resistances. There is also another way that is sometimes used to describe Radiation Resistance. Some textbooks will explain that "it all depends at which point along the aerial the current is measured and where the feeder is to be connected".

Many aerials are end-fed and hence, like the vertical, have a return to the source via the ground. If the aerial happens to be λ/2 and is fed at the end, the feed impedance is going to be high. But, nevertheless, the current squared times the resistance at that point ($I^2R$) will equal the energy radiated... just the same as with the centre-fed aerial. The resistance measured is sometimes called the radiation resistance.

# AN INTRODUCTION TO RECTIFICATION

**by Alan Betts, G0HIQ**

The mains power delivered to the home is 230V alternating current. It is AC because that allows the generating and distribution to be performed at the most efficient voltage and then stepped-down for domestic use by means of a transformer. We also will need a transformer to step the 230V down to, say, 12V for a power supply.

This AC must then be rectified to provide DC for the radio. This is done using a diode, which is a device allowing current to pass through it in one direction but not the other. Often four diodes are used to form a 'bridge rectifier'. The photograph shows two diodes of different construction and a bridge rectifier as a single device.

## SIMPLE RECTIFIER CIRCUIT

A simple transformer, diode and smoothing capacitor circuit is shown in **Fig 1**. The load is represented by the resistor, R. In reality, the load might be a radio. If it requires 12V and draws 0.5A, this is equivalent to a 24Ω resistor. When the top lead of the transformer secondary goes positive (the positive half-cycle of the AC), the diode is 'forward biased' and allows current to flow. A half-cycle of current flows through the load resistor.

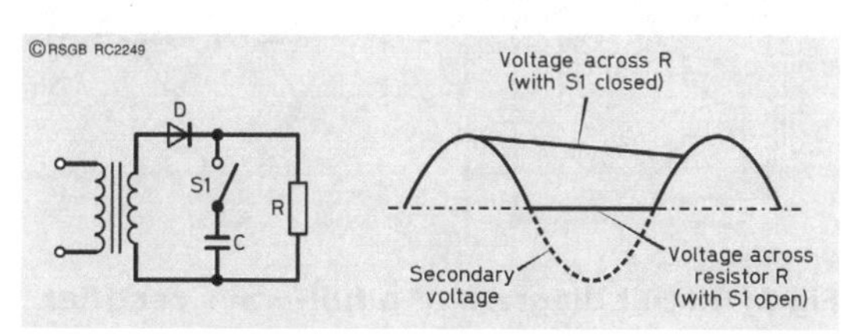


**Fig 1: Circuit diagram of a half-wave rectifier.**

When the top lead of the transformer goes negative (the negative half cycle), the diode is 'reverse biased' and no current flows. The supply to the radio consists of positive pulses of current and voltage. This is still not the nice, smooth, constant supply that is obtained from a battery, and the radio would not work properly. In fact it would hum very loudly and may well be damaged.

## ADD SOME SMOOTHING

If we now close switch, S1, the capacitor, C, is connected across the supply to the radio. Now, on the positive half-cycles, the capacitor will charge up almost to the peak voltage of the AC supply from the transformer's secondary.

After the first quarter-cycle, the voltage from the transformer will fall below the voltage on the capacitor. The voltage on the cathode end of the diode will then be more positive than the anode; the diode will be reverse-biased and will not conduct.

The capacitor will remain charged and will not discharge back through the diode. It will, however, discharge through the load resistor, so its voltage will slowly fall. On the next positive half-cycle, the capacitor will be charged back up to the peak voltage.

The voltage supplied to the radio (load resistor) is now reasonably constant. Our radio would work, but the 'ripple' in the supply voltage may still cause a hum loud enough to be objectionable, but it would not be as loud as before. The size of the ripple depends on how much the capacitor voltage falls during the short time between one positive half-cycle and the next. In fact, because the capacitor will only be topped up when the voltage from the transformer and diode is greater than the voltage on the capacitor, the time between charges is almost the time for a complete AC cycle. That is $^1/_{50}$ second, or 0.02s.

## RIPPLE VOLTAGE

There are three ways of reducing the ripple in the supply voltage to a reasonable level.

1. Top up the capacitor twice as often. This will approximately halve the voltage drop between successive cycles.
2. Use a larger value capacitor, so that the voltage drop is reduced.
3. Use electronic regulation. This was introduced in the March 1999 *RadCom*.

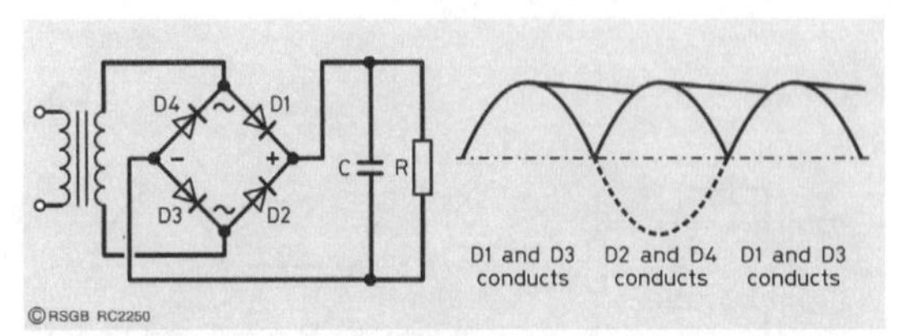


**Fig 2: Circuit diagram of a full-wave rectifier.**

## FULL-WAVE RECTIFICATION

Up to now, only the positive half-cycles have been used in supplying current to the load. Since only half of the waveform has been used, the circuit is known as a 'half-wave rectifier'. What we now want to do is to find a way of using the negative half-cycles as well - a full-wave rectifier. **Fig 2** shows such a circuit, which is also known as a 'bridge rectifier'. The four diodes can be made into a single component, as can be seen in the photograph.

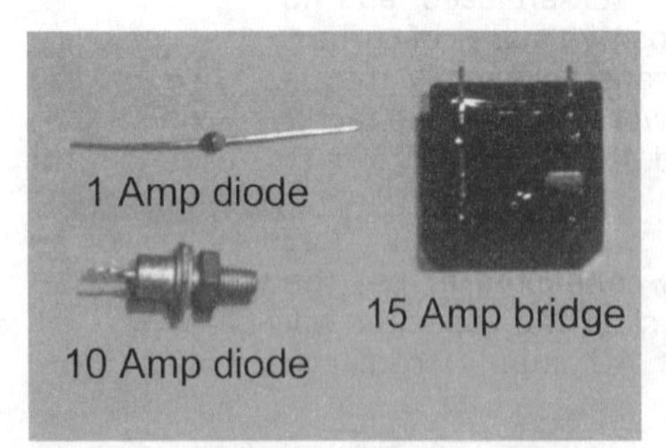


**Rectifier diodes and a bridge rectifier.**

Consider what happens when the top lead of the transformer secondary is positive. Current will flow out of this secondary lead, through D1, through the load and back to the bottom of the secondary winding via D4.

Now consider half a cycle later, when the bottom lead of the secondary winding is positive. Current will flow through D2, through the load and back to the transformer via D4. The important thing to notice is that the current flows through the load in the same direction each time. The negative half-cycles have been steered to become like positive half cycles.

The capacitor, C, performs exactly the same function as before, but is now topped up twice as often. The time between charges has been halved, and so has the ripple voltage.

## CHOOSING A SUITABLE CAPACITOR

If the mains power was switched off while the radio was turned on, the top-up charges into the capacitor would cease. It would continue to discharge until it was 'flat'. If the voltage continued to fall at the same rate then it would reduce to zero after a time set by the capacitance of the capacitor, C, and the resistance of the load, R. The formula is

$t = R \times C$,
where C is in farads, R in ohms and t in seconds.

For our radio, the resistance 24Ω. If the capacitor was 20,000μF, the time would be:
$t = (20,000 \times 10^{-6}) \times 24 = 0.48s$

For a half-wave rectifier working at 50Hz, 0.48s is the time for 24 tops-up; for the full wave circuit, there are 48 tops-up. So, with a half-wave rectifier circuit, the voltage will have fallen by about $^{1}/_{24}$ of its peak value before it is re-charged. For a 12V supply that is 0.5V. The drop of the full-wave circuit will be 0.25V.

## TOO BIG A CAPACITOR

The charge in the capacitor is topped up only when the voltage from the transformer exceeds the capacitor's voltage. If a very large capacitor is used, the capacitor voltage drop will be small and the time during which it is being topped up will only be a very small fraction of a cycle. For a 0.25V ripple in a 12V supply, this time is about 6.5% of the cycle. During this time, enough charge must be stored to run the radio for the entire cycle. Since the average discharge current is 0·5A, during the charges the current must be 7·7A. The diode(s) must be able to cope with these surges and a capacitor designed with a high 'ripple current' rating for smoothing circuits must be used.

An analogy may be useful to explain this effect. Regard the capacitor as a water butt, continuously watering the plants in a greenhouse. The water is draining out slowly at the water equivalent of 0·5A, say 0.5 litre/hour, and the level (voltage) is slowly falling. Every evening it rains. The rain lasts 1 hour 34 minutes, which, conveniently, is 6·5% of the day.

The question is, what is the rate of rainfall such that the water butt is just topped up? The answer is that it must be $^{1}/_{0\cdot065}$ x 0·5 litres/hour = 7·7 litres/hour.

This analogy can be taken further. A small water butt would see its level change markedly over the day. Indeed, if it were too small, it may even run out. A large butt would maintain a much more constant level.

## EFFECT OF A LARGE CHARGING CURRENT

Apart from the need for the diodes and capacitor to handle these large pulses of current, we must also consider the transformer. Part of the heating effect in the transformer is the current flowing in the resistance of its windings. This given by $I^2r$ (where *r* is the resistance of the copper winding).

This $I^2$ is a problem. During charging, the current is 7·7A, over 15 times greater than the average current drawn by the radio. The power dissipation will be about 237 times greater, but only for 6·5% of the time, giving an average increase of some 15 times. This is only a part the total heating effect in the transformer, but it is necessary to de-rate the transformer to prevent overheating. Typically a de-rating factor of 0.6 may be used so, for 500mA, the transformer must be rated at 833mA. This de-rating factor is for the full-wave rectifier; for a single diode half-wave rectifier the de-rating factor is 0.28 so, for 500mA DC, the transformer must actually be rated at 1.8A, a good reason for not using half-wave rectifiers!

# AN INTRODUCTION TO SCREENING

**by Peter Buchan, G3INR**

Screening (or shielding, as it is sometimes known) can be very much a practical exercise when it becomes necessary to restrict a field or fields, close to their source, or alternatively to prevent a field or fields from reaching a sensitive point in a circuit. However, unless the underlying principles are applied, the outcome may not be successful. Textbooks are vague on the subject, so perhaps the following few 'rules of thumb' may go some way toward a better understanding.

As far as this feature is concerned, there are two fields, magnetic and electric. They can and do exist independently but, the instant one of them changes, an electromagnetic field is produced which is able to reproduce itself and propagate into the surrounding space (as an electromagnetic wave), or possibly be constrained to travel along a transmission line of some description. For example, if a DC source is switched briefly into a transmission line of any length, the capacitance of the line causes a charging current to flow. This, along with the applied voltage, creates an electromagnetic wave which, willy-nilly, has to set off down the line.

Only changing or alternating magnetic or electric fields create electromagnetic waves. At low frequencies, eg at 50Hz (mains supply), the radiation is very small, as it is also at audio frequencies.

The field surrounding the secondary of a mains transformer of a modern solid-state transceiver would be almost entirely magnetic, whereas the field surrounding the supply terminals of a valve linear power supply (2kV @ <1A) would be almost entirely electric.

## MAGNETIC SCREENING

Now to the question of screening. There are basically two methods available:

(1) diverting the path of the field; and
(2) cancelling it out with an opposing field.

As an example of (1), consider **Fig 1**. Here, the magnetic flux of the coil finds an easy path through the high permeability material of the enclosure, leaving only a very small residual field outside. It should be remembered that the lines depicting the path of the magnetic flux do not mean that the flux is made up of lines... it is spread throughout

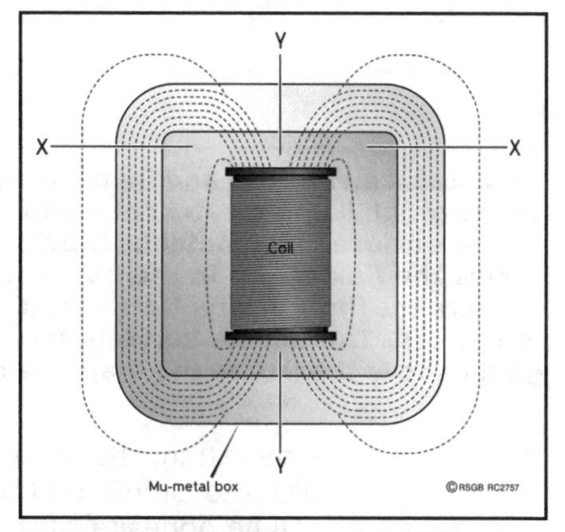


**Fig 1: The high-permeability box concentrates the flux within the walls. Remember not to join the box along the line X – X.**

the space, though beyond the enclosure only very very thinly.

For the enclosure to be effective yet reasonably light in construction, the material used must have a very high permeability. It is here that an alloy called *mu-metal* is employed, though whether this is readily available nowadays is uncertain. At one time, microphone transformers were all enclosed in mu-metal boxes, to prevent low and audio frequencies being picked up by the windings.

Note that this method *increases* the inductance of the coil. A further very important point is to realise that the join in the box should *not* be made along the line X-X; to do this would insert a high resistance (reluctance) path to the flux. The join should be made along a line Y-Y instead.

A mu-metal enclosure is effective from 0Hz to the higher audio frequencies but, as frequency increases, the permeability of the alloy diminishes and the resistance increases, so method (2) is employed.

For method (2), copper or aluminium is used. Each material has a low resistance to current flow, though the relative permeability is that of air, ie 1, hence it cannot offer a path for the magnetic flux. What it *does* do is act as a short-circuited secondary to the coil, and the current induced produces a field which tends to cancel out the field produced by the coil. This reduces the inductance of the coil, but providing the screen is made large enough the effect can be tolerated and/ or allowed for in the design of the coil and screen. The screen diameter should be twice the coil diameter, and the ends of the coil should not come within one diameter of the ends of the screen. **Fig 2(a)** shows a coil with the normal flux lines depicting a magnetic field. The letters a, b, and c indicate the falling-off of flux density as one moves away from the coil, the field being strongest at a and much lower at c, though assumed to be greater than wanted. We assume the current in the coil is alternating, therefore when the screen is placed around the coil, forming a closed circuit, currents will also be produced in it, though their direction will be opposite to those creating them (Lenz's Law).

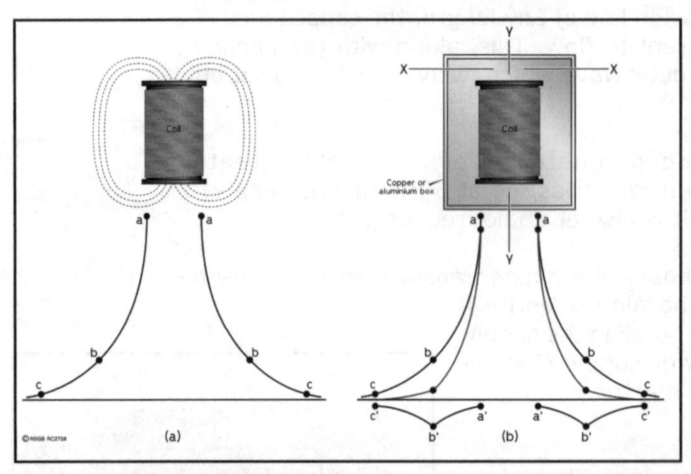


**Fig 2: (a) A simple coil and field. The flux density is shown by the graph being greatest at a and least at c. (b) Thecoil is now enclosed in a metal box. Below the line, the graph shows the flux set up in the box; the letters a', b' and c' indicate the flux density. Note that it is the reverse of that in the coil. The middle curve is the result of adding the graph above the line to the graph below the line.**

Provided the screen resistance is very low, the currents produced in it will tend to reduce the flux at b *almost* to zero, the small difference being that which is needed to create the reverse flux, b′, see **Fig 2(b)**. The letters a′, b′ and c′ below the axis of the graph show the level of the flux present due to the screen at these points - note that they are in the reverse direction to those causing them and very slightly less in amplitude at b′ and c′ and very much less at a′. Note also that the flux at a′ will only marginally affect the flux in the coil, therefore the inductance is not greatly reduced. It should be pointed out that the direction of the current in the screen is very different from the flux in Fig 1; it is in a continuous ring, parallel to the turns of the coil, hence the joint in this screen may be made in the direction X-X, but not in the direction Y-Y (which would prevent current flowing). As frequency is increased the depth of field penetration into the screen becomes less, so relatively thin copper or aluminium sheet may be used, but for best results the boxes or enclosures must be watertight (the ideal), requiring joints to be lapped and special care taken with lids, covers, etc. See [1] and [2].

Incidentally, method (2) is completely useless for static or DC fields, and is not successful at low frequencies, therefore use mu-metal for DC and low frequency and copper or aluminium for RF screening.

## ELECTRIC SCREENING

Perhaps you will have noticed that nothing has been said about 'earth' so far. This is because earthing has nothing to do with magnetic screening. However, it has everything to do with electric screening. Metallic screening is used for electric fields, but the permittivity is not considered in the same way as the permeability must be in magnetic screening (DC and LF fields) - it is the *conductivity* that is important.

To screen a DC or LF electric field all you need to do is to enclose the field in an *earthed* can, box or enclosure.

**Fig 3(a)** represents a high voltage point, such as may be found in a valve linear's power supply. Between this point and chassis/earth there will be an electric field, depicted by the dotted lines. This means that every point in space between V and chassis/earth is at some potential (and hence not zero). If this field is enclosed in a conducting screen connected to chassis/earth, then the screen is considered to be at the potential of the chassis/earth, all points on the screen are at chassis/earth potential, so there is no longer an electric field outside the screen. This is true for non-varying or LF fields, but if the potential at V is varying at a high frequency then there will be a capacitive current between V and the screen, which could cause a potential difference across the screen, so some loss is possible. In practice it is advisable to have the screen as far away as possible from V, to avoid too much capacitive loading of the screened high voltage point.

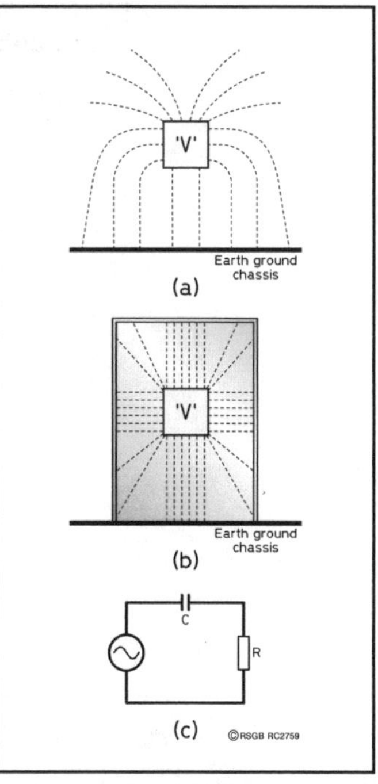


**Fig 3: (a) The point V is at high potential and surrounded by an electric flux or field. (b) The field or flux is contained inside the screen. Every part of the screen is at earth potential, consequently the field or flux is no longer present outside. (c) The AC generator represents an HF or VHF source of voltage. Due to the presence of the screen, some capacitive coupling is certain. Should the screen be of poor conductivity, a potential difference will develop across it, giving some loss.**

It is not necessary to have a continuous metal screen, it may be made up of individual conductors in the form of a cage.

Although important, the joints need not have the perfection of those in a magnetic screen, so lids etc may be just a push fit. The degree of screening is measured by taking the ratio of the field strength before and after the screen is fitted. Screen effectiveness below 20dB is poor; from 20 to 80dB, average; 80 to 120dB, above average; above 120dB, there are cost problems.

**REFERENCES**

[1] *Practical RF Handbook*, by Ian Hickman, Newnes, 1993. Good Appendix.

[2] *Circuit Designer's Companion*, by Tim Williams, Newnes, 1993. pp248 – 52.

# AN INTRODUCTION TO SOLAR INDICES

**by Gwyn Williams, G4FKH**

To gain a better understanding of the figures quoted each week on the RSGB GB2RS bulletins, some of the more important indices are explained below. This is not a comprehensive coverage of indices, rather a simple introduction. To retain simplicity, emphasis will be given to HF conditions. The indices that will be discussed are Solar Flux, the Ap and Kp indices, and how these interact and affect our ability to propagate signals around the earth.

## SOLAR FLUX

The 10.7cm (2800MHz) radio flux is the amount of solar noise that is emitted by the sun at these wavelengths. The solar flux is measured and reported at approximately 1700UTC daily by the Penticton Radio Observatory in British Columbia, Canada. The solar flux is used as a basic indicator of solar activity. It can vary from values around 60 to values in excess of 300 (representing very low solar activity and high to very high solar activity, respectively).

It is worth noting that it takes several days of high values for the ionosphere to improve, one high day will not improve communications. Values in excess of 200 occur typically during the peak of solar cycles.

The solar flux is closely related to the amount of ionisation (electron concentration) taking place at F2 region heights (heights sensitive to long-distance radio communication), **Fig 1**. High solar flux values generally provide good ionisation for long-distance communications at higher than normal frequencies. Low solar flux values can restrict the band of frequencies which is usable for long-distance communications. The solar flux is measured in 'Solar Flux Units' (SFU).

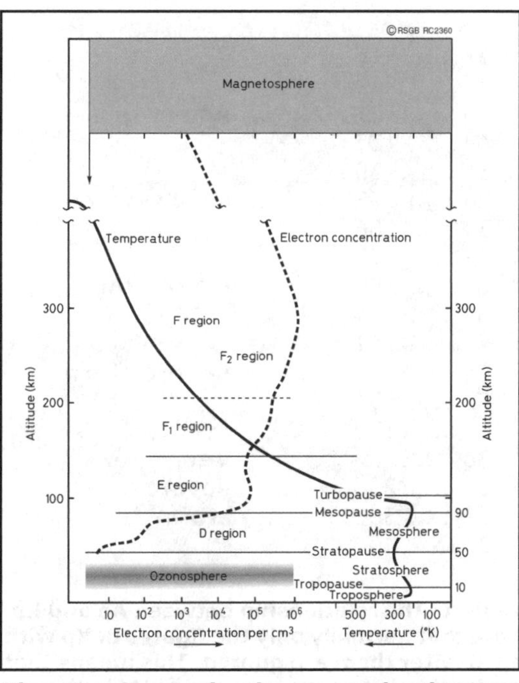


**Fig 1: General shape and electron concentration at the various region heights.**

### A-INDEX

The geomagnetic A-index represents the severity of magnetic fluctuations occurring at local magnetic observatories. During magnetic storms, the A-index may reach levels as high as 100. During severe storms, the A-index may exceed 200.

Great 'rogue' storms may succeed in producing index values in excess of 300, although storms associated with indices this high are very rare indeed.

The A-index varies from observatory to observatory, since magnetic fluctuations can be very local in nature. A quoted A-index will indicate a local reading, whilst an Ap-index represents a *planetary* A-index.

### K-INDEX

The K-index is related to the A-index. Comparison figures are given in **Table 1**.

| Kp | Ap | State |
|---|---|---|
| 0o | 0 | |
| 0+ | 2 | |
| 1- | 3 | **Quiet** |
| 1o | 4 | |
| 1+ | 5 | |
| 2- | 6 | |
| 2o | 7 | **Unsettled** |
| 2+ | 9 | |
| 3- | 12 | |
| 3o | 15 | |
| 3+ | 18 | |
| 4- | 22 | |
| 4o | 27 | **Active** |
| 4+ | 32 | |
| 5- | 39 | |
| 5o | 48 | **Minor storm** |
| 5+ | 56 | |
| 6- | 67 | |
| 6o | 80 | **Major storm** |
| 6+ | 94 | |
| 7- | 111 | |
| 7o | 132 | **Severe storm** |
| 7+ | 154 | |
| 8- | 179 | |
| 8o | 207 | **Very** |
| 8+ | 236 | **severe** |
| 9- | 300 | **storm** |
| 9o | 400 | |

**Table 1: The relationship between Ap and Kp. Note that normally only the figures of Kp with an 'o' after them are quoted. This means that each 'K' index covers a range of 'A' indices.**

Each UTC day is divided into eight 3-hour intervals, starting at 0000UTC. In each 3-hour period, the maximum deviation from the quiet day curve is measured and the largest deviation is selected. It is then put into a standard mathematical equation to yield the K-index for the period.

The K-index is useful in determining the state of the geomagnetic field, the quality of radio signal propagation and the condition of the ionosphere. Generally, K-index values of 0 and 1 represent *quiet* magnetic conditions and imply good radio signal propagation conditions.

Values between 2 and 4 represent *unsettled* to *active* magnetic conditions and generally correspond to less-impressive HF radio propagation conditions. K-index values of 5 represent *minor storm* conditions and are usually associated with fair to poor propagation on many HF paths. K-index values of 6 generally represent *major storm* conditions and are almost always associated with poor radio propagation conditions. K-index values of 7 represent *severe storm* conditions and are often accompanied by 'radio blackout' conditions (particularly over higher latitudes). K-indices of 8 or 9 represent *very severe storm* conditions and are rarely encountered (except during exceptional periods of solar activity). K-indices this high most often produce radio blackouts for periods lasting in excess of 6 to 10 hours (depending upon the intensity of the event).

**COMPLETE PICTURE**

Keeping the above explanations in mind, it is now time to put it together and get a more complete picture of the various interactions. Referring once again to Fig 1, the various regions of the ionosphere are ionised by the sun. Basically, the greater the received energy, the more capable the ionosphere is of bending radio transmissions back to earth. Also, the greater the received energy, the higher will be the Maximum Usable Frequency (MUF).

**Picture of a very active sun, taken during a solar maximum.**

Some of you may not be fully aware of the term MUF, so here's a brief explanation. It is the maximum frequency that allows reliable HF radio communication over a given ground range by ionospheric refraction (bending). Frequencies higher than the MUF penetrate the ionosphere and become useful for extra-terrestrial communications.

Ionisation is produced, as mentioned earlier, by the sun. By far the most important type is extreme ultra-violet (EUV) radiation. Bright areas called 'plage' can be seen surrounding sunspot groups, as shown in the photograph. Plage is the primary source of the sun's EUV radiation. As plage levels vary throughout the solar cycle, so too does the level of EUV radiation.

This leads to changes in the electron concentration of the ionosphere during the course of a solar cycle. The ionosphere's ability to refract signals is dependent upon the electron concentration - the higher the electron concentration and SFU, the better able the ionosphere is to refract radio signals. The MUF also rises with a greater electron concentration; which explains why radio communication conditions are enhanced during the peak of solar cycles. The second photograph shows the sun at the bottom of a solar cycle. Notice the distinct lack of plage regions and, therefore, the shortage of EUV. Also, the low solar activity will result in very low SFU levels.

**Picture of a spotless sun, taken during a solar minimum.**

The key points to remember are that a high SFU will result in superior communications capabilities, while a low SFU will result in inferior communications capabilities. Another important point is that it takes some time for the ionosphere to react to increasing SFU, so it requires quite some building-up before it is capable of supporting higher MUFs.

## THE DOWN SIDE

With increasing solar activity on the sun, as a cycle gathers momentum towards a peak and following that peak, geomagnetic activity also tends to build. This is the result of activity on the sun producing interactions with the earth and causing disturbances to the earth's magnetic field. As previously mentioned, this activity is measured, the 'A' and 'K' indices being the result.

The higher the A/K index, the more likely that there will be an interaction with the ionosphere, ie a high A/K level will depress the MUF. The amount by which the MUF is depressed will depend upon the severity of the storm, while the period of depression will depend upon the duration of the geomagnetic / ionospheric storm. Geomagnetic and ionospheric storms are intrinsically associated, however - a geomagnetic storm is a disturbance affecting the earth's geomagnetic field, and an ionospheric storm is a disturbance in the earth's ionosphere.

## CONCLUSIONS

For the best chance of HF DX, look for high levels of SFU, say >150, and low levels of K, say 0 - 2. It is when these indices have been around these levels for some days that the best communicating conditions will be observed.

# AN INTRODUCTION TO SPEECH PROCESSING

**by George Brown, M5ACN / G1VCY**

There is nothing like a demonstration to illustrate a point, so I would strongly recommend this experiment.

On the FM waveband, tune in BBC Radio 4, and adjust the volume to give a comfortable listening level on ordinary speech. Then, retune to a commercial 'pop' station. Whether the content is music or speech, you will probably reach for the control to turn down the volume.

The reason for this is the subject of this article, although it is its application to amateur radio that concerns us most. Just like us (but even more so), the broadcasting companies must not over-modulate or over-deviate, so why is speech on one broadcast so much louder than on another?

### FIRST STEPS

The waveform (as might be shown on a cathode ray oscilloscope) of male speech, taken from a discussion programme on BBC Radio 4, is shown in **Fig 1(a)**. **Fig 1(b)** shows the waveform of male speech from a commercial radio station, and **Fig 1(c)** the waveform of music taken from the same station as in (b).

All three recordings were made with the same audio gain setting (with no audio AGC), and are of the same length, so they are thus directly comparable. The equipment used was an Icom IC-PCR1000 receiver, a Soundblaster® PC sound card, and proprietary software for recording wave files. To analyse these waveforms quickly, I wrote a short program in *Visual BASIC 6.0*, which analyses the waveform and calculates the power (in arbitrary units) in each waveform. The results are summarised here.

**Radio 4 speech: Power = 138**
**Commercial radio speech: Power = 504**
**Commercial radio music: Power = 522**
**Note: Power in arbitrary units, constant audio gain, equal durations.**

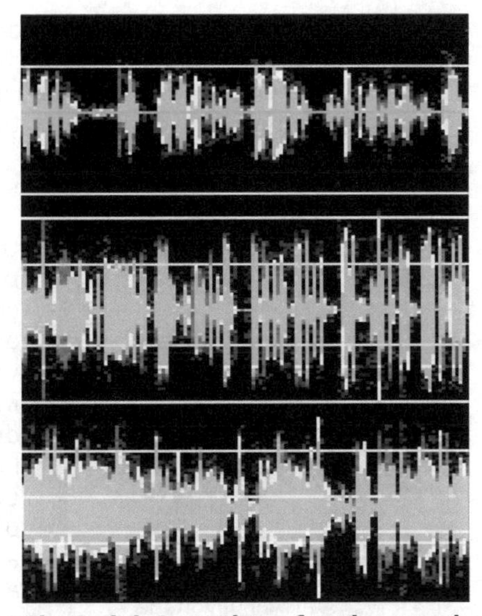

**Fig 1: (a) A section of male speech from BBC Radio 4. (b) A section of male speech from a commercial radio station. (c) Vocal pop music from the same commercial radio station.**

For these random selections of programme output, you can see that there is almost four times as much power in commercial radio speech as there is in Radio 4 speech, and that the music played on commercial radio is only marginally louder than the DJ's pearls of wisdom! This process of 'making everything approximately the same loudness' is called *dynamic range compression, contrast compression* or, simply, *compression*.

As we amateurs are concerned only with speech, we know it as *speech compression* or *speech processing* and, for reasons to be discussed later, we use it only when working on SSB.

**THE PROBLEM**

Speech is remarkably 'spiky' in nature, as Figs 1(a) and 1(b) show when viewed on a cathode-ray oscilloscope. This spikiness causes us problems when we have to set the modulation level on our transmitters.

On sideband transmissions, over-modulation causes splatter at the very least andon FM transmissions, it produces over-deviation. To avoid this, we must set the transmitter drive so that the *peaks* of our audio waveform do not over-modulate. To illustrate this, use Fig 1(a), and lay a ruler on it horizontally so that it *just* touches the biggest negative-going peak (we choose a negative-going rather than a positive-going peak, so that the ruler does not hide the waveform!). You will now notice one thing – that the average modulating voltage (compared with our self-imposed maximum) is very low. In practical terms, our transmission will lack 'punch', and will not be heard very well under difficult conditions.

**SOME SOLUTIONS**

In very basic terms, what we need to do is to turn up the audio gain when the speech is quiet and turn it down when the speech is loud. Unfortunately, these variations of loudness occur very quickly, often between syllables, and the technique is something which cannot be done manually.

Two basic methods are used, one of which involves using an amplifier of non-linear transfer characteristic, the other uses automatic control of the audio gain. It should be understood at the outset that whatever technique is used, it *distorts* the speech waveform. This distortion is (hopefully) controlled so as to improve the intelligibility, even if the result is sometimes unnatural.

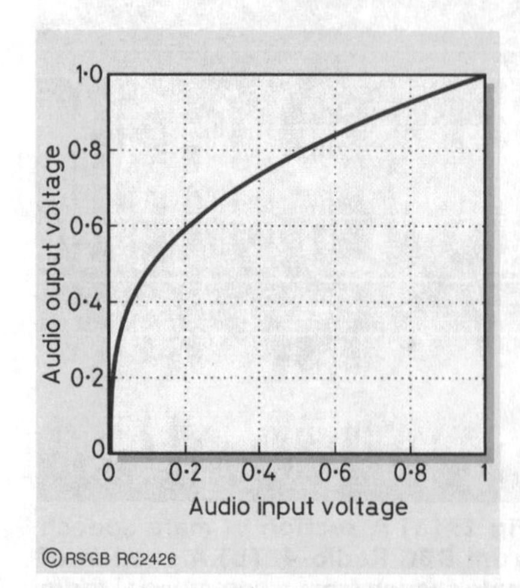


**Fig 2: Using a non-linear transfer characteristic as the basis of a speech processor.**

Using a non-linear transfer characteristic is, in theory, probably the most attractive solution. A transfer characteristic is the graphical representation of the output voltage from a device compared with the input voltage. Normally this should be a straight line, indicating that the output and input voltages are directly proportional. To achieve the desired result, look at the characteristic of **Fig 2**; notice that, for example, a small input of 0.1V will produce an output of about 0.5V (a gain of about 5), 0.2V will produce an output of about 0.6V (a gain of about 3), whereas a large input of 1V will produce an output of 1V (a gain of unity). Provided the input voltage never exceeds 1V, the circuit will produce an increasing gain at lower voltages. The main problem with a technique such as this is that a circuit to implement it is quite difficult to design. It does have the advantage (see later) that nothing *physically* changes as a result of changing input levels. The other techniques are simpler to implement, but have their own disadvantages.

Manually-altering the audio gain control may be completely impractical, but circuits can be made which will achieve the same result automatically. The operations involved are straightforward.

(a) Rectify the audio voltage. The average value of an audio signal is zero, so rectification is needed to 'lop off' the lower half of the signal, so that it *does* have a mean value.

(b) Feed the rectified signal into a low-pass filter, a simple connection of one resistor and one capacitor (see **Fig 3**). The output from this filter is the mean value of the signal over short time periods, the audio components having been short-circuited by the capacitor.

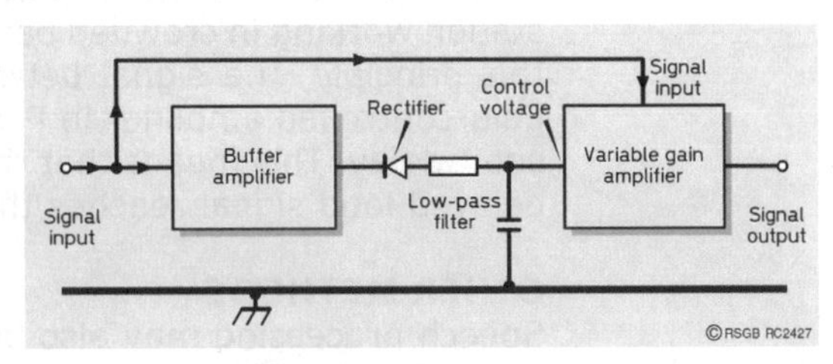


**Fig 3: Block diagram of a simple audio AGC circuit.**

(c) This varying mean value is then used to control the gain of another audio amplifier, the gain being reduced as the mean value increases. A block diagram of the circuit is shown in Fig 3. A buffer amplifier is a device which helps to separate the rectifier and filter from preceding circuits and to supply enough gain to operate the rectifier cleanly.

A simple gain control circuit, eminently suitable as the basis for experimental work, is shown in **Fig 4**. An N-channel JFET is used as one of a pair of feedback resistors in a standard non-inverting Op-Amp circuit. Connected this way, the resistance between the JFET drain and source depends upon the voltage applied to its gate. The overall gain [1] is adjustable between 1 and 1000 by changing the voltage at the control input.

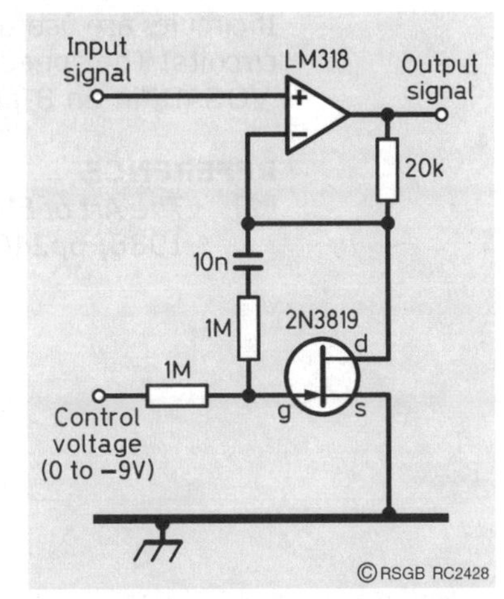


**Fig 4: A simple amplifier circuit where the gain is dependent upon the control voltage. The circuit supply voltage should be around ±15V.**

## ATTACK & DECAY

While Fig 4 is quite a simple and acceptable circuit to use for automatic audio gain control, the derivation of the control voltage to operate it can be quite tricky.

The simplest circuit, comprising the rectifier, capacitor and resistor, shown in Fig 3, is a good starting point. The values of the resistor and capacitor need to be varied in order that the circuit will act quickly to reduce the gain when a loud sound suddenly appears (this is known as 'attack'), yet will take a little longer to return to its initial gain (known as 'decay') in case another loud sound follows-on quickly. Getting the attack and decay right is quite an art, and a circuit more complex than that of Fig 3 is usually needed.

### THE PROBLEMS

No circuit like this is without its problems. Correct attack and decay are obvious candidates, and the solution is somewhat subjective, because no two people speak in exactly the same way. However, the overriding problem with all AGC-based speech processing circuits is that a loud sound *must* get through to the output *before* the circuit can react to reduce the gain. This means that, in unfortunate cases, transient peaks, which are louder than they ought to be, will still get through to the output and could still cause transmitter over-modulation.

The overall drive must be backed off a little to allow for this. Nevertheless, speech processing, when properly used, is of great benefit to the SSB station working in crowded band conditions. In professional circuits using this principle, the signal between the main input and the input to the gain-controlled amplifier in Fig 3 is subject to a short delay. This means that the gain *can* be turned down *just before* the delayed loud signal reaches the amplifier!

### OTHER METHODS

Speech processing may also be achieved (and even more effectively so) by performing the control operation on the RF signal, or by using DSP. Both these techniques are beyond the scope of this article, however.

### FINALLY

We have not discussed speech processing in the context of FM. FM is essentially a short-distance mode, and interference from adjacent stations is not a problem. This removes the need for processing, hence the restriction of our discussions to SSB.

If circuits are useful, you can bet that they have been made into integrated circuits! The speech processor is no exception, and one such device is the 'VOGAD' in an 8-pin DIL chip, the SL6270 (Maplin order code UM73Q).

### REFERENCE

[1] *The Art of Electronics*, Horowitz and Hill, Cambridge University Press, 1988, pp240 - 41.

# AN INTRODUCTION TO SIDs

**by Gwyn Williams, G4FKH**

Sudden Ionospheric Disturbances (SIDs) are well-documented phenomena associated with solar flares. You cannot discuss one without involving the other. Thus, SIDs can be introduced in the following way.

The solar flare serves as a useful indicator of solar activity, since more occur when the Sun is more active. They also have direct consequences for the ionosphere.

In 1937, J H Dellinger recognised that fadeouts in high-frequency radio propagation were the result of abnormally strong absorption in the ionosphere, occurring at the same time as a solar flare. The fadeout had a rapid onset and a typical duration of tens of minutes, like the visible flare. Because they begin suddenly, all the immediate effects of a solar flare are known as Sudden Ionospheric Disturbances. For a long time, the absorption effect discovered by Dellinger was called the 'Dellinger fade', but is now generally termed the Short-Wave Fadeout (SWF) [1]. SWFs only effect the sunlit side of the earth.

The nature and timing of the SWF immediately provide two clues to its nature. The simultaneous nature of the fadeout and the visible flare shows that the cause is electromagnetic and the occurrence of radio absorption indicates that the electron density in the D-region has been increased. When it became possible to measure hard X-rays from space rockets, it was observed that they intensified by several orders of magnitude during a flare; thus the SWF is now attributed to the X-rays emitted from the flare.

In fact, there is a wide range of ionospheric disturbances, as summarised in **Table 1**. Though the effects are most marked in the D-region, E- and F-region, effects can also be detected. The electron content, governed mainly by the F-region, is increased by a few percent. The range of effects shows that a considerable band of the spectrum is enhanced. Moreover, there is some difference of timing: extreme ultraviolet (EUV) radiation and the hardest X-rays tend to be enhanced early in the flare, whereas the softer X-rays last longer and correspond more closely with the optical flare.

All SID effects cover the whole of the Earth's sunlit side, and are essentially uniform except for a dependence on the angle of the Sun to the horizon (ie when the Sun is directly overhead it is at 90° and when at the horizon 0°). Even flares that are not observed visually can be detected by their ionospheric effects and some properties of the flare deduced.

| | Technique | Effect | Region | Radiation |
|---|---|---|---|---|
| SWF (Shortwave Fadeout) | HF radio | Absorption propagation | D | Hard X-rays 0.5-0.8Å |
| SCNA (Sudden Cosmic Noise Absorption) | Riometer | Absorption | D | Hard X-rays 0.5-0.8Å |
| SPA (Sudden Phase Anomaly | VLF radio propagation | Reflection height reduced | D | Hard X-rays 0.5-0.8Å |
| SEA (Sudden Enhance-ment of Atmospherics) | VLF Atmospherics | Intensity enhanced | D | Hard X-rays 0.5-0.8Å |
| SFE ({magnetic} Solar Flare Effect) | Magnetometer | Enhanced ionospheric | E | EUV & soft X-rays |
| SFD (Sudden Frequency Deviation) | HF Doppler | Reflection height reduced | E & F | EUV |
| - (Electron content enhancement) | Faraday effect | Content enhanced | F | EUV |

**Table 1: SID phenomena.**

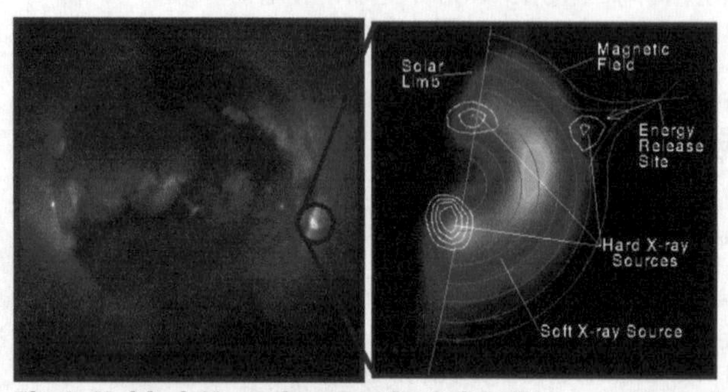


**Fig 1: Yohkoh X-ray image of a Solar Flare, combined image in soft X-rays (left) and soft X-rays with hard X-ray contours (right).**

## SOLAR FLARES

A solar flare is a sudden brightening of a small area of the Sun's photosphere, which is the part of the Sun that can be seen (see **Fig 1**). This may last between a few minutes and several hours. Flares are classified on a scale of 1 to 4 according to the area of the brightening when the Sun is viewed in $H\alpha$ light (656.2nm wavelength). In addition, very small brightenings are designated 'S' for 'subflare'. Flares tend to develop near the long lasting dark areas, called sunspots, and the regions of enhanced Calcium-K and Hydrogen-$\alpha$ emission, known as 'plages'.

Flare activity and its resulting radiation may vary extensively from hour to hour. The short term activity, which may be related to short-term HF effects, is correlated with SIDs, and are closely related to other solar features which are responsible for certain geomagnetic substorm phenomena, enhanced auroral activity and ionospheric storms.

| Class of X-ray flare | E, flux level at 1AU |
|---|---|
| C | 10-6 < E < 10-5 |
| M | 10-5 < E < 10-4 |
| X | E > 10-4 |

**Table 2: Classification of X-ray flares. The X-ray flux is referred to that value which would be observed at 1AU (the Earth's distance from the Sun). The units are in watts per square metre.**

Flares are classified similarly, but another optical designation provides an indication of the brightness: 'F' is faint, 'N' is normal, and 'B' is bright. The most important flare classification for association with ionospheric effects is the flare strength as measured in the X-ray band. **Table 2** shows the X-ray classification in the 1 - 8Å band (0.1 - 0.8nm).

The number of X-ray flares observed during solar cycle 21 totalled 172. About 10 times more optical flares were observed than X-ray flares. More X-ray flares were observed following a sunspot maximum than at the peak itself. **Fig 2** shows that flare activity and storm occurrence were not strongly correlated with sunspot number, at least during solar cycle 19. Definitive patterns linking sunspot activity with ionospheric storms or solar flares (and thus SIDs) have been elusive, since observable relationships seem to vary from cycle-to-cycle [2].

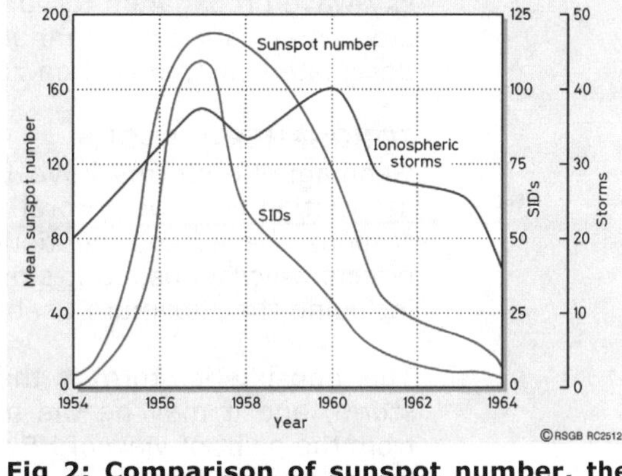


**Fig 2: Comparison of sunspot number, the number of ionospheric storms, and the number of SIDs for solar cycle 19.**

## IONOSPHERIC RESPONSE

SIDs constitute those events which arise as a result of the atmospheric interaction with electromagnetic flux from solar flares. We recognise that the Sun is the ultimate source for a large variety of ionospheric and magnetospheric effects. Many of these are of some importance in HF working, while others are only of marginal interest. There are many types of SID which are observed, and many are of importance only to long-wave systems or have limited diagnostic application. At HF, the most important form is the SWF, although a relatively small and sudden enhancement in the F2 layer critical frequency ($f_{oF2}$) may also be observed. The SWF that affects the entire sunlit ionosphere is by far the most important form of SID. An X-ray flare generates a significant increase in D-layer ionisation content with a temporal pattern, which mimics that of the flare itself.

This results in an increase in the product of the electron density and the collision frequency. Put simply, the definition of collision frequency is just the number of collisions that one particle makes with others in one second. It is the growth of this product which accounts for the absorption of HF signals.

## GEOMAGNETIC STORM

A geomagnetic storm develops as the direct result of the energy transferred from the solar wind to the magnetosphere in a series of substorm events. We can regard the geomagnetic storm as being composed of two parts:

1. an initial (short-lived) positive phase associated with an increase in the horizontal component of the magnetic field, which is followed shortly by enhanced auroral displays;
2. a main negative phase (or bay) in the horizontal field intensity which may last for several days.

The initial phase is associated with short-lived enhancements in electron concentration at atmospheric heights, while the main phase is associated with large-scale decreases in electron concentration. By monitoring the total electron content of the ionosphere we can see how an ionospheric storm is almost a one-to-one mapping of the geomagnetic storm-time profile. Since $f_{oF2}$ is of fundamental importance in sustaining ionospheric sky-wave propagation for long-haul HF communications, an ionospheric storm may have a substantial negative influence on performance within a designated communication zone.

**IONOSPHERIC STORM**

Geomagnetic storms have their origins in the magnetosphere. Their association with the auroral zone and some insight into the ionospheric response have been mentioned. What we now need to mention are the effects which ionospheric storms have on certain HF system parameters, including the Maximum Usable Frequency (MUF).

The ionospheric storm is the ionosphere's response to a geomagnetic storm, and it may be the most important solar-induced phenomenon from the point of view of HF propagation impact. Ionospheric storms are known to introduce substantial and lasting decreases in F-region electron concentrations, which have the effect of reducing the availability of the upper portion of the HF spectrum. This is true even during the daytime.

At mid-latitudes, the ionospheric storm signature is one in which the F-region ionisation momentarily increases in the dusk section following Storm Commencement (SC), after which it decreases dramatically.

The initial short-lived enhancement is observed in $f_{oF2}$ records and is correlated with the initial positive phase of the geomagnetic storm. The main phase of the geomagnetic storm is correlated with an attendant $f_{oF2}$ decrease. This reduction in $f_{oF2}$ may last for a day or longer.

It is thought that the initial enhancement in $f_{oF2}$ is a result of electrodynamic forces and the long-term $f_{oF2}$ reduction with ionospheric heating.

An enhancement in the MUF is observed to occur within the first 12 hours of a Sudden Storm Commencement (SSC), in step with the initial positive phase of the associated geomagnetic storm, but it lasts for only a few hours. The major effect is associated with the main phase condition, during which time the decrease in MUF may exceed 10% of the undisturbed MUF for about 24 hours, depending upon the season. This effect starts roughly 12 hours after a SSC and may persist for several days.

**CONCLUSIONS**

A SID is a phenomenon that is a direct consequence of a solar flare. The effects are numerous and have far-reaching consequences on HF propagation. These effects include SWF (of prime importance and most disruptive), sudden enhancement of atmospherics and sudden frequency deviations, to name just a few. The lengths of disruptions vary, as do the number of occurrences per solar cycle. Research into this cause and effect cycle is still taking place, particularly the correlation of SIDs to other solar phenomena.

**REFERENCES**

[1] J K Hargreaves, *The Solar-Terrestrial Environment*, Cambridge University Press, 1992.

[2] John M Goodman, *HF Communications – Science and Technology*.

# AN INTRODUCTION TO THE CRO

**by George Brown, M5ACN / G1VCY**

In 'An Introduction To The CRT', (see elsewhere in this book), the principle of the Cathode-Ray Tube was discussed. Now it is time to look at a particular piece of equipment which uses the CRT. In the radio amateur's shack, this equipment is usually the Cathode-Ray Oscilloscope, or CRO.

In that article, the functions of the heater, cathode, grid, anodes, deflector plates and phosphor were explained, together with the operation of the brilliance (or intensity) and focus controls. It is now instructive to see how signals external and internal to the CRO are fed to the CRT, and to understand why the oscilloscope is so useful as a measurement and test tool.

## X AND Y

For 90% of its everyday uses, the CRO presents a real-time display of voltage (along the y-axis) against time (along the x-axis). The varying voltage on the y-plates is usually an amplified version of an external signal applied to the **y-amplifier** through a socket on the front panel (see **Fig 1**). This amplifier has a large, fixed gain, but is preceded by a wide-range **frequency-compensated attenuator**, in order to accommodate the display of signals of widely-varying magnitudes. The calibration scale on the y-attenuator enables the amplitudes of complex signals to be measured. The signal present on the x-plates varies uniformly with time, and is generated internally by circuits known as the timebase and the x-amplifier.

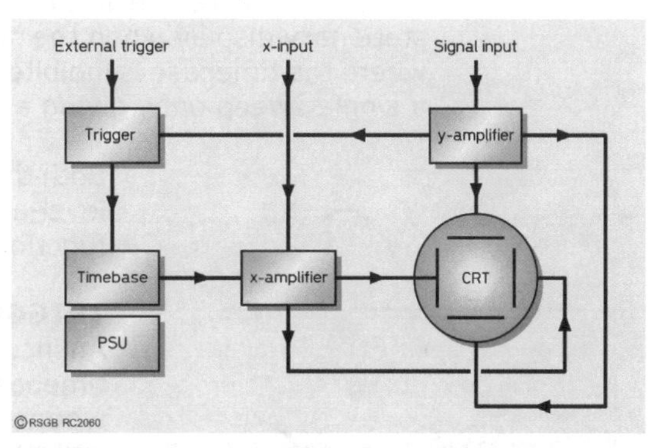


**Fig 1: Block diagram of the basic cathode-ray oscilloscope.**

## TIMEBASE

In order to 'draw' the spot at a constant speed across the face of the tube, a **sawtooth** waveform is used, as shown in **Fig 2(a)**. The waveform consists of two parts: the **sweep** is a rising voltage which is linear with respect to time and deflects the spot from left to right across the tube screen; the **flyback** is the falling section, which should occur very quickly, returning the spot to the left-hand side of the screen.

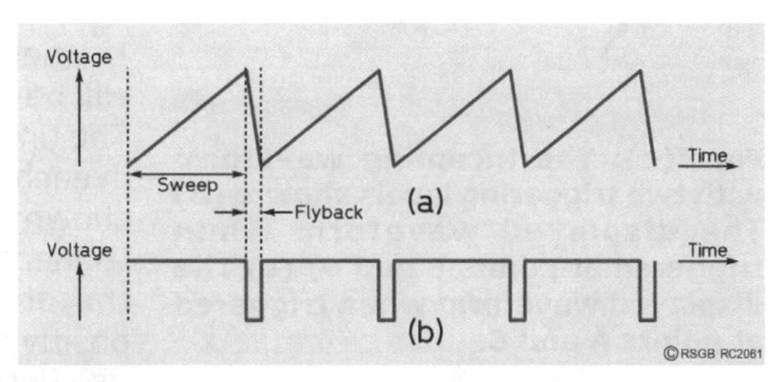


**Fig 2(a): The voltage waveform applied to the x-plates of the CRT; (b) The blanking pulses applied to the CRT grid to suppress the appearance of the flyback on the screen,**

This process is repeated, to give the continuous display of the input waveform. At very high sweep speeds, the flyback may occur in a time comparable with the sweep; when this occurs, the flyback becomes visible and clutters the display. To overcome this problem, **flyback blanking** is used. When the flyback occurs, the negative-going **blanking pulse**, shown in **Fig 2(b)**, is applied to the grid of the CRT; this effectively switches off the beam, and no flyback is visible. Referring to the tube bias circuit given in Fig 3 of the original article, the blanking pulse would be applied to the top end of R1 which is connected directly to the grid of the CRT. The speed at which the timebase runs is governed by a **calibrated rotary switch** (coarse) and an **uncalibrated potentiometer** (fine), both on the front panel. These enable the speed of the timebase to be matched to the frequency of the incoming signal, to display a convenient number of cycles on the screen. The calibration on the timebase control allows the measurement of time intervals to be made from the display, from which signal frequencies
can be calculated if required.

There are three basic modes of timebase operation: **free-running**, where the timebase runs freely at the speed set by the controls; **synchronised**, where the free-running timebase is given 'kicks', derived from the incoming signal, which return it to the left side of the screen, and will give a stationary display when the sweep controls are correctly set; **triggered**, where the timebase is inhibited (stopped) and can be triggered to produce a single sweep only, giving a stationary display under all circumstances.

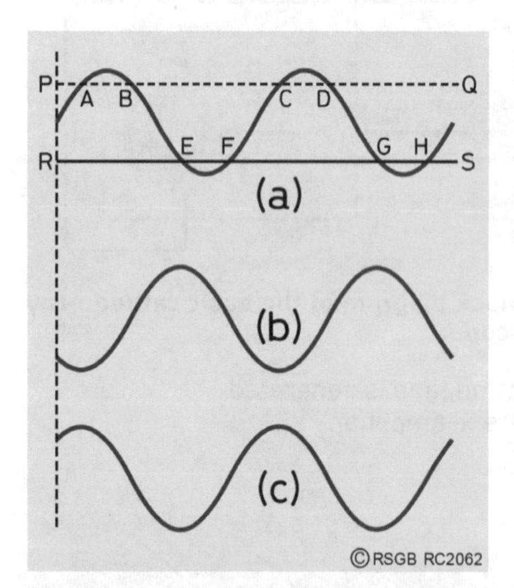


**Fig 3(a): The incoming waveform with two triggering levels shown; (b) The displayed waveform when triggered at points E and G; (c) The displayed waveform when triggered at points A and C.**

Both the synchronised and triggered modes require level-and slope-detecting circuits to operate. These functions are illustrated in **Fig 3**.

## TRIGGERING

An incoming sine wave is shown in **Fig 3(a)**. If the timebase were to be triggered (or synchronised) whenever the voltage level crossed the value indicated by the line RS, it would begin the sweep at two points in every cycle, such as E and F or G and H. If the display is to remain static (the aim of triggering), this is unsuitable. In order to isolate only one point per cycle, the detection of the waveform's slope is also needed.

If a negative slope is selected, only points E and G will be found at the level RS, ie one point per cycle. The timebase will begin its sweep at the same point in each cycle, and thus will produce a static display, shown in **Fig 3(b)**. A positive slope selection would trigger the timebase at F and H. **Fig 3(c)** illustrates the display that would be obtained by using a positive slope at level PQ. As Fig 1 shows, it is usual to derive the triggering signal from the incoming signal via the y-amplifier, but most oscilloscopes have a socket for feeding in a trigger signal from an external source. This is most useful when viewing a complex waveform. There is a simple test to check if a timebase is being synchronized or triggered: if the timebase stops when the signal is removed, it is being triggered. Many types of CRO do not distinguish between synchronising and triggering on the front panel controls.

Finally, what about the 10% of cases where the x-axis is not time, but is another signal generated externally? One such application is illustrated in **Fig 4**. It uses the y-against-x capability of the CRO to plot the characteristics of a single component, in this case a Zener diode. The voltage signal is fed to the x-amplifier, and generates the x-axis; the voltage developed across R is proportional to the current flowing, and generates the y-axis. In this case, the x-axis scale is 5V per division; the display shows a 10V reverse Zener breakdown, together with the standard forward characteristic of a silicon diode. The voltage source can be derived from a transformer or from a function generator, preferably using a triangular wave. The measurements are not source waveform- dependent.

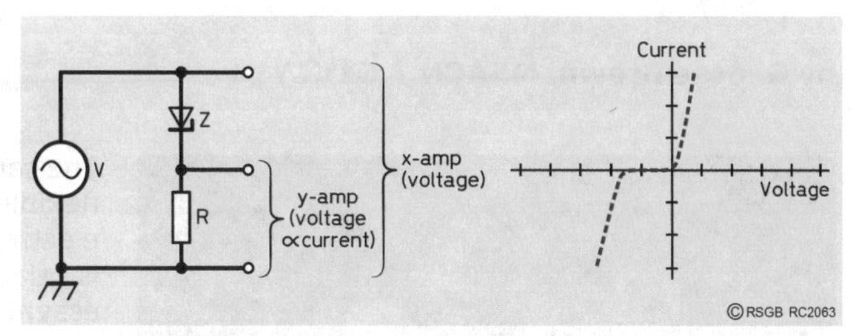


**Fig 4: Displaying Zener diode characteristics using a CRO.**

Other non-timebase applications include the measurement of frequency using Lissajous Figures, but this has little direct application in amateur radio.

## SPECIAL MODELS

Oscilloscopes made specifically for amateur radio are usually called Station Monitors; typical examples being the Kenwood SM-220 and SM-230. These are conventional oscilloscopes but have direct HF/VHF access to the y-plates, which allows a radiated RF waveform to be displayed, and trapezoidal displays of HF waveforms to be produced. They also act as panoramic displays with the TS-950 range of transceivers.

# AN INTRODUCTION TO THE CRT

**by George Brown, M5ACN / G1VCY**

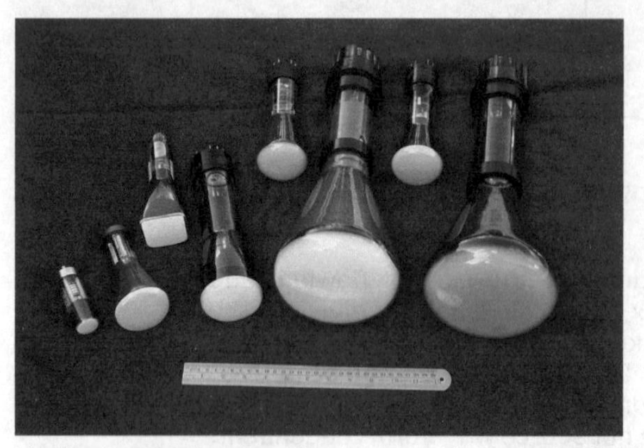

**Various cathode-ray tubes commonly to be found at rallies, and suitable for use in the amateur station. Left to right: DH3-91, DG7-5, D7/231GH, 3BP1, 23D, VCR517, VCR139A, and VCR97.**

The cathode-ray tube (CRT) is the most flexible measuring instrument in the electronics laboratory. Its construction is relatively simple, and has remained essentially constant since its invention by Braun in 1897. A selection of those commonly available to the amateur is shown in the photograph.

## DESCRIPTION

The basic oscilloscope CRT is shown in **Fig 1**, using its standard circuit symbol, which is very similar to its actual construction. Very much an electronic 'ship-in-a bottle', all its electrodes are contained inside an evacuated glass envelope.

**Fig 1: Circuit symbol for the cathode-ray tube.**

## ELECTRON EMISSION

The *cathode* emits a copious supply of electrons when raised to red heat by the *heater*. Only a minute fraction of these are allowed past the *grid,* a metallic cylinder capped at one end with a plate containing a circular aperture. This electrode is maintained at a potential which is negative compared with that on the cathode, and tends to repel the electrons back towards the cathode.

The number of electrons which pass beyond the grid is thus controlled by the potential on the grid, which is varied by the 'brilliance' (or 'intensity') control on the front panel of the oscilloscope.

## ANODES

Beyond the grid (at a distance much greater than that between the cathode and the grid) are three more circular plates with small apertures in their centres. These are called *Anodes 1*, *2* and *3* (working away from the grid) and have a very special role to play.

Anode 1 and Anode 3 are usually connected together, and are maintained at a potential of between +600V DC and +5000V DC relative to the cathode; the exact voltage depends upon the design of the equipment operating the CRT and its purpose. This Extra High Tension (EHT) attracts those electrons which the grid has allowed to pass beyond it, and accelerates them to very high speeds (comparable with the speed of light - see later). The beam of electrons diverges slightly as it approaches the anode structure, and is travelling so fast that it passes through the holes in the three electrodes, rather than colliding with them. Anode 2 is maintained at a lower positive potential than Anodes 1 and 3 (usually around +150V DC), and its purpose is to 'bend' the electron beam so that it converges rather than diverges. The amount of convergence is controlled by the potential on Anode 2, this being varied by the 'focus' control.

The electrodes considered so far form two groups within the cathode-ray tube. The heater/cathode/grid combination is known as the *electron gun*, as its purpose is to 'fire' electrons towards the anodes.

The three anodes are known collectively as the *electron lens*, and are required to focus the electron beam on to the screen. Turning the focus control changes the focal length of the lens.

## DEFLECTION

Beyond the electron lens are two parallel pairs of *deflector plates*, as shown in **Fig 2,** which is *not* to scale. First come the *Y-plates*, two horizontal plates between which a potential can be established; this will exert a transverse force on the beam passing between them, and thus will deflect the beam up or down (ie along the Cartesian y-axis), depending upon the polarity of the voltage applied. The *X-plates* then perform an identical function, except that the deflection produced is horizontal (along the x-axis).

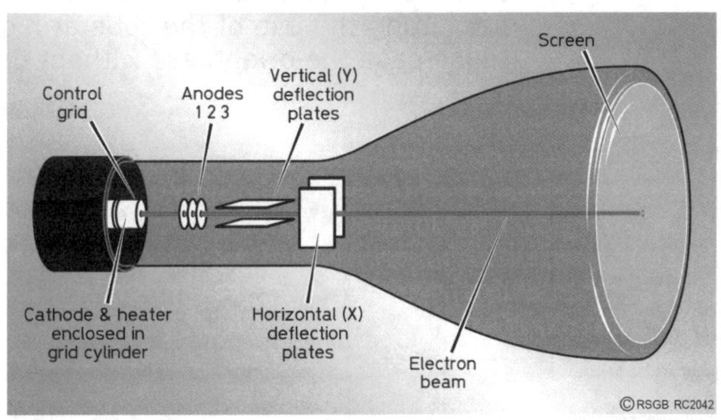


**Fig 2: The electrode structure within the cathode-ray tube.**

## DISPLAY

Once clear of both pairs of deflector plates, the beam continues on its way to the screen at a constant speed. The screen is simply a coating of phosphor powder on the inside of the glass envelope at the end of the *flare*, where the tube increases in diameter between the neck and the screen, as Fig 2 shows.

The purpose of the phosphor is to glow under the bombardment of electrons - the more intense the bombardment, the brighter the glow. Different phosphor materials give different colours, and the simplest of phosphor materials is zinc sulphide, which glows green.

## CONNECTIONS

The cathode ray tube is a form of thermionic valve and, in common with its more numerous brethren, has the connections to its electrodes brought out to a series of pins in the tube base. This mates with a socket, to which the connections are made from the rest of the equipment of which the CRT is a part.

If it is assumed that no electrons strike the electron lens on their way through, then there is no circuit present, as there is no conventional anode to collect the electrons.

To overcome this problem, and to prevent the screen charging up, the insides of the neck and flare are usually coated with colloidal graphite which, in turn, is connected to Anode 3. This collects the secondary electrons emitted when electrons strike the screen, thus completing the circuit.

## REFINEMENTS

There are two main changes to the structure which are found in more sophisticated CRTs.

Firstly, to minimise the capacitance between the pairs of deflector plates, their connections can be brought out on the neck of the tube, thus permitting the use of the tube at higher frequencies. Secondly, in order to increase the brightness without degrading the deflection sensitivities, another electrode after the deflector plates is used to provide additional acceleration to the beam. This is known as Post-Deflection Acceleration (PDA).

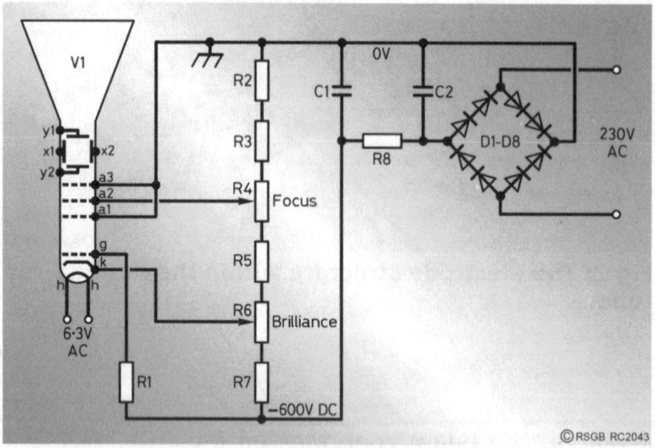

The deflection of the beam, as measured in millimetres horizontally or vertically on the screen, is directly proportional to the voltage between the corresponding deflector plates; a doubling of deflection indicates a doubling of voltage. It is this property which makes the CRT such a valuable tool for the measurement of very complex waveforms. **Fig 3** shows how the CRT is biased.

**Fig 3: A simple circuit for biasing the electrodes of a cathode-ray tube.**

## SPEED

Finally, it is worth mentioning the speeds at which the electrons travel, even in modest systems, such as may be constructed for amateur use. For an EHT of only 600V, the electrons strike the screen at around 1/20 the speed of light, and at 5000V, this becomes 1/7 the speed of light!

# AN INTRODUCTION TO THE FOURIER TRANSFORM

**By George Brown, M5ACN / G1VCY**

**[Editor's note: *This article was written several years ago, when computing power and the programmes available were somewhat pedestrian by modern standards. The programme SbFFT, which was used extensively in this work may be successfully replaced by* Spectran, *obtainable via ww.weaksignals.com/*]**

Ignored by many, revered by some, the techniques pioneered by Fourier form the basis of much of the theory of communications. We all use them (perhaps unwittingly) when we discuss the bandwidth of a signal. Here, just enough of the specifics of the subject are presented in order that those interested may pursue the concept in more detail.

## HISTORY

Jean Baptiste Joseph Fourier, born in 1768, fought for Napoleon I in Egypt, and was an expert on Egyptian antiquities. He was made a Baron by Napoleon in 1808 and was elected to the Academy of Sciences in 1816. It was in 1822 that he published *The Analytical Theory of Heat* and, from this arguably obscure treatise, came the mathematics which were to change the evolution of science. In it, he used a trigonometrical series (later to become known as the *Fourier Series*) by which a complicated mathematical function can be expressed as a sum of an infinite series of sine and cosine functions. He was, justifiably, elected to the French Academy in 1827, three years before his death.

## INTRODUCTION

From these unlikely roots have evolved the techniques which are so fundamental to our appreciation and understanding. With the common availability of the computer, non-mathematicians can now enjoy realising the potential of Fourier's work without needing to understand the intricacies of the equations involved. Just as electronics is becoming more biased towards systems as being assemblies of 'black boxes', with no knowledge being necessary of what goes on inside them, so the worlds of mathematics and communications are being opened-up by the use of software packages which can be used as tools to solve specific problems. Such problems can now be addressed (and solved) by those who, ordinarily, could never understand the underlying complexities. This article was prompted by the description of the use of digital signal processing (DSP) by Peter Martinez, G3PLX, for the study of long-delay echoes, reported in 'TT' in the January and February 1998 issues of *RadCom*.

Because the Fourier transform and DSP are very closely linked, a simple discussion of the principles involved was thought to be timely. AF9Y has also reported using the FFT to identify moonbounce signals, and to detect the 70cm beacon from the Mars Global Surveyor (complete with 30Hz earth-rotation Doppler shift) at a distance of 5 x $10^6$km! [1].

Unfortunately, it is still true that Fourier Series are quickly dismissed in A-level maths courses, with little or no mention of their application in *any* area of science, thus consigning them at the first opportunity to the student's waste paper basket. Fourier's work, complex though it may be, is readily understandable in a general sense, provided it is acknowledged at the outset that its basic premises are true. We can paraphrase these for our own purposes as:

- Any complex waveform can be *synthesised* by the addition of a number of sine waves of different frequencies, amplitudes and phases. And its converse statement:
- Any complex waveform can be *analysed* into a number of sine waves of different frequencies, amplitudes and phases.

Strictly speaking, the words 'a number' should be replaced by 'an infinite number', but since we are restricting our descriptions to practical cases the word 'infinite' is not in our vocabulary, and we are restricted to using finite numbers. This statement, as will be realised later, is simply recognising the very practical restriction that the bandwidths of our signals (at audio and RF) must be limited! The two statements also introduce the concepts of *analysing* a signal *into* its components, and *synthesising* a signal *from* its components; thus we talk about Fourier Analysis or Fourier Synthesis, depending upon the *direction* in which we are interested. In general, the expression 'Fourier transform' is used, and is applicable both to the synthesis and analysis processes. The transform enables processes which occur in the *time domain* (eg changes of any parameter with *time*) to be examined in the *frequency domain* (eg changes of an equivalent parameter with *frequency*) and *vice versa*.

## ANALYSING A RTTY SIGNAL

By far the most common application of Fourier's work to amateur radio lies in the *analysis* of an existing signal. This can be a live signal, extracted from the audio stages of a transceiver, or a recorded version of the same signal.

The two should be the same, provided the recording process does not introduce artefacts of its own (bandwidth reduction, wow and flutter, etc). Computer programmes exist which will perform the analysis process very rapidly, depending upon the speed of the particular processor. One such, if a little long in the tooth, which is extremely easy to use, is *SbFFT* by Kevin McWilliams, KW5Q [2]. It uses a digital technique known as the *Fast Fourier Transform* (FFT), and displays on the screen an intensity (or colour) plot of frequency horizontally and time vertically. It requires a Soundblaster SB16 (or 100% compatible) sound card, into which the audio waveform is fed from the signal source. The display is in real-time, and can be saved as a file for future study.

An example of the display of the *SbFFT* programme is shown in **Photo 1**. The legend on the top five lines need not concern us; it relates to the parameters that can be used to customise the display. The sixth line is

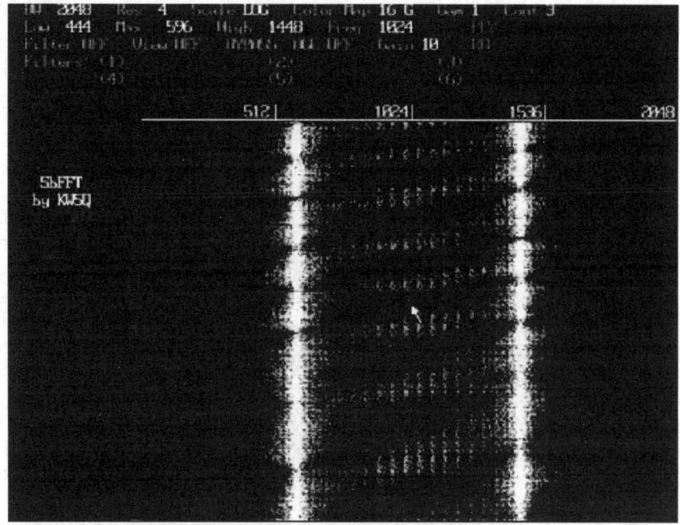


Photo 1: SbFFT display of a RTTY signal, showing the frequency shift of the keying, and fading of the signal.

the frequency axis calibration, in this case extending from 0 to 2048Hz. The horizontal axis is frequency, and each horizontal line of the display is a single Fourier transform. The power (ie a number proportional to the voltage squared) at any frequency is represented by the brightness of the display along a single line. The process is repeated, and the next transform is displayed along the next lowest line of the screen, and so on. This enables 370 lines of FFT to be displayed on a low-resolution screen (640 x 480 pixels). On a higher-resolution display (1024 x 768 pixels), 658 lines (transforms) are available.

The vertical time axis, which starts at the top and progresses downwards, is not calibrated, as it is seldom used for measurement purposes. In this particular case, the line rate was four per second, thus completing the 370 active-line display in 370 / 4 = 92.5 seconds. The signal being analysed is a normal, two-tone commercial RTTY transmission.

Purists should note that the signal is not tuned-in as it would be for normal reception, but instead to centre it in the display. No information is lost in this way. The two RTTY tones are the two prominent vertical lines. The 'graininess' of the picture indicates the presence of noise (of which there was a significant amount), and the effects of frequency- switching transients. The cursor can be seen at the centre of the display. This can be moved about to any frequency of interest, and the value of that frequency is given beside the word 'Freq' on line two of the legend.

The display in question has a horizontal frequency resolution of 4Hz per pixel, meaning that any frequency up to 2048Hz can be identified within 4Hz. The two tones can be measured in this way to be (592 ± 4)Hz and (1436 ± 4)Hz, showing the frequency shift of the RTTY signal to be (844 ± 8)Hz, the standard value being 850Hz.

The second feature of interest is the series of vertical stripes between (and outside) the two RTTY tones. These are caused by the frequency switching which is used to produce the binary codes corresponding to the characters being transmitted. All modulation, of whatever sort, causes extra frequencies to be added to the basic spectrum. The frequency spacing of these subsidiary lines will tell us the frequency of the switching. Moving the cursor to a peak near the upper tone gives a frequency of 1300Hz; repeating the measurement 11 lines lower (near the lower tone) gives

750Hz, showing a frequency separation of (550 ± 8)Hz. Because of the 11 subsidiary lines between these measurements, the spacing of adjacent lines is 550/11 = (50.0 ± 0.7)Hz. This particular station uses RTTY at 50 baud (50 bits per second), hence the appearance of the harmonics of 50Hz!

The third feature of this photograph was not intended at the outset, but appeared as the data capture proceeded - the diagonal 'empty' areas between the two tones. This is a classic example of frequency-selective fading. Anyone who has listened to HF RTTY signals knows that, from time to time, one or other of the two tones disappears, and the decoder (depending upon its design) will either carry on working or produce gibberish until the missing tone returns.

This behaviour is very interesting to observe on the photograph. It begins at high frequency, almost removing the upper tone, and then reduces its frequency quite steadily, taking out the 50Hz harmonics as it does so, then removes the lower tone. By this time, another round of fading has begun at the upper tone, and so it continues. As time passes, the time interval between successive rounds of fading increases.

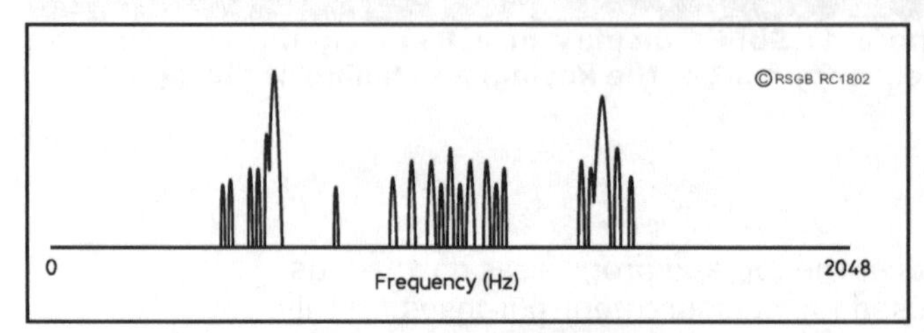


**Fig 1: A trace of power spectral density, from one line of Photo 1. This can be compared directly with the graph in Photo 2, which averages all 370 lines of the display.**

This photograph shows the intriguing nature of the FFT, especially when displayed as a time-dependent feature, as with the *SbFFT* software. It is easy to spend hours tuning through various signals and studying their spectra.

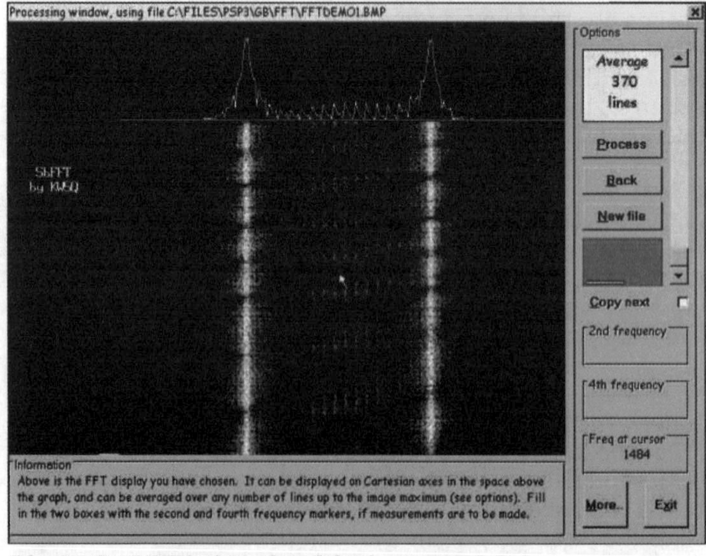


**Photo 2: RTTY signal, with the averaged spectrum displayed above it.**

Few signals can be received noise-free, as the photograph shows. If a single line of this FFT is displayed as a simple graph of power against frequency (**Fig 1**), it means surprisingly little, because of the noise. However, for signals like this one where the noise is random, and where the spectral information of the wanted signal is not radically changing, there is a way in which science can come to the rescue of noisy FFTs. It uses the fact that noise is random in nature and will average (given infinite time) to zero. The 'wanted' signal spectrum remains essentially unchanged. Consequently, if the FFT is averaged over all 370 lines, the noise will reduce and the signal will be seen more clearly. My own programme takes the FFT display from *SbFFT* and averages it, line by line, over the 370 lines of the original FFT. The result is shown in **Photo 2**.

The averaged spectrum is displayed in the area above the original data, on the same frequency axis, and can be compared directly with the single-line graph of Fig 1. Averaged signals can usually be measured more accurately than unprocessed data, although time-dependent displays do show artefacts that averaged data can remove. Frequency-selective fading is a good example of this.

### FFTS AND DOPPLER SHIFTS

One of the most eye-catching and interesting uses of programmes like *SbFFT*, is to display the Doppler-shifted signal received from a polar-orbiting satellite. First, a few words about the equipment and the receiver settings are in order. A Kenwood R-5000 receiver with a VHF adapter was used, the aerial being a standard 2m colinear, mounted in the house loft. The 136.770MHz beacon on the NOAA-12 weather satellite provided the signal source. The R- 5000 was set for USB reception, the idea being to use the carrier insertion oscillator of the receiver to mix with the incoming carrier, and produce an audible beat note in the speaker. The beat note will be the absolute difference between the frequencies of the carrier insertion oscillator and the incoming signal.

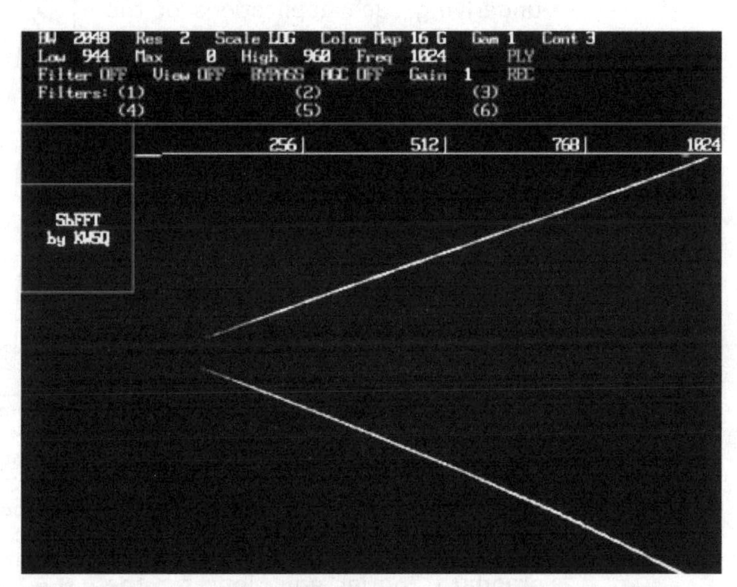

**Photo 3: The Doppler shift of a passing weather satellite.**

With the receiver tuned to the source frequency of 136.770MHz, the beat note is exactly equal to the magnitude of the Doppler shift! What is heard is a high-pitched whistle which slowly decreases in frequency, until it is finally removed (at around 200Hz) by the receiver filtering. The satellite is at its closest approach when the two frequencies are at zero-beat. Unfortunately, this crucial moment is missed. However, the R-5000 has the facility to remove the normal sideband filtering so that heterodynes are heard from frequencies in what would normally be the 'other sideband'. Switching the filters out in this way allows the Doppler shift to be heard as it comes down in frequency, is lost around 200Hz due to the RC-coupled audio stages, and then reappears at 200Hz again, this time as a rising frequency. **Photo 3** shows just what happens.

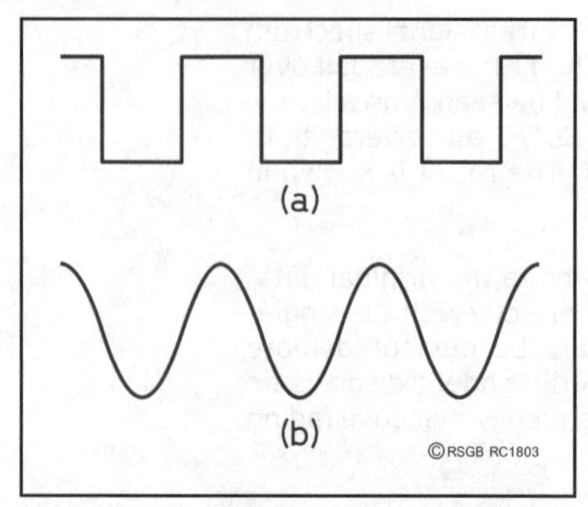


Fig 2: (a) The desired waveform. (b) The fundamental frequency sine wave (the first term of the series).

An equation exists [3] which allows the minimum slant range of the satellite to be calculated from the maximum rate of change of the Doppler shift, and I have written a programme which automatically calculates this range from the display shown in the photograph. The results agree very well with theory. It is also useful, and instructive, to compare the FFT displays for near-overhead satellite passes, and other more distant passes. The characteristic 'V' shape has a very narrow apex angle for an overhead pass, the angle increasing for passes at greater distances, as the rate of change of received frequency decreases.

A final (practical) note about the *SbFFT* programme. It also allows six independent digital filters to be simultaneously applied to the signal, and the result fed to the speakers attached to the sound card. Each filter can be toggled in or out of the signal chain at will. These filters all have -3dB to -50dB skirts of 25Hz, and pass-band filters down to 1Hz bandwidth are possible. CW signals only a few Hertz apart can be resolved in this way, making it a truly remarkable package!

**FOURIER SYNTHESIS AND ANALYSIS**

Some explanation of the processes underlying such applications of the work of Fourier now helps to build an appreciation of the techniques involved. Let us start with the process which is the reverse of that which we have just considered.

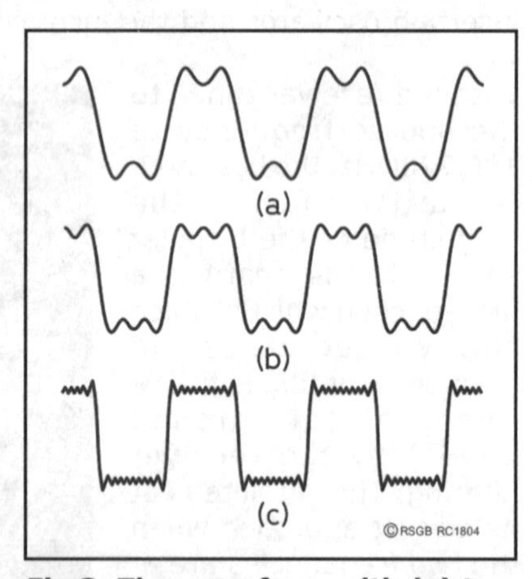


**Fig 3: The waveform with (a) two components (the fundamental and the 3rd harmonic), (b) three components (the fundamental, 3rd and 5th harmonics), (c) the first 10 terms (ie with harmonics up to the 19th).**

Fourier synthesis concerns the creation of a complex signal from simple sine waves. Consider the rectangular wave of **Fig 2(a)**. There is little comparison between its shape and that of a pure sine wave of the same frequency, **Fig 2(b)**, except that each has an equal positive and negative excursion, once per cycle. Despite this, Fourier says that Fig 2(a) *can* be produced by the sum of sine waves alone. This can be proved very simply on a computer [4], using the standard Fourier equations for the components of a rectangular wave [5]:

$$V = {}^{4E}/_{\pi}[\cos(f) - {}^{1}/_{3}\cos(3f) + {}^{1}/_{5}\cos(5f) - {}^{1}/_{7}\cos(7f) + \ldots],$$

where: $^{4E}/_{\pi}$ allows a particular amplitude of the resultant waveform to be specified, and f is the frequency of the synthesised wave.

Notice the following aspects of the equation:

- the signs of the terms alternate, positive and negative;
- the terms represent *only odd harmonics* (3, 5, 7, etc) of the fundamental frequency, f;
- each term is preceded by a decreasing factor ($^1/_3$, $^1/_5$, $^1/_7$, etc) showing that the amplitudes of the harmonics decrease quite rapidly;
- the dots at the end show that the equation can be extended as far as is meaningful in any particular context.

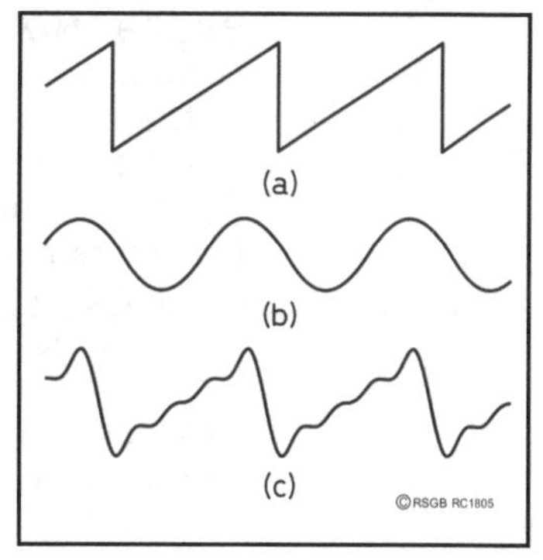


**Fig 4:(a) The sawtooth wave to be synthesised. (b) The fundamental frequency sine wave. (c) The fundamental frequency sine wave with the first four terms (ie with harmonics up to the 4th).**

Using this equation, the shape of the resultant waveform can be derived for any number of terms in this expression. **Fig 3** shows the resultant waveform for 2, 3 and 10 terms. It is easily seen that the first term gives rise to the basic frequency of the final waveform, while the harmonics serve to add the fine detail to the wave - the rapid rise and fall and the sharp corners - both being attributes which we intuitively associate with a high frequency response. The more terms we use in the evaluation, the better the final wave approximates to the shape of Fig 2(a).

The same can be done with a sawtooth wave, shown in **Fig 4(a)**. This wave can be synthesised using the Fourier Series:

$$V = 2/\pi E[\sin(f) - {}^1/_2 \sin(2f) + {}^1/_3 \sin(3f) - {}^1/_4 \sin(4f) + \ldots]$$

This series has *all* harmonics, both even and odd; the terms still alternate their signs, and the denominator of each fraction matches the harmonic number. **Fig 4(b)** shows the fundamental frequency used as a basis for the synthesised wave. **Fig 4(c)** shows the wave synthesised with four terms only, ie using the expression as it stands, with no additional terms. Notice that the peak of the fundamental does not coincide with the peak of the sawtooth; its amplitude is less than that of the sawtooth; as before, the smoothness of the rise and the sharpness of the fall are both due to the harmonic content of the wave.

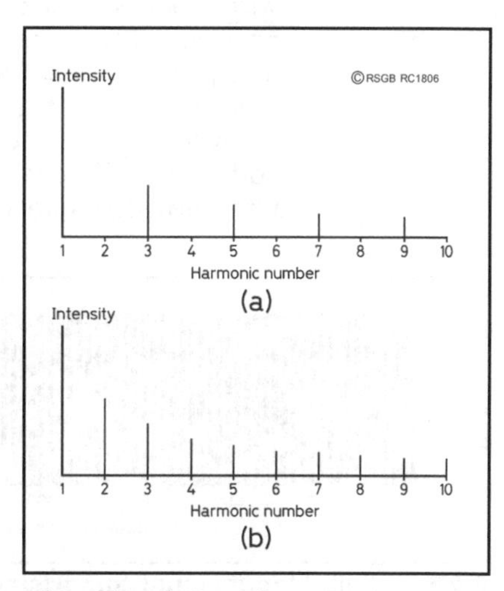


**Fig 5: (a) The component frequencies and intensities for the rectangular wave up to the 9th harmonic. (b) The component frequencies and intensities of the sawtooth wave up to the 10th harmonic.**

All the diagrams shown so far are illustrations of processes taking place in the *time domain* (characterised by the use of *time* as the horizontal axis). The frequencies needed to synthesise the two waveforms are shown in **Fig 5**. Note that, as the equations suggest, the higher harmonic terms ecome smaller and smaller in intensity.

## BASEBAND FREQUENCIES AND BANDWIDTH

For our purposes, we can define *baseband frequencies* as those present in the modulating signal *before* they are used to modulate the carrier. **Fig 5(a)** shows the baseband frequencies of a rectangular wave, and **Fig 5(b)** those of a sawtooth wave. The range of baseband frequencies, from the fundamental to the highest harmonic present, defines the *bandwidth* of the modulating signal. These diagrams show that, for practical purposes, there comes a point beyond which adding higher harmonics becomes inefficient, as their presence changes the waveform only slightly at the expense of greatly increased bandwidth. This illustrates the well-known compromise between acceptable intelligibility at narrow bandwidth and acceptable fidelity at wide bandwidth.

Fig 5 shows two examples of processes occurring in the *frequency domain*. They can be transformed into their rectangular-wave and sawtooth-wave time-domain equivalents (and back again) by repeated use of the Fourier transform.

Fourier synthesis performed in this way requires the availability of a mathematical equation which specifies the frequencies and amplitudes to be used. In the case of amateur radio, this is not very common. However, the way in which a well-known waveform is built up from its component parts, as in Fig 2 and Fig 3, gives an intuitive 'feel' for the way in which other waveforms will be affected.

By now, you should have no real problem in accepting that any wave can be synthesised from a set of sine-wave components. Performing the reverse process is computationally more complex, but should not represent a conceptual problem. Any 3-year old will know that he can build a house with wooden blocks, and then demonstrate with great glee the reverse process! The reverse process in our case is called *Fourier analysis*, analysing a wave into its building blocks - a number of sine waves. We have looked at the use of the *SbFFT* [2] freeware programme for the analysis of RTTY and Doppler signals in Photo 1 and Photo 3. Mathematical software packages, such as *Mathcad* [6], can also perform the Fourier transform on digital functions, allowing theory and practice to be compared.

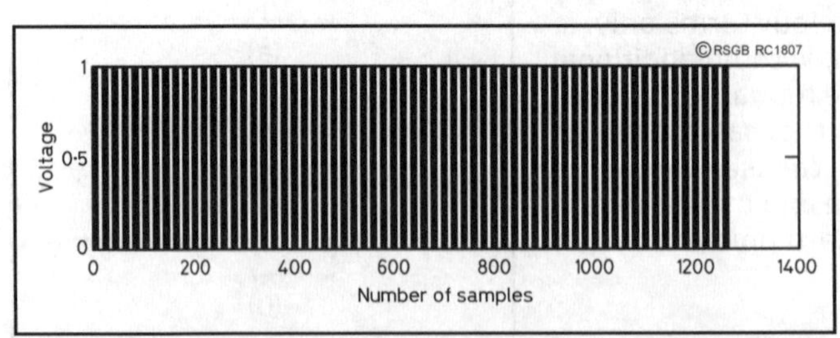


**Fig 6: A digital input signal comprising a rectangular wave of 10 samples of 1V and 10 samples of 0V, repeated 64 times.**

Some of the graphs used later on were derived using this software. The interpretation of the output of a standard FFT requires a little mental agility. Take an example, shown by **Fig 6**, of an input signal consisting of a rectangular wave. This wave is represented digitally in the form of 1280 samples, which means that the voltage of the wave has been examined at 1280 equal increments of time, and the results plotted. The wave is a repetition (64 times) of 10 samples of 1V and 10 samples of 0V. The 1V areas are blocked in, so that it is visually obvious that the wave is rectangular, and has a 1:1 mark-space ratio.

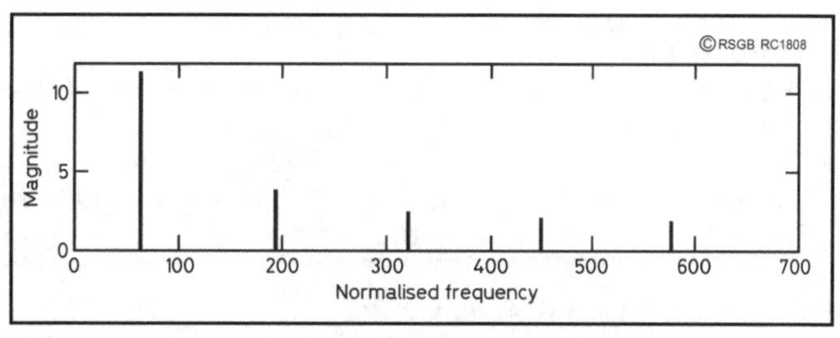


**Fig 7: Fourier transform of the waveform of Fig 6. See the text for an explanation of the frequency axis.**

The unique part of the FFT output (now in the frequency domain) is shown in **Fig 7**, and is taken from the *Mathcad* programme. There is little doubt that it is identical with Fig 5(a), but the labelling of the horizontal axis is not at all clear. I shall try to keep the explanation brief! The maximum frequency that can be represented by our 1280 samples is represented by one in which the sample values are '10101010 . . .'; all other combinations of zeroes and ones give lower frequencies (which is a first-principles statement of the famous theorem attributed to Shannon and Nyquist, that the lowest possible sampling frequency must be at least twice that of the highest frequency present in the signal). It is not necessary to know the time period of the 1280 samples; this is quite arbitrary. To make things simple, suppose this time period is one second.

Our maximum frequency has a wavelength of two samples, since it is the signal '10' repeated 640 times to give the 1280 samples. 640 cycles in the one second signal interval correspond to a frequency of 640Hz. Our hypothetical input signal (Fig 6), occupies 20 samples, therefore it must have a frequency which is 1/10 that of the highest frequency, making it 640/10 = 64Hz. The graph in Fig 7 has 640 entries, corresponding to frequencies from DC to 640Hz, in 1Hz increments. If the graph is examined closely, there are entries *only* at 64Hz, 192Hz, 320Hz, 448Hz and 576Hz, corresponding to the fundamental frequency together with the 3rd, 5th, 7th and 9th harmonics!

The amplitudes of these signals are in the ratio 1: $^1/_3$: $^1/_5$: $^1/_7$: $^1/_9$, as we would expect from our original synthesis of the rectangular wave.

We need not concern ourselves about the absolute magnitudes of these individual entries, as it is the relative magnitudes in which we are interested. Our allocation of one second to the 1280 samples is also quite arbitrary, as the fundamental frequency will always be $^1/_{10}$ of the highest frequency for this particular waveform; all the other frequencies will maintain the same relationship to each other.

We have now demonstrated the two converse activities of breaking a rectangular wave into its sine-wave components, and reconstituting it by the algebraic addition of these components. This Jekyll and Hyde existence of waves permits great flexibility in their processing, because this can occur in the time domain or in the frequency domain. Although digital processing in the frequency domain is much more complex, the results *can* be quite remarkable, as shown by the examples given here. Perhaps this article may stimulate anyone interested in waves and their significance to experiment further, using packages such as *SbFFT* and *Mathcad*.

## APPENDIX

THIS SIMPLE THEORY of Amplitude Modulation (AM) serves to illustrate the principle of Fourier Analysis, but by a different route.

Suppose we have a carrier at a frequency $f_c$ and a single modulating frequency $f_m$. These are represented by the formulæ:

$V_c = \sin(2\pi f_c t)$ and $V_m = \sin(2\pi f_m t)$.

The modulation process effectively multiplies these two together, and introduces the depth of modulation, m [m = 0 for zero modulation, and m = 1 for 100% modulation]. The resultant modulated signal thus becomes:

$V = [1 + m.\sin(2\pi f_m t)] \sin(2\pi f_c t)$.

For simplicity, let us assume that m = 1. The expression reduces to:

$V = \sin(2\pi f_c t) + \sin(2\pi f_m t) \times \sin(2\pi f_c t)$.

Recourse to an A-level book of trigonometrical formulæ will produce an equivalent form for the product of two sines:
$\sin(A) \times \sin(B) = {}^1/_2\cos(A-B) - {}^1/_2\cos(A+B)$.

Applying this to the previous equation, we have:

$V = \sin[2\pi(f_c)t] + {}^1/_2\cos[2\pi(f_c - f_m)t] - {}^1/_2\cos[2\pi(f_c + f_m)t]$.

This gives the same result that Fourier theory would derive, that the radiated signal has *three* discrete frequency components, at frequencies $(f_c - f_m)$, $f_c$ and $(f_c + f_m)$. These frequencies have been put in round brackets in this expression to facilitate their identification.

## REFERENCES

[1] 'A Conversation with Mike Cook, AF9Y', *QST*, January 1998, p9 and pp 56, 57. The FFT software is available from http://www.webcom.com/af9y

[2] The *DOS* freeware programme *SbFFT*, by Kevin McWilliams, KW5Q, is available by anonymous ftp from ftp.funet.fi/pub/ ham/. Alternative sources have been mentioned by Roger Cooke, G3LDI, in his 'Data' column (*RadCom*, July 1997 and February 1998).

[3] *The Satellite Experimenter's Handbook*, by Martin Davidoff, K2UBC, ARRL, 1st ed, p10-2, 2nd ed, p13-2.

[4] The programme *FOURIER.EXE*, which runs under *DOS*, was written some years ago by the author for use with sixth form students. It shows how rectangular, triangular and sawtooth waves are made up from their individual components, using any number of terms.

[5] *Radio Engineers' Handbook*, by Frederick E Terman, McGraw-Hill (New York) 1943, p20.

[6] The *Windows* programme *Mathcad* has been available for several years now, and is currently supplied by Adept Scientific (see current computing press for details).

# AN INTRODUCTION TO VARIABLE TUNED CIRCUITS

by Stewart Revell, G3PMJ

***This article necessarily involves a little mathematics, but you can ignore the mathematical description and download a spreadsheet from the RSGB Members-Only website, which will do all the hard work for you.***

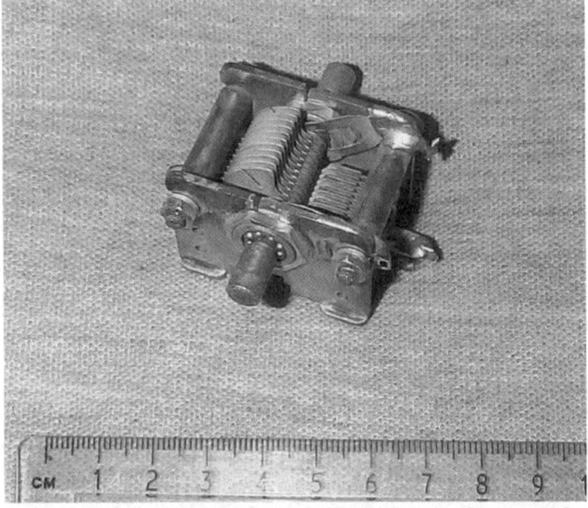


**A variable capacitor. The capacitance is changed by varying the area of overlap of the fixed and rotating vanes.**

Of fundamental importance to our leisure time, either listening or transmitting, is the humble tuned circuit at resonance. Our very existence as radio amateurs, the whole of the RSGB, the vast systems of world-wide radio communications, are totally dependent on the fact that when an inductor and a capacitor have equal and opposite reactances at a certain frequency, that circuit exhibits a phenomenon called resonance.

A capacitor and inductor when used together are called a 'tuned circuit'. A fixed value of inductance (L) and a fixed value of capacitance (C) will produce a fixed resonant frequency. For obvious reasons, this is called a fixed-tuned circuit. However, there are many cases where we need to make our resonant frequency variable, so that it can be made to 'tune' over a band of frequencies.

**Typical small inductors.**

This can be achieved in three ways - change the value of C, change the value of L, or both. Although it is possible to change the value of an inductance, it is not as easy as using a variable capacitor, particularly when you are new to the business of circuit construction. The rest of this article will show you how to derive the value of inductance to use with a given variable capacitor, knowing the frequency band you want to cover.

It necessarily involves a little mathematics, but you can ignore the mathematical description and download a spreadsheet from the RSGB Members-Only website which will do all the hard work for you.

### THE MATHS

It is relatively easy to calculate the frequency of resonance given the values of capacitance and inductance of a tuned circuit from the wellknown formula which you will find in most radio textbooks:

$$f = \frac{1}{2\pi\sqrt{LC}},$$

where $\pi$ is a constant (3.141), $L$ is the inductance (in henries) and $C$ the capacitance (in farads).

From this fundamental formula, we can see that the frequency is obviously dependent, in quite a complex way, on $L$ and $C$, a fact which was mentioned earlier.

Things become more involved when we try to bring in our variable capacitor and the range of frequencies over which we want to tune. Changing two variables in an equation at the same time is not recommended practice!

First of all, let us make the equation simpler by combining all the constants into one number and, at the same time, make the units of $f$, megahertz, $L$, microhenries and $C$, picofarads, as these will be the units we use 'on the bench'. Doing this, the equation becomes:

$$f = \frac{159.2}{\sqrt{LC}}.$$

Things are still confused by the square-root sign, which we can remove by squaring both sides of the equation, as follows:

$$f^2 = \frac{159.2^2}{(\sqrt{LC})^2} = \frac{25330.3}{LC}.$$

We are almost in a position to use our equation specifically for our problem; the only thing left to do is to rearrange it to make $LC$ the subject (ie put $LC$ on the left-hand side). This simply involves swapping over the $f^2$ and the $LC$, thus:

$$LC = \frac{25330.3}{f^2}.$$

This gives the $LC$ product for any frequency, $f$. How this is used to design a tuned circuit is shown in the two examples to follow.

Bear in mind that the calculations can be used with a varactor diode as the tuning element, using the upper and lower limits of its capacitance in the formulae.
A program performing the necessary calculations has been designed by Harry Lythall, SM0VPO / G4VVJ, and this is available (with other radio-

type calculations) at a small cost. Contact Harry by e-mail to sm0vpo@home.se or the author for details.

**EXAMPLE 1: A VFO is required to cover 5.0MHz to 5.5MHz using a high-quality variable capacitor of 50pF.**

**METHOD**

We shall make the assumption (which will be discussed later) that, at its extremities of motion, the capacitor has values of zero and 50pF. First, calculate the *LC* values at each end of the band. At the low-frequency end, we use $f$ = 5.0MHz and at the high-frequency end, $f$ = 5.5MHz. In this case we obtain the following:

$$LC_{(LF)} = \frac{25330.3}{5.0^2} = 1013.2, \qquad (1)$$

and

$$LC_{(HF)} = \frac{25330.3}{5.5^2} = 837.6. \qquad (2)$$

To give a slight extra margin at each end of the band to be tuned, it is best to round off the figures in the opposite directions; for the LF figure use $LC$ =1014, and for the HF figure use $LC$ = 837.

As the capacitor is turned from being fully meshed (the highest value of $C$, giving the lowest frequency), to being fully open (the lowest value of $C$, giving the highest frequency), the value of $LC$ swings from 1014 to 837, a range of 177.

This range of 177 must be produced by our chosen capacitor, which has a 'swing' of 50pF, so we can divide the *LC* range by the swing in order to derive the value of $L$ to use in our circuit, thus:

$$L = \frac{LC}{C} = \frac{177}{50} = 3.5\mu\text{H}. \qquad (3)$$

At this point, we must think very clearly. Having derived the correct value of inductance, we must examine the two frequency limits and work out the values of capacitance corresponding to these frequencies, which we shall call $C_{LF}$ and $C_{HF}$.

$$C_{LF} = \frac{LC_{(LF)}}{L} = \frac{1014}{3.5} = 286\text{pF}, \qquad (4)$$

and

(5)
$$C_{HF} = \frac{LC_{(HF)}}{L} = \frac{837}{3.5} = 236\text{pF}. \qquad (5)$$

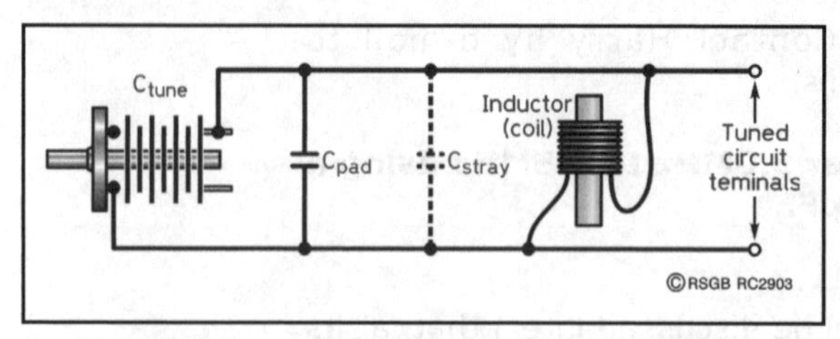


**Fig 1: Diagram of the elements of a variable tuned circuit.**

It can be seen from this that the 'swing' of capacitance is 50pF, which is what we want, but the actual values of calculated capacitance are somewhat larger than this.

If you now recall some of the theory of the Advanced examination, you will know that when capacitors are connected in parallel, their values add. Look at **Fig 1**. This shows the circuit of the components necessary to implement our example. For the moment, ignore $C_{stray}$.

You will see immediately that there are two capacitors in the circuit: $C_{tune}$ and $C_{pad}$. $C_{tune}$ is our 50pF variable. We can infer the value of $C_{pad}$ from the last equation. The minimum calculated capacitance is 236pF, and this occurs when $C_{tune}$ is zero (see the first sentence of 'Method'). This makes $C_{pad}$ = 236pF. When $C_{tune}$ is fully meshed, its value is 50pF, which adds to the 236pF of $C_{pad}$, producing 286pF, the calculated value of $C_{LF}$, as shown above.

When fully open, most capacitors have a residual capacitance of around 5pF. To account for this, subtract 5pF from the 'swing' used in equation (3). This will then produce slightly different values for $L$, $C_{LF}$, $C_{HF}$ and $C_{pad}$, but the frequency swing will be unaltered! The next example will include this.

**EXAMPLE 2: The front end of a short-wave radio is required to tune from 12 to 16MHz using a 30pF variable capacitor.**

**METHOD**

As before, we first need to calculate the $LC$ values at each end of the frequency range:

$$LC_{(LF)} = \frac{25330.3}{12^2} = 175.9\text{, and } LC_{(HF)} = \frac{25330.3}{16^2} = 98.9.$$

To give working margins, we round the two results in opposite directions, thus giving $LC_{(LF)} = 176$ and $LC_{(HF)} = 98$.

The range, or 'swing' of $LC$ is the difference of these two numbers, 78.

This range of $LC$ must be produced by our 30pF capacitor. Most variable capacitors have a residual capacitance of 5pF. This means that, although the maximum capacitance (with the vanes fully meshed) is 30pF, its minimum value is not zero, but 5pF. This gives rise to the figure of 25pF for the 'swing' of the capacitor, which will now be used.

To calculate the inductance for our circuit, the *LC* swing must be divided by the *C* swing, as follows:

$$L = \frac{LC}{C} = \frac{78}{25} = 3.1\mu\text{H}.$$

We now need to derive the values of capacitance needed to produce our stated frequency limits using the value of *L* we have just found:

$$C_{LF} = \frac{LC_{(LF)}}{L} = \frac{176}{3.1} = 56\text{pF}, \quad \text{and} \quad C_{HF} = \frac{LC_{(HF)}}{L} = \frac{98}{3.1} = 31\text{pF}.$$

As before, you can see that the swing in capacitance is 25pF, but that an extra capacitor (pad) of 31pF is needed in parallel with the variable, in order to bring the minimum capacitance up to 31pF.

Try as you may, it is never possible to eliminate the effects of 'stray' capacitance. This is capacitance due to the mutual proximity of circuit elements, and can never be accurately predicted.

Unfortunately, you cannot go to a shop or component supplier and ask for a 31pF capacitor; instead, a fixed capacitor plus a very small variable capacitor (called a trimmer) are used to make up $C_{pad}$. Not only can the correct value be set by varying the trimmer, but it will also enable the effects of the strays (unique to your construction) to be compensated.

In this particular case, a fixed capacitor of 25pF would be used with a trimmer of 20pF. The setting of the trimmer would be accomplished by using a Dip Meter. Measure the resonant frequency at the upper band edge (with $C_{tune}$ fully un-meshed) and vary the trimmer until the frequency reaches the wanted value of 16MHz. Then confirm with the meter that the LF limit (with $C_{tune}$ fully meshed) is 12MHz.

Not only have you designed a tuned circuit to cover a specific frequency range, but you have calibrated it in practice and have shown that it works to specification!

# AN INTRODUCTION TO VOLTAGE REGULATION

**by George Brown, M5ACN / G1VCY**

Nothing is perfect; the AC mains, transformers, resistors, transistors, integrated circuits; we know how they ought to behave, but how they really behave can be rather different.

The AC mains is not constant, particularly if you live out in the country and are served by several miles of overhead cable. Your next-door neighbour switches on her electric oven, and your lights dim! Transformers have losses in the resistance of their windings, and these losses vary depending on the load at any given time.

Resistors produce a voltage drop when a current flows through them. The current may cause a resistor to heat up unduly, change its resistance, and produce a different voltage drop. Transistors are also sensitive to changes in temperature, and may cause major problems in a poorly-designed circuit.

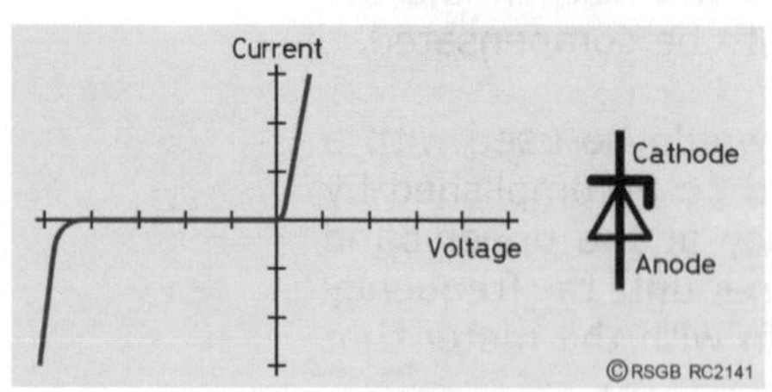


**Fig 1: The circuit symbol and characteristic of a typical Zener diode.**

One problem is that the TTL family of integrated circuits requires a supply of 5.00 ± 0.25V. Whatever happens in the rest of your equipment must not cause more than 0.25V variation of the 5V supply.

### THE ZENER DIODE

For voltages below about 30V, the heart of almost all voltage stabilisation (or regulation) circuits is the Zener diode. This is a discrete silicon device, which has the normal diode characteristic when forward-biased (ie when the anode is positive with respect to the cathode); but, when reverse-biased, it passes negligible current up to a well-defined point, at which the current increases sharply. This is illustrated in **Fig 1**, which also introduces the circuit symbol.

The point at which the reverse current starts to flow can be closely controlled in manufacture between about 3V to 200V. If such a Zener diode is intentionally biased into this steep region, the voltage across the diode varies very little for large changes in current; the device is serving to stabilise (or regulate) the voltage across it. A typical circuit is shown in **Fig 2**.

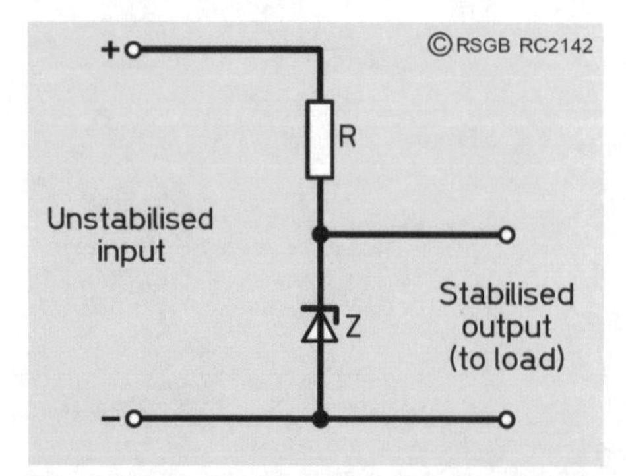


**Fig 2: A simple circuit that produces a stabilised output from an unstabilised input.**

The value of R is given by the equation:

$$R = \frac{V_{out}(V_{in} - V_{out})}{P},$$

where *Vin* is the unstabilised input voltage, *Vout* is the stabilized output voltage (the Zener voltage), and *P* is the Zener dissipation in watts (see text).

Zener diodes are specified by two parameters – the Zener voltage and the maximum power dissipation of the device. For example, all Zener diodes of the types BZY88C and BZX55C have 0.5W dissipation, and are by far the commonest. A 15V Zener would be marked BZY88C15V; you are expected to know it is a 0.5W device!

**SEE-SAW ACTION**

A see-saw is a simplistic but very useful analogy of how the Zener works in a circuit. In Fig 2, the current from the unstabilised supply has two paths - through R and Z to earth, or through R and the load connected across Z. Suppose we have chosen a 10V Zener of the BZY88C series. It is a 0.5W device, therefore it will draw a current, I, given by:

$$I = \frac{P}{V} = \frac{0.5}{10} = 0.050A, \text{ or } 50\text{mA}.$$

Suppose no load is connected. All 50mA will go through the Zener, as there is no other route. Suppose the load now takes 5mA; here the current see-saw comes in. With 5mA going through the load, 45mA now goes through the Zener. If the load takes 10mA, then 40mA flows through the Zener. This keeps the total current constant (from the unstabilised supply); this helps to keep the voltages constant also.

This see-saw action is not perfect, particularly if the load takes more and more current. As the current through the Zener approaches zero, the see-saw action fails and the stabilisation becomes inoperative. This circuit is simple and works well, provided the demands of the load are small, but it has one big disadvantage - your power supply must deliver the maximum current at all times, whether or not the load takes any.

Notice in the example that the Zener takes 50mA when there is no load. Notice also that the unstabilised voltage must always be significantly greater than the stabilised voltage, because of the voltage drop across R. It is inadvisable to try to construct a 12V stabilised supply from a 12V unstabilised input! As the range of currents demanded by the load increases, so also should the difference between the supply voltage and the regulated voltage.

**A REFERENCE VOLTAGE**

Removing the need for the power supply to supply the maximum power at all times can be achieved by keeping the current through the Zener constant (and relatively low). In this way, it does not need to dissipate maximum power, and the regulation it provides is much better. It is used

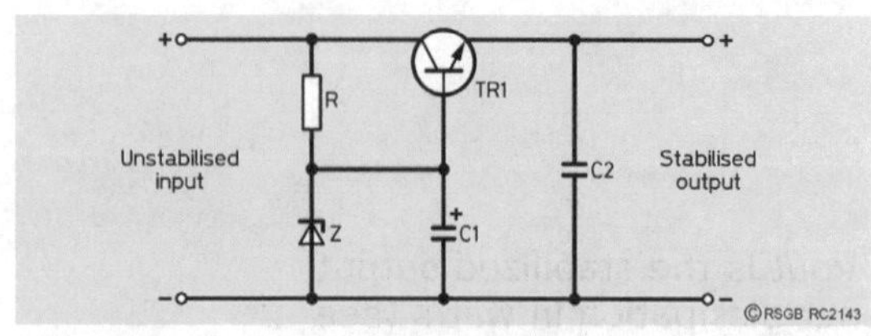

**Fig 3: An improved circuit.**

more as a reference voltage than as a type of voltage source. This idea is illustrated in **Fig 3**.

The stabilised voltage is produced at the base of the transistor, which acts as an emitter follower, producing about 0.6V less than the voltage across the Zener, but at a current limited only by the capacity of the power supply and the characteristics of the **series** or **pass transistor**, TR1, which usually requires a heat sink. There is a shunt-stabilising version of this circuit, but it is little used because of its inefficiency.

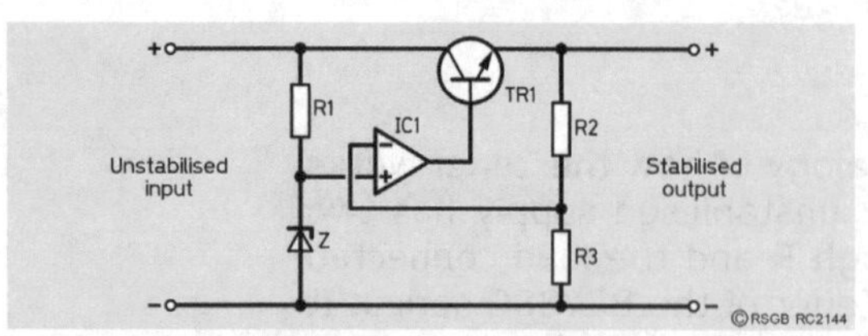

**Fig 4: Using an operational amplifier to improve the circuit.**

To take this circuit to its logical conclusion and produce a well-stabilised supply, the circuit of **Fig 4** is common. This produces, by means of the integrated circuit Op-Amp, a very sensitive comparison between the Zener voltage, VZ, and the output voltage divided down by R2 and R3. This allows the output voltage to be greater than the Zener voltage and, depending upon the voltages and currents involved, can obviate the need for extra smoothing capacitors. Readers interested in refinements such as foldback current limiting and crowbar protection are referred to [1] and [2].

Needless to say, integrated circuits have been produced that possess all the qualities (and some refinements, too) discussed so far, but for the lower output current ranges only. These are typified by fixed positive-voltage devices such as the LM78**CT series of 1A regulators and their negative voltage equivalent, the LM79**CT series. The LM338K is a 5A regulator for output voltages between 1.2V and 32V. The LM317T is a popular 1.5A variable regulator.

## HIGHER VOLTAGES

When stabilisation is required at high voltages, solid-state regulation gives way to the regulation afforded by ionised gases at very low pressures. The use of neon and hydrogen discharge devices is beyond the scope of this article, but basic details may be found in [3].

## REFERENCES


[1] *ARRL Handbook*, 1998 edition, p 11.15.
[2] *Radio Communication Handbook*, RSGB, 6th edition, p 3.13.
[3] *The Services' Textbook of Radio*, HMSO, Volume 3, 1963, pp180 - 198.

# Chapter 2 SIMPLE TEST EQUIPMENT
# A JUNK-BOX SPECTRUM ANALYSER

**by Pat Hawker, G3VA, from 'Technical Topics'**

Desmond H Vance, GI3XZM, has for many years used a simple, improvised piece of test gear that he finds indispensable. He jokingly refers to this as his "junk box spectrum analyser", although it utilises a multimeter rather than a CRO display and is not frequency-selective. He writes: "The function of this device is to give a rough indication of the purity of the output of any oscillator or single-ended RF power amplifier while monitoring the output level. Such circuits sometimes have an asymmetrical output with one voltage peak higher than the other. Not always, but sometimes, this is important. When we try to measure an asymmetrical RF waveform using a single diode probe (as often the case), we obtain a value for either the positive or the negative peak, depending on the polarity of the diode.

"There are several consequences.

- The reading will be inaccurate, perhaps grossly.
- If the reading is not what we expected, we may waste time trying to adjust or alter the circuit to achieve a 'correct' reading, unaware that we have chosen one of two inaccurate readings.
- A spectrum analyser would show the signal to have harmonic content (principally second harmonic), possibly to a significant degree.

"The fairly common voltage-doubler probe (peak-to-peak reading) gives a better indication of the fundamental output, but it does not provide warning of harmonic content. A probe that measures the positive and negative peaks *separately* yields more useful information:

- by adding the two readings we have the peak-to-peak output, a reasonable indication of the (required) fundamental;
- if the readings agree (single-ended circuits, remember) we are likely to have a fairly clean sine wave; if not, we have been warned;
- comparison of the sum and difference readings of the two peaks can be used to give a rough estimate of second harmonic content, as shown in **Fig 1**.

"No exhaustive tests have ever been made on this simple device, although I have been guided by it for about 20 years! Initially, I built an oscillator (about 5MHz) with two outputs, one partially filtered, and then increased

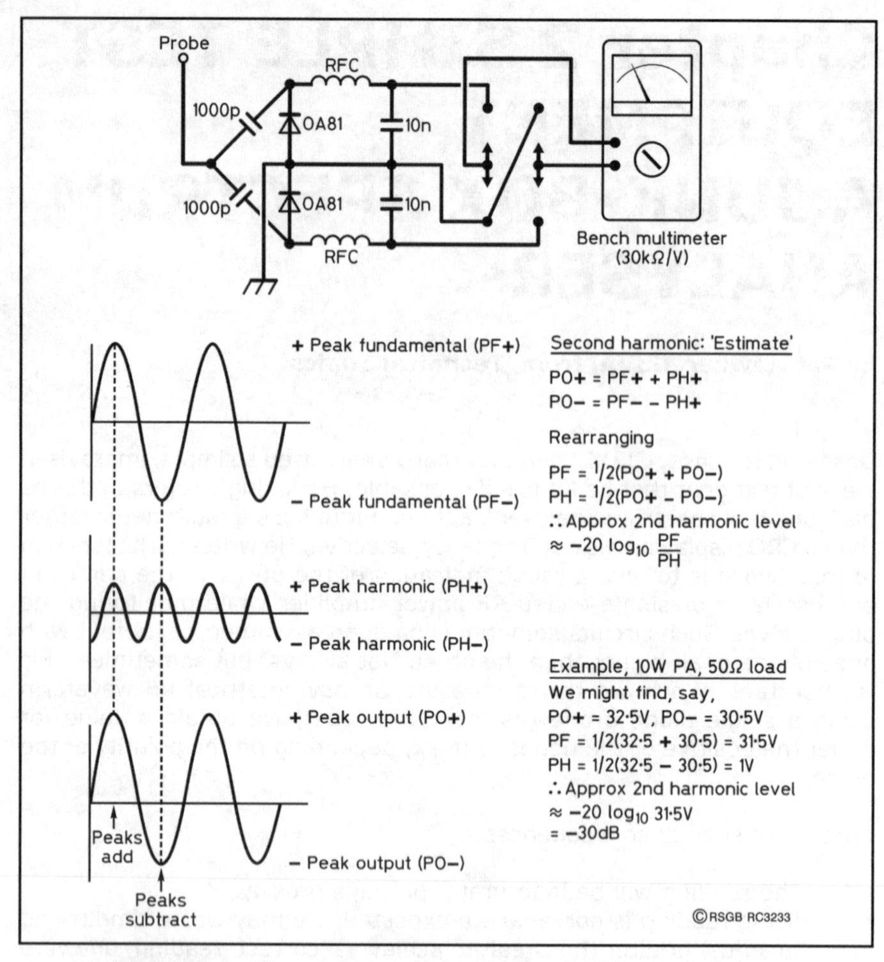


**Fig 1: GI3XZM's 'Junk box spectrum analyser' or, perhaps more accurately, a simple RF wave analyser useful for estimating the degree of second harmonic content in the output of an oscillator or amplifier.**

the feedback until my method suggested a second harmonic output of about -22dB and -28dB. Ian Kyle, GI8AYZ, measured the outputs on an HP spectrum analyser and (from memory) found things to be about 3dB worse than I had estimated from my readings. This still puzzles me because I believed I was 'measuring' total even-order harmonic-content and therefore obtaining a conservative figure. I would stress that I make no claim to expertise in this field, although I still find the device of practical use. Perhaps readers might like to comment?"

GI3XZM notes my suggestion ('TT' December 2001, p62) that dissatisfaction with 'the modern easy-to-use black-box concept' is linked to the growth of QRP operation. He writes: "While I much enjoy QRP and have the highest regard for the encouragement given by G3RJV and the G QRP Club, I have long thought that there must be a place for a form of similar 'sister' club. This would be committed to simple home-built equipment.

# AN npn TRANSISTOR TESTER

**By David Clark**

A frequently-used method of checking that an npn transistor is not faulty involves measuring the resistance of its p-n junctions, ie between the base and emitter and the base and collector. These resistance values should be low when the junction is forward-biased (anode positive, cathode negative), and high when reverse-biased (anode negative, cathode positive). Additionally, the measured resistance between the collector and emitter should always be high since, whatever the polarity of the test voltage, one of the two 'back-to-back' junctions will always be reverse-biased. This means a minimum of five checks must be made with a multimeter, for example, before a transistor can be considered good.

An 'ideal' p-n junction would have zero resistance when forward-biased and infinite resistance when reverse-biased. Real p-n junctions, however, have a significant forward resistance and a far-from-infinite reverse resistance and so, when checking a transistor by this resistance method, it is often unclear whether the transistor is functioning or not. A more conclusive check would be to see if the device performs its correct function as a current amplifier, and this npn transistor tester circuit does exactly that. The tester also gives an LED indication if the transistor is working, thus avoiding the necessity of having to make a decision based on several meter readings. In this way, multiple checks and reconnections of the transistor leads are not required - only one test is needed after having connected the transistor to three test leads.

**HOW IT WORKS**

The 'detector' part of the circuit works by making the transistor under test part of an astable multivibrator (a free-running oscillator) circuit via flying test leads. If the transistor operates correctly, the output of the multivibrator will be approximately a square wave; otherwise it will be a DC voltage fixed at some point between zero and the supply voltage.

The signal from the multivibrator is connected via a DC blocking capacitor to straightforward rectification and smoothing circuits. The output voltage of this section will be zero if the signal from the multivibrator is a DC voltage (ie if the test transistor is faulty). However, if the transistor is switching correctly and the multivibrator is oscillating, the output will be a positive voltage.

The final stage is the 'indicator' section. The voltage from the rectification and smoothing circuits is connected to a transistor that drives the indicating LED; if this voltage is positive (greater than around 0.6 to 0.7V) the LED lights, indicating that the transistor under test is operating. If it is zero, the LED remains unlit.

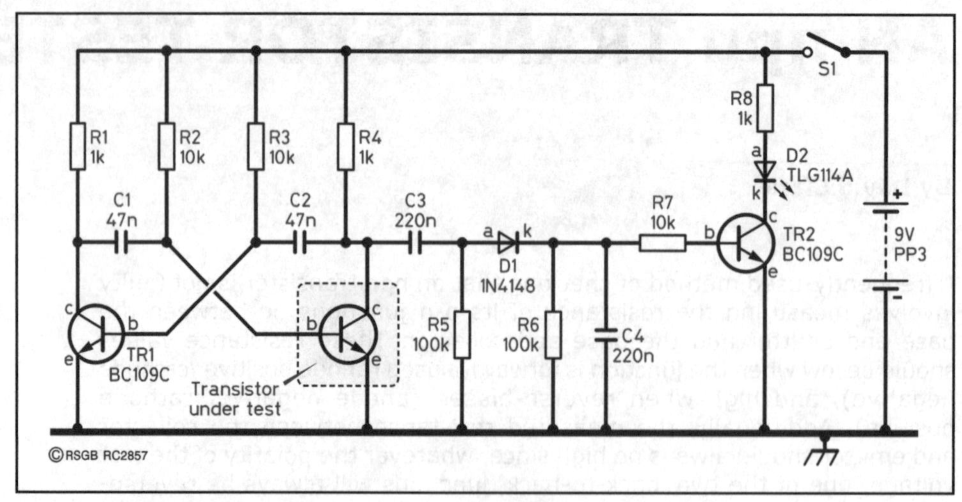


Fig 1: Circuit diagram of the npn transistor tester.

## THE CIRCUIT

The circuit diagram of the device is shown in **Fig 1**. Resistors R1 to R4, capacitors C1 and C2, transistors TR1 and the transistor under test, form the astable multivibrator. The basic operation of a transistor-based multivibrator involves each transistor alternately saturating (switching completely on) and cutting off as the output of each transistor is fed back to the input of the other. The rate of switching is primarily determined by the rate of charging of C1 and C2 and hence on the values of C1 and C2, R2 and R3 (the resistors through which they charge). If C1 = C2 and R2 = R3, the frequency of oscillation, *f* (in Hz), is given by

$$f = \frac{0.7}{R2 \times C1},$$

when R2 is in ohms and C1 in farads. For this project, these values have been chosen to give a frequency of around 1.5kHz in order that both audio- and radio-frequency transistors can be checked. For correct operation, the ratio of R2/R1 and R3/R4 must be less than the current gain ($h_{FE}$) of the transistors. If this is not the case, the transistor will not switch completely on and the voltage across the associated capacitor will not fall enough to switch off the 'opposite' transistor. By making this a ratio of 10, all transistors likely to be tested are covered. Although a few transistors can have an $h_{FE}$ as low as 25, most have a value of between 100 and 500, some rising to as much as 10,000 (unless faulty of course).

C3 is the DC blocking capacitor that links the multivibrator output to the rectification and smoothing sections made up of D1, C4 and R6. R5 provides a discharge path for C3, hence the voltage on the 'right hand side' of this capacitor can fall to zero if there is no oscillation present. The time constant of C3 and R5 which, at around 22ms, is much longer than the 0.7ms period of the 1.5kHz signal, allows the AC signal (if present) to pass through C3 to the rectifier diode, D1. The time constant of C4 and R6 is also around 22ms, so C4 and R6 smooth the rectified signal to provide a

DC voltage to the indicator section (R7, R8, TR2 and D2), thus lighting LED D2 when the transistor under test is working correctly.

R7 and R8 limit the current values to those required to drive TR2 and switch on D2. R6 provides the discharge path for C4 and so, as well as being part of the smoothing circuit, it also ensures that the voltage at the base of TR2 can fall to zero when there is no AC signal. In this way, the LED remains unlit when the transistor under test is faulty.

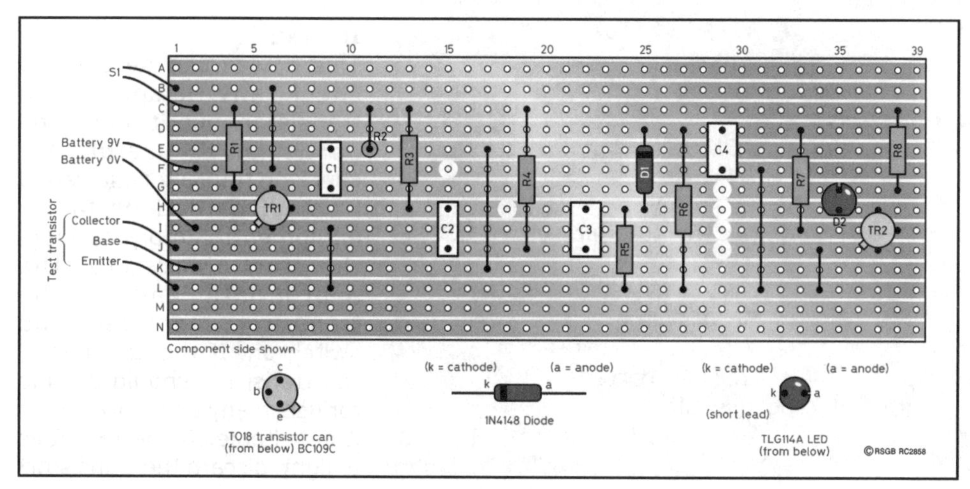


**Fig 2: Stripboard layout and wiring diagram for the npn transistor tester.**

## CONSTRUCTION

A suitable stripboard layout for this project is shown in **Fig 2**. The correct orientations of TR1, TR2, D1 and D2 need to be observed; details of the appearances of these components are also given. They are all general-purpose devices.

To facilitate the connection of the transistor under test to the device, it is useful to terminate the test leads with small crocodiles or test clips. It is helpful to colour-code the leads as reminders of their functions. For example, green can be chosen for the lead to connect to the emitter, since it links to 0V. Orange might be chosen for the lead to connect to the transistor base, and red for the lead to the collector, the colours suggesting the increasing positive voltages applied to the emitter, base and collector. The use of coloured test clips or coloured covers for the crocodile clips would be a suitable alternative.

**The completed tester, laid out on matrix board.**

The device could, of course, be run from a

**COMPONENT LIST**

**Resistors, metal film, 0.6W, 1%**

| | |
|---|---|
| R1, R4, R8 | 1kΩ |
| R2, R3, R7 | 10kΩ |
| R5, R6 | 100kΩ |

**Capacitors**
(all polyester film)

| | |
|---|---|
| C1, C2 | 47nF |
| C3, C4 | 220nF |

**Semiconductors**

| | |
|---|---|
| TR1, TR2 | BC109C |
| D1 | 1N4148 |
| D2 | TLG114A (or any standard green LED) |

**Miscellaneous**

| | |
|---|---|
| S1 | SPST switch |
| Stripboard | |
| Battery clip | (for PP3 battery) |
| Test clips or small crocodile clips | |

power supply instead of a battery, if preferred. The supply voltage is not critical, but naturally should not exceed the maximum rated voltages of any of the transistors used. Values between 6V and 15V should be satisfactory.

**IN USE**

This device can be used to check any suspect npn transistor, but it would be particularly useful for checking transistors salvaged from discarded or faulty equipment. Simply connect the test leads to the appropriate leads of the transistor under test and switch on. If the LED lights, the transistor should be fine for use in any circuit for which it is suitable. If the LED fails to light, discard the transistor!

**WARNING!**

*This device should not, under any circumstances, be used to test a transistor already in another piece of equipment, whether this equipment is switched on or off, or whether it is mains- or battery-powered. Damage could occur to your transistor tester, or even worse, to your other equipment, should you disregard this warning.*

# A TIME DOMAIN REFLECTOMETER

**by Martin Beekhuis, PA3DSC, and Klaas Robers, PA0KLS**
**(translated by Erwin David, G4LQI)**

Did you ever have a cable with a bad contact in one connector? Which one to squeeze? Solution: send a very short pulse into the cable and measure on an oscilloscope how long it takes for the reflection to come back. You then know how much cable there still is connected to the sending end.

Ever wanted to know the surge impedance of a piece of coax? Terminate it with a (non-wirewound) variable resistor, tune-out the reflection with it, then measure its DC resistance with an ohmmeter. Its reading is the wanted impedance.

The velocity factor of that coax? Measure its physical length with a tape measure and its electrical length with the reflectometer. The quotient of the two is the velocity factor.

## THE PULSE GENERATOR

If you plug a BNC T-connector into the Y-input of your oscilloscope and feed a very short pulse into one of its arms, you are ready to measure whatever you connect to the other arm. That pulse can be generated by the circuit shown in **Fig 1**, which will work with any standard oscilloscope. The pulse width is adjustable to the narrowest your instrument can display, which coincides with the condition for maximum sensitivity.

The pulse repetition frequency is chosen high enough to get adequate brightness, but low enough to measure cables up to 200m. Two inexpensive TTL ICs are used. The 74LS132 comprises four Schmitt-trigger NAND

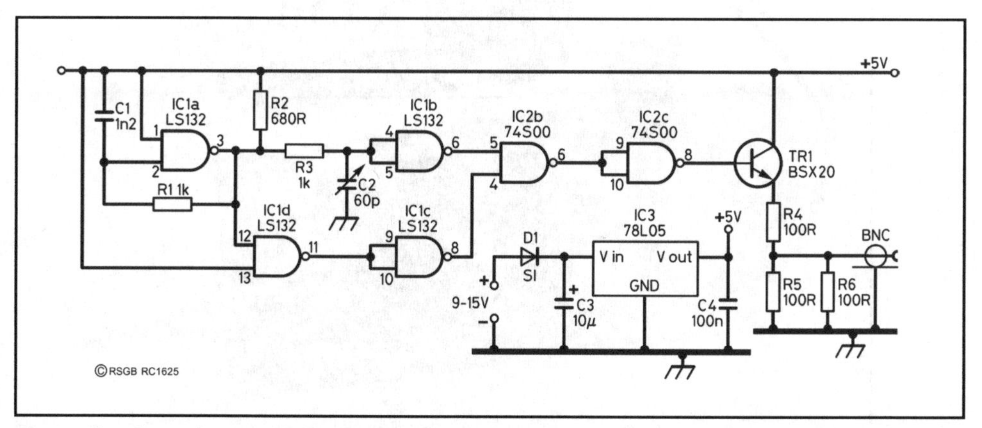


**Fig 1. The time-domain reflectometer described here consists of this pulse generator and a general-purpose oscilloscope.**

gates. One of these, IC1a with R1 and C1, forms a square-wave oscillator running at approximately 500kHz. R3 and C2 delay the square wave by an amount which can be adjusted with C2. IC1b, IC1c and IC1d serve to restore the square wave shape. They also invert the voltage and delay the pulse by 10ns per NAND gate. The difference in arrival time between the signals on the two gates of IC2b determines the width of the output pulse, which is adjustable with C2 between zero and 40ns.

After TR1 and its emitter voltage divider, the output pulse height is 1V and the output impedance 50Ω. If you have a fast oscilloscope, IC2 should be a 74S00. If your oscilloscope is no faster than 10 or 15MHz, a 74LS00 is good enough. IC3, a 7805, is a 5V regulator. Total current drain is approximately 35mA. The input voltage can be anything between 9 and 15V.

**TUNE-UP**

Connect the pulse generator directly or through 50Ω coax to the T-connector on the oscilloscope's input, terminate the open socket of the T-connector with 50Ω, set C2 to maximum capacity and the oscilloscope timebase at its fastest.

You will see a more-or-less square pulse. Now reduce C2. First you will see the pulse narrow, then lose its flat top, then decrease in height. The optimum setting is at 70% of its full height. This setting of C2 will be different for every oscilloscope.

With some oscilloscopes you may find it difficult to keep the trace still on auto-triggering. If so, switch to external triggering and take a trigger

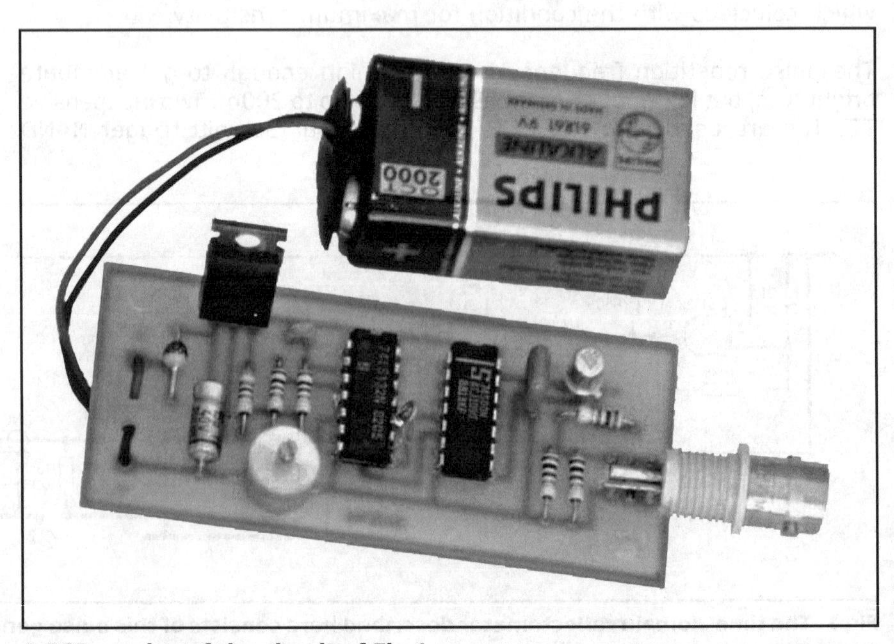


**A PCB version of the circuit of Fig 1.**

input through a 10:1 probe from the junction of pins 6, 9 and 10 of IC2. You can see a small wire loop to hook the probe into in the photograph.

**NO CALIBRATION**

The calibrated sweep of your oscilloscope automatically calibrates the time-domain reflectometer. Signals in air travel at close to 300,000km/s. With the oscilloscope sweep set to 0.1µs/cm, each 1cm division represents an electrical path length of 30m; that is the path from the oscilloscope to the end of the cable and back. Thus 15m of air-dielectric coax would show the reflection 1cm to the right of the input pulse. Solid-dielectric coax such as RG-213 has a velocity factor near 0.66. That means that an electrical length of 15m corresponds with a physical length of 10m (15 x 0.66). For semi-air-dielectric 75Ω TV coax with velocity factor of 0.8, each division would represent 12m (15 x 0.8). You will not have to unroll that reel of coax any more to see how much there is left!

**FAMILIARISATION**

First, a note. A very strong return pulse will bounce back from the capacitive oscilloscope input into the cable, creating, beyond the first reflection, a series of equally spaced reflections. Now try these experiments:

- Remove the 50Ω termination from the T-connector and connect a few metres of coax in its place. With the other end of the coax open or short circuited, you will see a reflection; but, if terminated with the cable's characteristic impedance, the reflection disappears.
- See how short you can make a piece of open-ended coax and still distinguish between the outgoing pulse and the reflection. With the sweep of my 150MHz oscilloscope at 10ns per horizontal division, I still see a distinct separate reflection, delayed by one whole division, from a one-metre length of coax.
- Try pieces of coax of equal length and impedance but with different dielectric, eg solid and foam. Correlate this with the difference in the specified velocity factor.
- Couple a few lengths of coax together with connectors and terminate the far end. Do the connectors reflect?
- Try the same with mixed cables of different impedance, eg 50Ω and 75Ω.
- Connect an aerial and see how wildly the broadband pulse reflects from the narrowband load.

# A USEFUL AUDIO LEVEL INDICATOR

**by RG (Danny) Dancy, G3JRG**

***Many amateurs, when wanting to try the many digital modes on offer, find that connections must be made from their transceivers to their computers. In the January 2004 RadCom, Ian White, G3SEK, explained how to make these connections, and gave advice on how to set the transmit audio levels so that overdriving and non-linearity were avoided.*** [This article is also in this book – *Ed.*] ***Here, Danny Dancy, G3JRD, provides an excellent yet simple piece of ancillary equipment to enable you to monitor the levels you set, thus ensuring complete repeatability when setting up.***

In the January 2004 *RadCom*, G3SEK, in his 'In Practice' column, described the problems of setting transmit audio levels for PSK31 operation; a similar problem exists on receive for the NOAA and other weather satellites requiring software for the computer sound card. Several years ago, a useful level indicator was designed and built, which has been in use in my shack ever since, for several modes including PSK31 and weather satellites.

**WHAT DOES IT DO?**

Connection of the indicator to the input terminals of the final piece of equipment in the chain enables the level for correct operation to be achieved easily. Receive calibration should be by the traditional hit-and-miss method, reviewing the results of each test until the optimum is found, the procedure being quite straightforward.

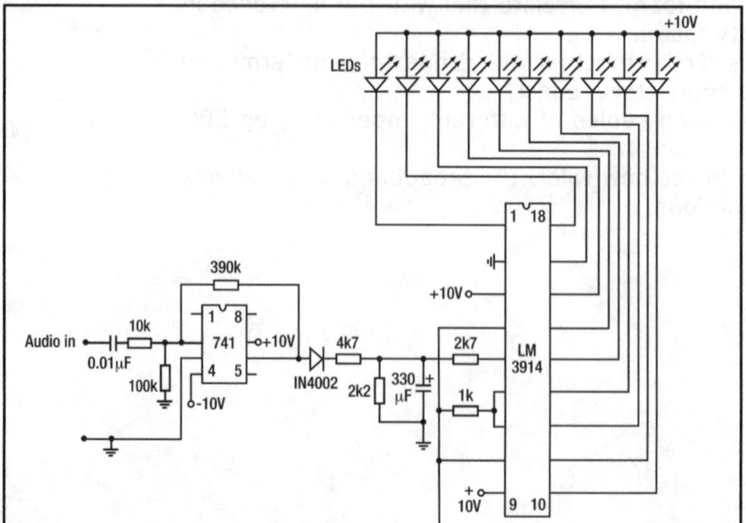

Then you can make a note of the indicated level for future use. When using the device to monitor your audio transmit level, particularly for PSK31, use the technique described by G3SEK to set the level, and then use the indicator to register its magnitude,

Fig 1: The level indicator circuit.

enabling you to set it precisely again when you have been using other modes.

## THE CIRCUIT

The original indicator, the circuit of which appears in **Fig 1**, was built on a piece of stripboard about 5cm x 12cm (2in x 5.75in), and drew its power from the RIG RX2 receiver, but a simple power supply, utilising a 12 – 0 – 12V transformer, two diodes and an electrolytic capacitor could be used. Note that the 741 op-amp requires ±10V. This balanced supply is derived partly from the circuit of **Fig 2**, which provides +10V regulated from a +15V unregulated supply, and **Fig 3**, which produces -10V from the single +10V supply.

In the circuit, the audio signal is applied to the inverting input of the 741 op-amp, the output of which is rectified and, after integration, an LM3914 LED bargraph display driver is used to provide visual level indication by means of 10 discrete LEDs or a single bargraph module.

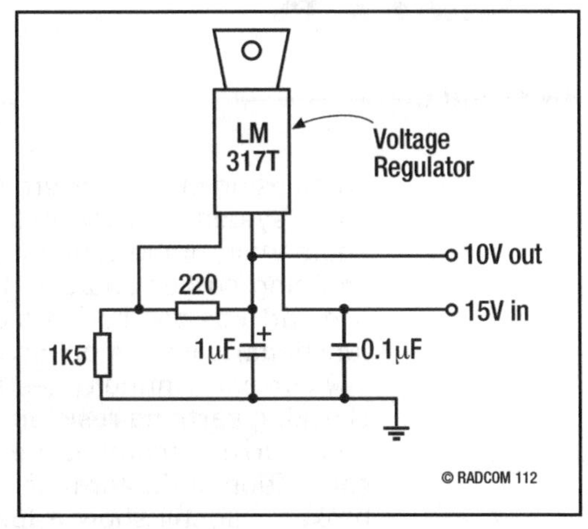


**Fig 2: Deriving the +10V regulated supply from a +15V unregulated source.**

## AN ALTERNATIVE

A sensitive moving-coil meter could easily be used here, in place of the LM3914 and LEDs, although I have not tried it.

The LED option is probably less expensive and is certainly very robust. No claim is made that the design is an optimum one, but it worked as soon as it was switched on, and has proved to be completely reliable. The unit could probably be used as a training project for potential radio enthusiasts, as well as being a useful piece of gear around the shack.

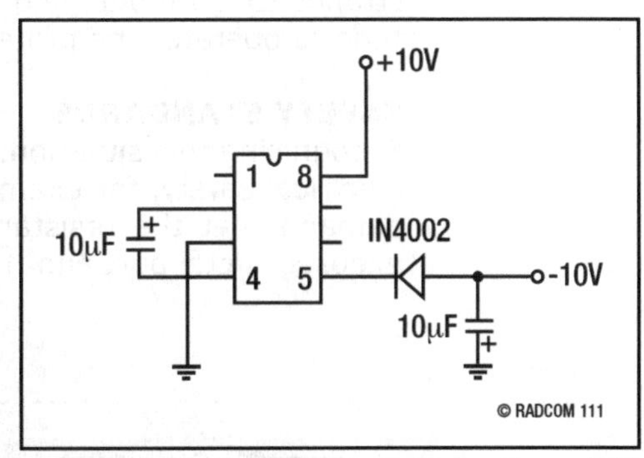


**Fig 3: -10V is generated from the +10V supply. The device shown is an SI7661CSA, available from Farnell (http://uk.farnell.com) and RS (http://rswww.com).**

# AN EARTH CONTINUITY TESTER

by David Clark

When using mains-powered electrical equipment, a good quality protective earth system is very important for safety. Good earth connections are additionally important for radio operation, both for protection against lightning strikes and also for the greater effectiveness of antennas that use earth as one half of a dipole. In situations where the earth path is a functional earth as opposed to a protective earth, a simple low voltage low current continuity tester or resistance meter is usually sufficient for checking earthing resistance, but for a proper test of a protective earth a high current tester is needed. This is because a deteriorating earth connection in the form of a stranded wire where many of the strands are broken will still show a low resistance to a low current tester, but in a fault situation when the earth path needs to pass a high current to ground and thus trigger a protective device, the high current causes the remaining strands to 'burn out', ie go open circuit, before the protective device has time to operate; the protection is then non-existent.

## SAFETY STANDARDS

Recognising this situation, the British and European safety standards for electrical safety, for example BS EN 60335-1 for household equipment, demand that the resistance of the protective earth path between an exposed metal part and the protective earth pin is less than 0.1Ω.

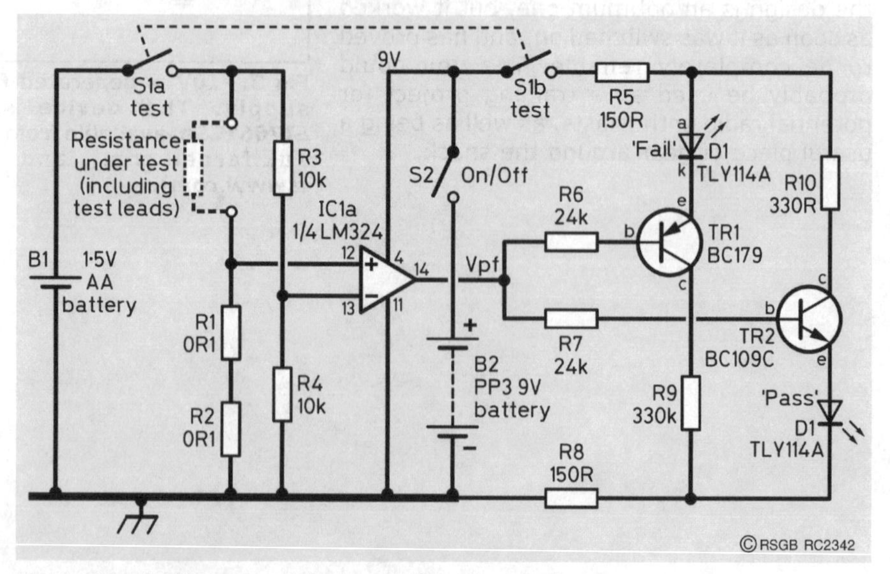


**Fig 1: Circuit diagram of the Earth Continuity Tester.**

The equipment needed for checking to this standard is specialised and expensive, but this simple project provides a low-cost alternative and will check resistance at two to three amps if good quality batteries are used. To simplify use, the circuit gives a pass/fail indication instead of a resistance value.

## WHEATSTONE BRIDGE

The circuit can be considered in three parts; *test*, *detector* and *output indicator*. See the circuit diagram in **Fig 1**.

The *test* part of the circuit is based on a Wheatstone Bridge, where the earth resistance path forms one of the resistance 'arms'. See **Fig 2** for the principle behind a Wheatstone Bridge. As a consequence of the values of resistance chosen (the test leads are assumed to have a resistance of 0.1Ω); if the earth resistance is less than 0.1Ω, the voltage between the midpoints of the two halves of the Wheatstone Bridge will be positive, and if it is greater than 0R1 it will be negative. This is fed to the detector part of the circuit.

The *detector* is an Op-Amp, wired as a comparator. In this way it has such a high gain that its output is roughly equal to either the positive or negative supply rail voltage, depending on whether the PD between its '+' and '-' inputs is positive or negative. It doesn't matter whether the PD is large or small, the output will always be at either extreme. This means there will always be definite pass or fail indication from the detector, no matter how large or small the output from the Wheatstone Bridge. This is important, as it means correct operation of the circuit doesn't depend on the voltage of the high-current battery, particularly as it is a chemical type whose output voltage can fall dramatically when a high current is being drawn. The pass/ fail voltage ($V_{pf}$) from the detector then passes to the output indicator circuit.

The *output indicator* circuit consists of two LEDs, driven by transistors to provide sufficient current, which indicate either a pass or a fail for an earth path resistance of less than or more than 0.1Ω. TR2 is npn, which switches on when its input is high, TR1 is pnp, which switches on when its input is low. A separate supply voltage is needed for the Op-Amp and LED circuit, since the test battery voltage will drop under a heavy load current.

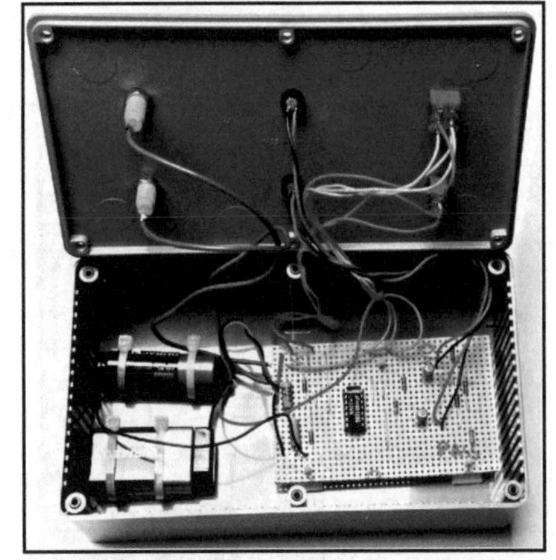

**Inside a completed tester. Note that in a 'cased' project, the LEDs are removed from the stripboard and brought out to the front panel.**

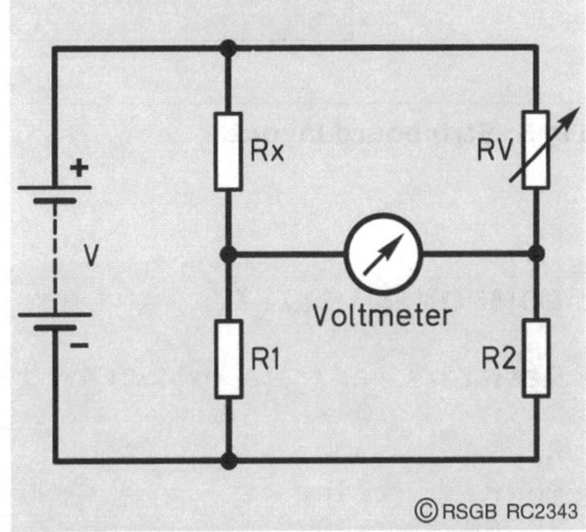


**Fig 2: In a conventional Wheatstone Bridge circuit, the value of an unknown resistance (Rx) is determined by adjusting a variable resistor (RV) with a calibrated scale, until the reading on the voltmeter is zero. At that point,**

**Rx/R1 = RV/R2,**

**so the value of Rx is then given by:**

**Rx = (RV x R1) / R2.**

**The advantage of this method is that at the balance point no current flows through the voltmeter, so the resistance of it doesn't affect the measurement. This is a sensitive method for detecting small changes in resistance, as a small change causes a large meter reading.**

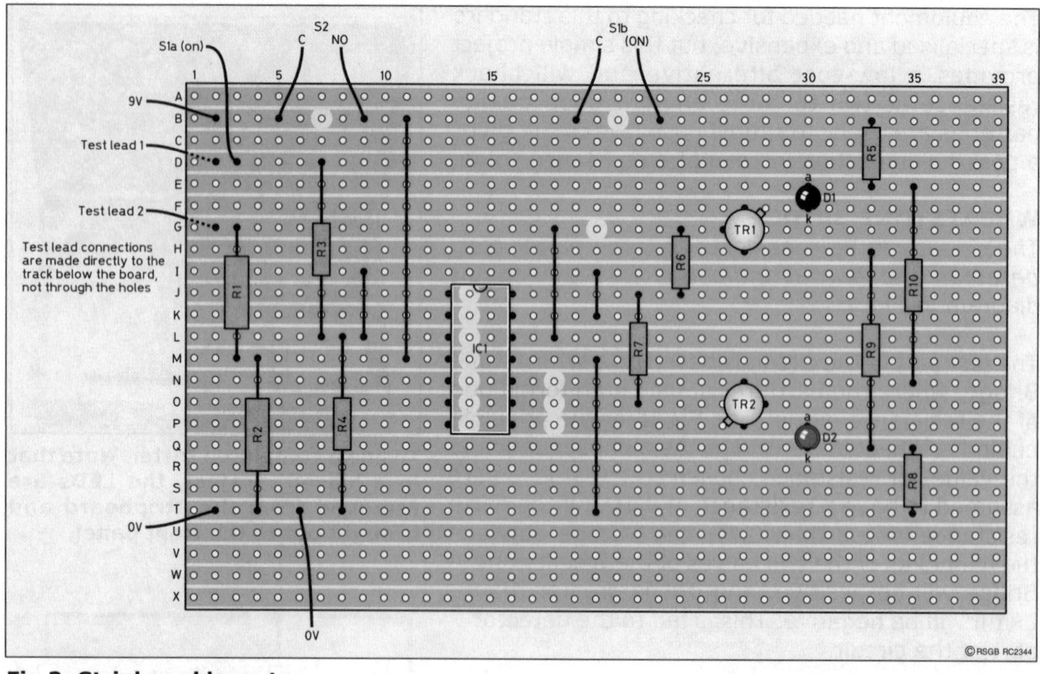


**Fig 3: Stripboard layout.**

## COMPONENTS LIST

**Resistors** - all metal oxide 0.4W 1%, except R1 & R2

| | |
|---|---|
| R1, R2 | 0R1, 2.5W |
| R3, R4 | 10k |
| R6, R7 | 24k |
| R5, R8 | 150R |
| R9, R10 | 330R |

**Semiconductors**

| | |
|---|---|
| IC1 | LM324 |
| D1, D2 | TLY114A yellow, or TLR114A red and TLG114A green |
| TR1 | BC179 (general purpose pnp) |
| TR2 | BC109C (general purpose npn) |

**Miscellaneous**

| | |
|---|---|
| B1 | 1 x AA Duracell |
| B2 | PP3 |
| S1 | double pole, momentary on, or push button |
| S2 | single pole on/off |

Battery clips/holders
Stripboard
Plastic case (only required if you are building the project in a case)
2 x 4mm plugs & sockets (only required if you are building the project in a case)
2 x crocodile clips

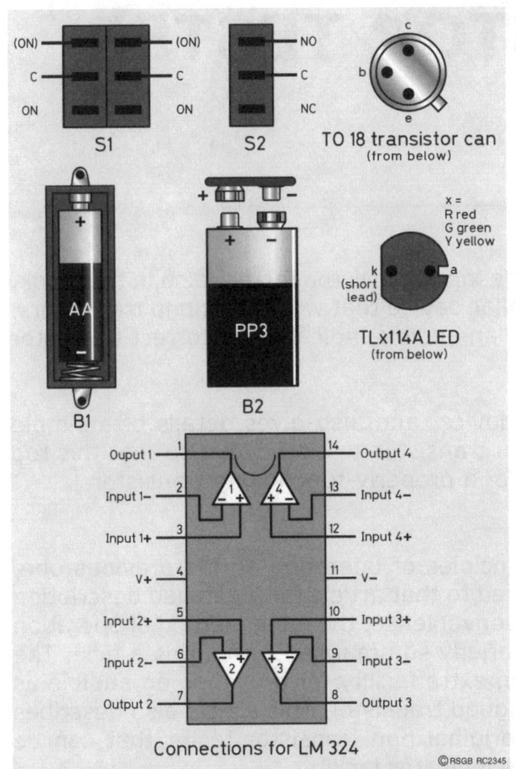


**Fig 4: Orientation (and pin connections) of the batteries, IC1, LEDs and transistors.**

Because the output of the Op-Amp does not swing completely to the positive (Vs) and negative (Vo) supply rails, measures need to be taken to ensure that the LED driver transistors switch off correctly.

## CONSTRUCTION

A suitable stripboard layout is shown in **Fig 3**, and **Fig 4** shows how to identify and orientate several of the components.

Use thick wire for the test leads!

## HOW TO USE

Using flying leads with suitable connectors, eg crocodile clips, connect the circuit to each end of the earth path to be tested. This would usually be the mains plug earth pin and any metal part meant to be earthed. Then press the test button. Release the test switch as soon as a pass/fail indicator lights (certainly within 5 to 10 seconds, to lengthen battery life and prevent possible overheating of R1 and R2).

**SAFETY NOTICE**

The project described here may be used to test the resistance of appliance earth connections, but it is *not* intended to conform to any *legal* requirements for the testing of electrical safety.

The RSGB and the author accept no responsibility for any accident or injury caused by its use. - *Ed*

# A pnp TRANSISTOR TESTER

**by David Clark**

This is a follow-up project to the 'npn Transistor Tester', also in this book. The requirement was for a similar device that would test pnp transistors, and would preferably also give an audible indication of correct transistor operation.

This article describes such a device, and also gives details of a simple modification to the original npn transistor-testing circuit so that this too will give an audible indication of a properly-functioning transistor.

## INTRODUCTION

This device uses the same principles of operation as the previous one, and so the constructor is referred to that article for a detailed description of 'how it works'. However, for convenience, the main method of operation of the npn transistor tester is briefly summarised within this article. The new tester also incorporates an extra facility for providing an audible as well as a visual indication of a good transistor. This article also describes a simple modification to the original npn transistor tester that can be 'retro-fitted' to add the audible indicator facility.

## THE pnp TRANSISTOR

The pnp transistor is a complementary device to the npn transistor. Both can be considered structurally as a 'sandwich' of two types of semiconductor, the npn bipolar transistor being a p-type semiconductor between two slices of n-type and the pnp type being an n-type semiconductor between two slices of p-type. This structure can be visualised as a pair of diodes connected 'back-to-back' (see **Fig 1**), which gives rise to one way of checking whether a transistor is undamaged or not. This involves taking measurements of the resistance between all combinations of the base, collector and emitter terminals in both directions, a total of six measurements. The situations where a p-n junction is forward-biased (conducting) should give a low, ideally zero, resistance. A reverse-biased junction (non-conducting) should give a high, ideally infinite, resistance, as should the collector-to-emitter test in both directions.

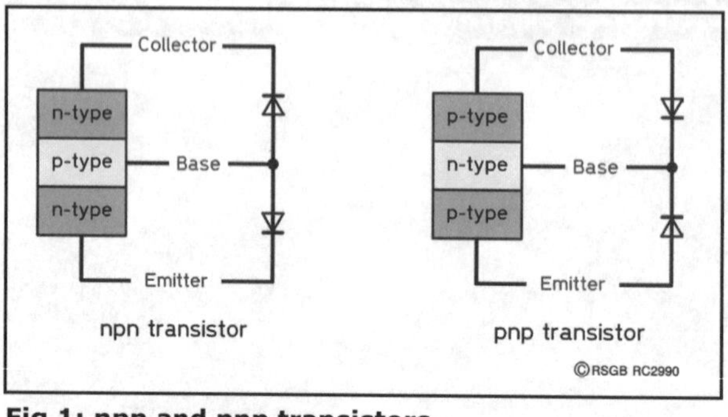

This can be a pretty haphazard method of testing a transistor, however. The resistance values for a 'real' device are far from the ideals of zero or infinite, hence giving a possibly ambiguous result.

**Fig 1: npn and pnp transistors.**

Furthermore the test itself, apart from being inconvenient in requiring multiple connections and reconnections, can destroy the device it is testing if care is not taken.

The reason for this is that the forward voltage for the test must be greater than the junction voltage of around 0.6V before the junction will conduct at all. Once it is conducting, the current flow must be limited because a high current can destroy the junction. Similarly, too high a voltage across a reverse-biased junction will destroy it - usually the reverse bias between the base and emitter must not exceed about 5V. This is because of the very small thickness of the junction (necessary for transistor operation). Even a small voltage appearing across a very thin junction gives a very high electric field strength within that junction.

A better method, therefore, is to test whether the transistor does what it is supposed to do - amplify current. This approach also avoids having to make numerous connections and disconnections in order to take all the necessary resistance measurements.

## HOW IT WORKS

The pnp tester makes the transistor under test part of an astable (free-running) oscillator circuit. This is then linked to a detector and an indicator circuit (see **Fig 2**). If the transistor is working correctly, the astable circuit will oscillate and the oscillations can be rectified and smoothed, and the resulting voltage used to switch appropriate transistors on, causing an LED to light and a buzzer to sound. If the transistor is faulty, oscillation will not occur and the LED will remain unlit and the buzzer silent. **Fig 3** shows the circuit diagram for the pnp Transistor Tester.

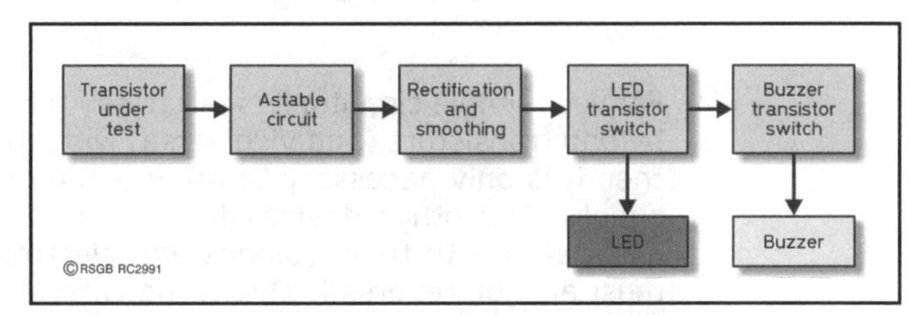


**Fig 2: Block diagram of the Transistor Tester.**

In fact, the transistor, TR3, controlling the buzzer, is switched via the action of TR2, controlling the LED, D2. When TR2 is switched off, there is no current through R8 (and the LED is unlit). There is therefore no voltage between the base and emitter of TR3 and it remains switched off, hence there is no current through the buzzer (ie it is silent). If TR2 is switched on, however, current flows through R8 and a voltage is dropped across it (and the LED lights). A voltage of around 0.6V then appears across the

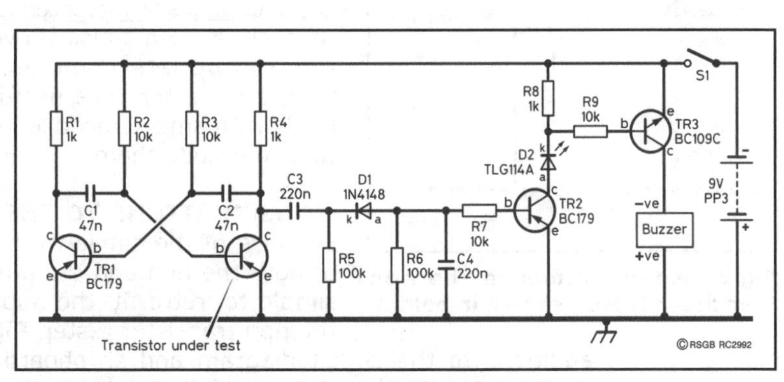


**Fig 3: Circuit diagram of the pnp Transistor Tester.**

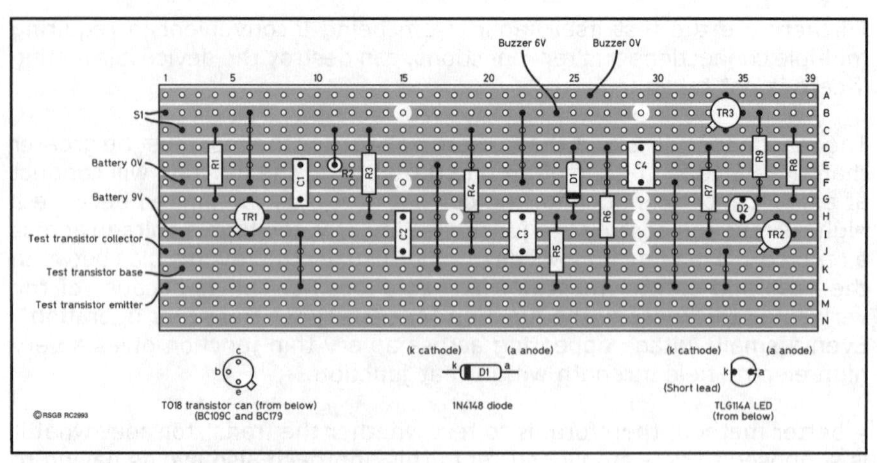


**Fig 4: Stripboard layout and wiring diagram.**

base-emitter junction of TR3, the current being limited by R9, as TR3 begins to conduct. Current thus flows through the buzzer which then sounds.

For very simple circuits, it is fair to say that if npn transistors are replaced by pnp transistors (and vice versa) with equivalent operating parameters then it is only necessary to reverse the polarity of the battery or power supply. Any other devices in such a circuit that also have a polarity associated with them (diodes and electrolytic capacitors, for example), must also be reversed. This is the case for this circuit, and in fact, apart from the modifications and additions needed to give an audible indication of a good transistor as well as a visual indication, the stripboard layout for this device is the same as that for the npn transistor tester. **Fig 4** shows the strip-board layout for the pnp Transistor Tester.

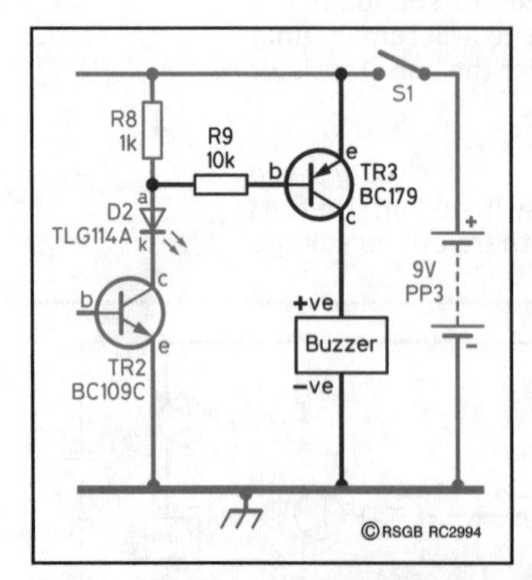


**Fig 5: Modifications to the npn Transistor Tester, shown in bold.**

## CONSTRUCTION

Constructing the tester is straightforward, but care in ensuring the correct orientation of the transistors and diodes is important. The correct polarity of the buzzer also needs to be observed if the recommended buzzer is used (see the note in the component list); the positive and negative connections are indicated by red and black leads respectively. Fig 4 also shows the configurations for the appropriate components. As with the npn transistor tester, it is useful to terminate the test leads with small crocodile or test clips, and it helps to colour code them.

## MODIFICATIONS TO THE ORIGINAL

Because of the similarity of the stripboard layouts of both the npn and the pnp transistor testers it is simple to 'retro-fit' the audible indicator facility to the npn transistor tester. **Fig 5** and **Fig 6** show the additions to the circuit diagram and stripboard layout. The additional components, track cuts and link are shown in bold.

**IN USE**
Connect the test leads to the transistor under test and switch on. A correctly-working transistor will cause the LED to light and the buzzer to sound.

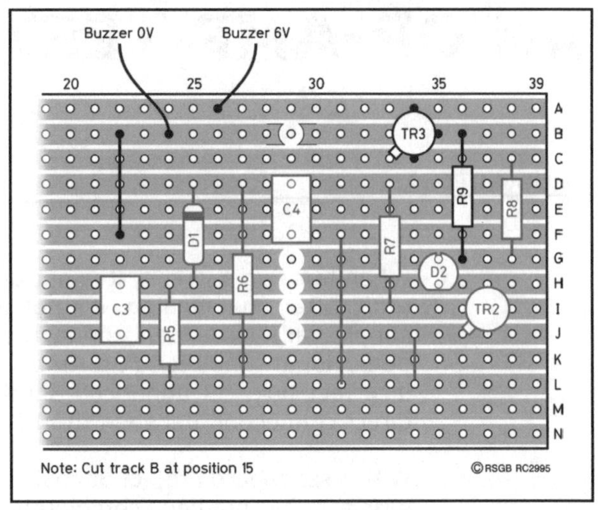


**Fig 6: Modified stripboard layout for the npn Transistor Tester.**

**Note**: The buzzer used was a 6V type obtained from Maplin Electronics (part number FL39N), but any buzzer operating from a nominal 6V supply and with a nominal impedance of around 250Ω, ie with a current consumption of up to a few tens of milliamps, should work satisfactorily.

**COMPONENTS LIST**

**Resistors (all metal film, 0.6W, 1%)**

| | |
|---|---|
| R1, R4, R8 | 1kΩ |
| R2, R3, R7, R9 | 10kΩ |
| R5, R6 | 100kΩ |

**Capacitors (all polyester film)**

| | |
|---|---|
| C1, C2 | 47nF |
| C3, C4 | 220nF |

**Semiconductors**

| | |
|---|---|
| TR1, TR2 | BC179 |
| TR3 | BC109C |
| D1 | 1N4148 |
| D2 | TLG114A green LED |

**Miscellaneous**

| | |
|---|---|
| S1 | SPDT switch |
| 6V buzzer | |
| Stripboard | |
| Battery clip (for PP3 power supply battery) | |
| Test clips or small crocodile clips | |

**npn tester modifications**

| | |
|---|---|
| R9 | 10kΩ |
| TR3 | BC179 |
| 6V buzzer | |

# AN OP-AMP TESTER

**by David Clark**

When building a circuit, it is not unusual to find that it doesn't work first time. After fault-finding, it's not unusual either to find that one of the supply rails had been connected to either an input or output of an operational amplifier (Op-Amp). This could be either because of an incorrect link, an unnoticed short circuit between copper tracks, or a direct connection between the IC pins due to a 'whisker' of solder.

With a stripboard project, it could also be due to an intended break in one strip that has not been completely cut through, or a burr of copper that is shorting onto an adjacent track. It might also be due to a required break that has been omitted.

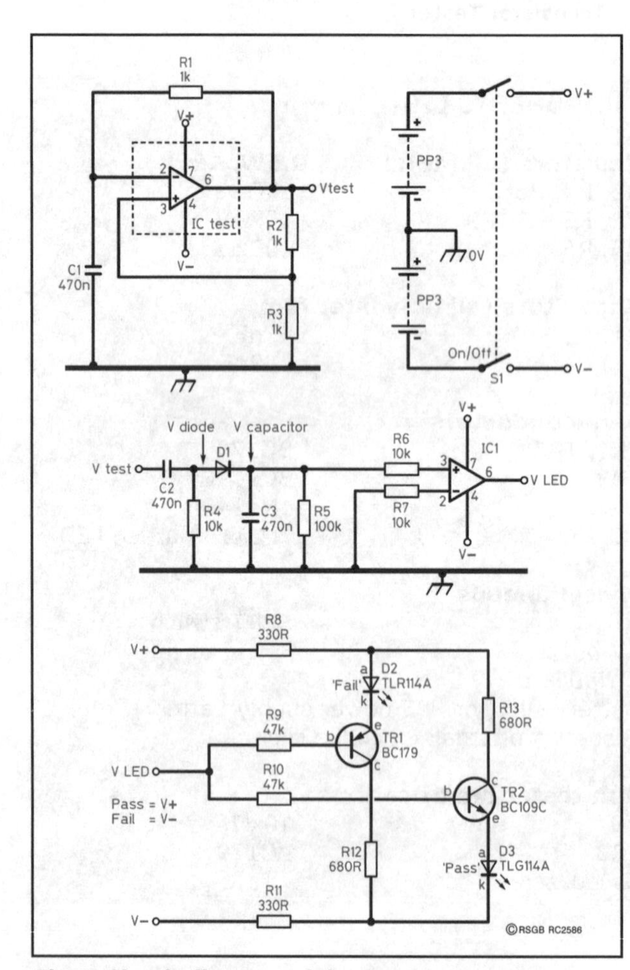


**Fig 1: Circuit diagram of the Op-Amp tester.**

When the fault has been rectified and the circuit still doesn't work, it then isn't possible to say whether this is now due to there being another fault present or whether it is because the Op-Amp was damaged by the initial fault. The best solution to this dilemma is to check independently that the Op-Amp is working correctly. This simple project will perform just such a check. It can also be used to check that an Op-Amp salvaged from unwanted equipment, for use in another project works correctly. It will prevent time being wasted in checking for construction faults when it is, in fact, the component that is faulty.

## HOW IT WORKS

When a component fails, particularly a semiconductor device, it usually fails catastrophically; ie it fails completely and doesn't just work half-heartedly.

In the case of an Op-Amp, this usually means that the output goes to the value of one supply rail or other and stays there. Another possibility is that the output goes to some fixed DC value between the two supply rail limits.

This circuit works by incorporating the Op-Amp under test into an astable oscillator circuit. If the circuit oscillates, the Op-Amp is fine; if not, it is damaged. So that an oscilloscope is not needed to observe the output waveform, the Op-Amp output is connected to a detector circuit and the output from this is connected to an indicator section. The indicator section drives two LEDs, a green one to indicate that oscillation is present, ie the Op-Amp passes the test; and a red one to indicate no oscillation, ie the Op-Amp fails the test.

## CIRCUIT

Components R1 to R3 and C1, in combination with the Op-Amp under test, form an astable oscillator (see **Fig 1**). When this is operating correctly, the output, Vtest, is a square-wave with a frequency of about 1kHz. Otherwise Vtest will be a DC voltage.

If Vtest is a DC voltage, it will be blocked by C2 and V diode will fall to 0V as the right-hand plate of C2 discharges via R4 (with time constant C2 x R4). D1 then isolates IC1 from this part of the circuit and both inputs of IC1 are connected via resistors to 0V. The non-inverting (+) input, however, is connected via 110k (R5 + R6), whereas the inverting (-) input is connected via 10k (R7). This means that the bias current entering the - input will be greater than that entering the '+' input, deliberately creating a differential input offset voltage.

Because of this, and the high gain of an Op-Amp connected without any negative feedback resistors, the output of IC1, V LED, will swing to very

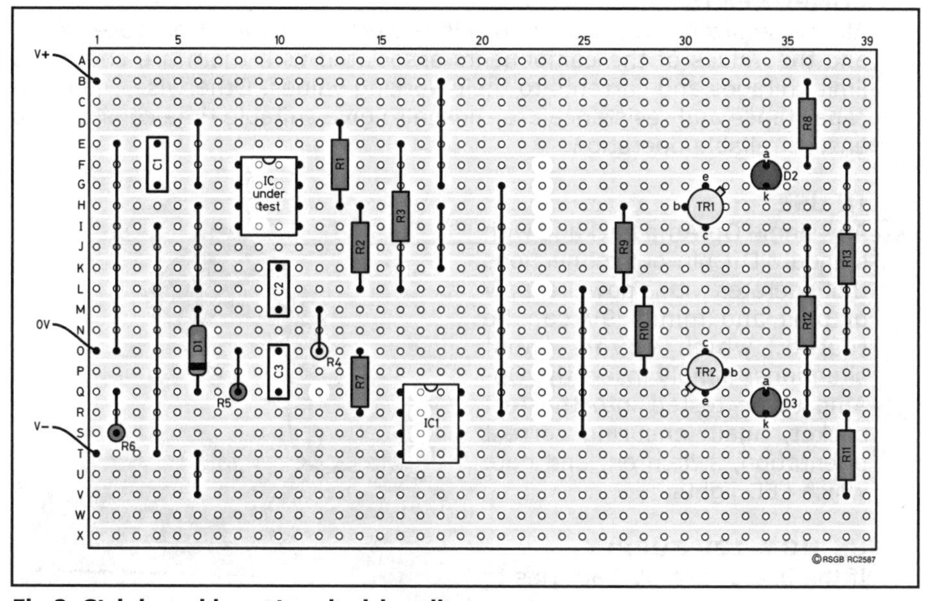


Fig 2: Stripboard layout and wiring diagram.

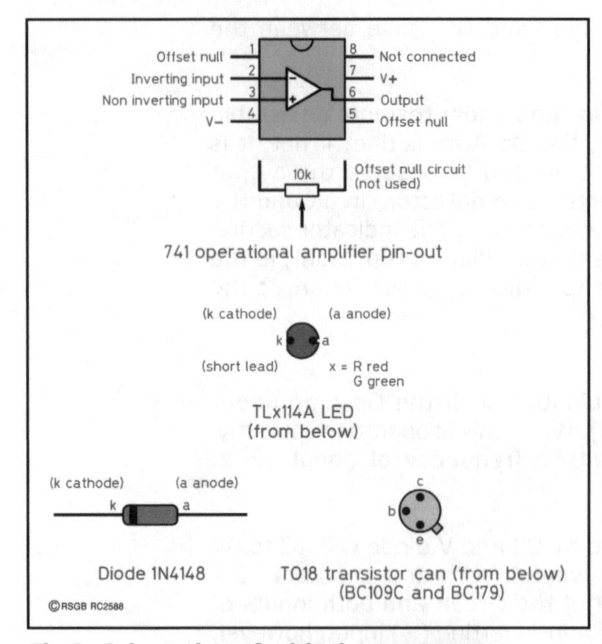


**Fig 3: Orientation of critical components.**

nearly the negative supply voltage. This switches TR1 on and TR2 off, and so the 'Fail' LED, D2, lights.

If, however, Vtest is a 1kHz square-wave, C2 will not block the signal and D1 will conduct during the positive part of the waveform. This means C3 will charge up, as R5 is now acting as the discharge path for C3 and the time constant C3 x R5 is long compared to the period of the 1kHz signal. The resulting positive value of V capacitor causes the current flowing into the + input of IC1 to exceed the current flowing into the – input, and V LED swings to nearly the positive supply voltage. This switches TR1 off and TR2 on, and so the 'Pass' LED, D1, lights.

## CONSTRUCTION

The stripboard layout for the Op-Amp tester is shown in **Fig 2**. An 8-pin DIL socket is needed for the Op-Amp under test, but IC1 can be inserted into a socket or soldered directly into place as preferred. It is always a good idea to install ICs in sockets though, since as well as eliminating the possibility of damage due to soldering, it makes removal for testing much simpler. Care needs to be taken to ensure that the Op-Amp, transistors, diode and LEDs are connected the right way round, and **Fig 3** shows how to determine the correct orientations.

## COMPONENTS

All the semiconductors used in this project are general-purpose types, and the values of the components associated with them are chosen to limit voltages and currents to their working values. Otherwise, values of capacitance and resistance are chosen to give suitable time constants to ensure reliable operation.

## IN USE

Any single Op-Amp package with standard DIL pin-out connections that operates with a power supply of plus and minus 9V can be tested, which covers most situations. Simply insert the Op-Amp to be tested into the test socket, again ensuring correct orientation, switch on, and note which LED lights.

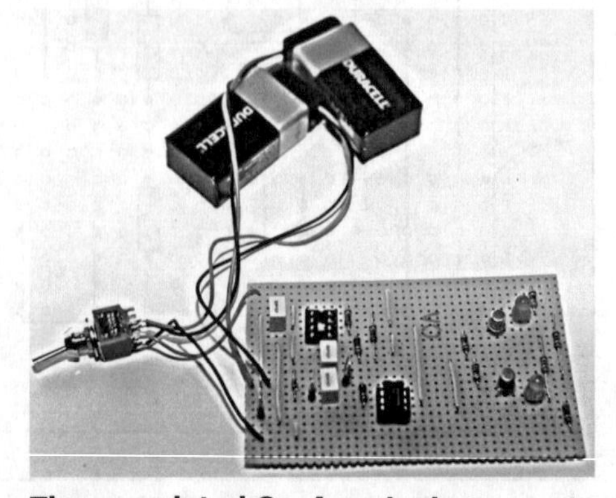

**The completed Op-Amp tester.**

## USING A 12V SUPPLY

If the use of two 9V batteries to power the Op-Amp tester doesn't

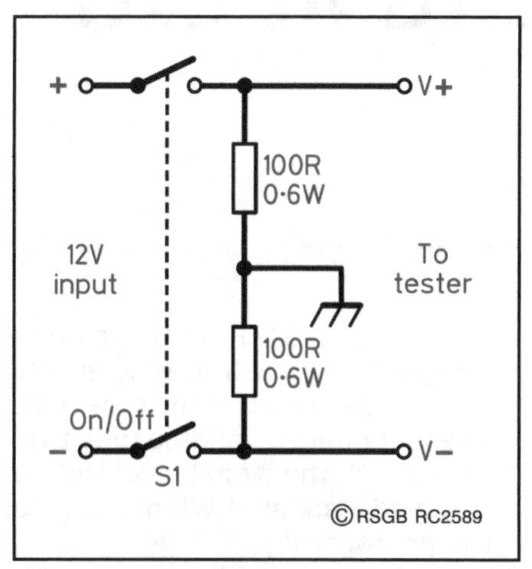


**Fig 4: To power the Op-Amp tester from a 12V supply, substitute the top right hand part of Fig 1 with this circuit. If you do this, remember that the 'chassis' of the tester (denoted by the chassis symbols) will not be at ground potential, so you will need to keep the unit isolated from ground.**

appeal to you, it is possible to adapt it to run from a 12V power supply. To do this, simply take two 100Ω resistors, make a potential divider (as shown in **Fig 4**), and use this instead of the top right hand part of Fig 1.

**COMPONENTS LIST**

**Resistors**

(all metal film, 0.6W, 1%)

| | |
|---|---|
| R1 - R3 | 1k |
| R4, R6, R7 | 10k |
| R5 | 100k |
| R8, R11 | 330R |
| R9, R10 | 47k |
| R12, R13 | 680R |

**Capacitors**

(all polyester film)

| | |
|---|---|
| C1 - C3 | 470nF |

**Semiconductors**

| | |
|---|---|
| IC1 | LM741CN Op-Amp |
| TR1 | BC179 |
| TR2 | BC109C |
| D1 | 1N4148 |
| D2 | TLR114A red LED |
| D3 | TLG114A green LED |

**Miscellaneous**

| | |
|---|---|
| S1 | DPDT on/off switch |
| | Stripboard |
| | Battery clip x 2 |
| | PP3 battery x 2 |

# SWR MODULE USING AUDIO TONES

**by David Berry, G4DDW**

**Front panel of the SWR tone module.**

This audio SWR module was designed and constructed to enable a blind operator to adjust his ATU. The unit emits a variable tone that varies with the signal level from a modified standard SWR meter (see the photograph).

The module provides a number of reference tones that have been preset by a sighted operator, to indicate reference audio meter readings. The circuit of the module is shown in **Fig 1** and the front panel is shown in the photograph. The rectified forward or reflected voltages from the SWR meter are amplified in IC1 and then used to control the frequency of the audio oscillator IC2. The circuit comprising S4 and RV1 to RV5 produces reference voltages that produce reference tones representing percentages of meter FSD.

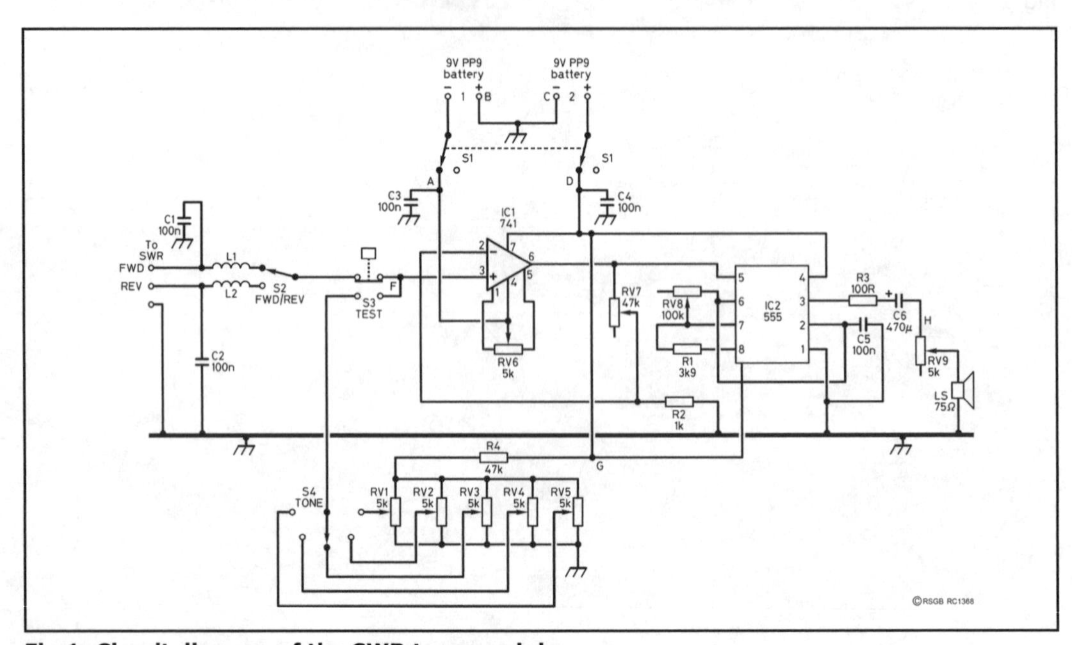


**Fig 1: Circuit diagram of the SWR tone module.**

## CONSTRUCTION

The first job is to modify the SWR meter to provide the signal for the SWR module. The modification will work with a single- or a two-meter system. I modified my meter by fitting a three-pin DIN connector to the case; the three leads were connected to the forward voltage, reflected voltage and earth points, respectively.

**SWR meter with DIN plug modification.**

The module is constructed in a diecast box and the layout is shown in the photograph. All the controls are mounted on the lid of the box and comprise the following:

- A two-pole on/off switch, S1.
- A single pole changeover push-button switch to select input or reference tones, S3.
- rotary switch to select the reference tones, S4.
- single pole switch to select forward or reference voltages, S2.

The speaker is glued to the hole in the lid with Araldite. Legends attached to the controls are used to enable sighted operators to set the equipment up.

The reference tone potentiometers were fitted to a section of PCB, mounted vertically in the slots of the diecast box, so that adjustments could be made easily.

The chokes L1 and L2 in the prototype were made from 7mm-diameter by 10mm-long ferrite cores, with four turns of insulated wire threaded through the 3mm hole. There is nothing critical about these, a few turns of wire on a ferrite rod will do.

**Internal layout of the components on the SWR tone module. The volume control is fitted to the side of the box**

## CALIBRATION

Connect a DVM or a high impedance multimeter across point F and earth. With the power meter / reflectometer in circuit, transmit RF power and note the reading on the DVM. Alter the transmitter power and note five positions on the SWR / power meter scales and their relative voltages at point F. Note voltages at the 0, 25, 50, 75 and 100% meter FSD; the 100% voltage should be a maximum of around 130mV.

Set S4, labelled 'tone', fully anti-clockwise and adjust RV1 for a voltage equal to 0% FSD. Repeat at all other switch positions, setting RV2, RV3, RV4 and RV5 for 25, 50, 75 and 100% FSD voltages, respectively.

You could calibrate the voltages differently, so that the tones represented specific ratios of SWR, such as 1:1, 1.2:1, 1.5:1 and 2:1, etc. If the SWR meter has a different voltage for FSD then increase, or decrease the gain resistor RV8 to suit.

Set the voltage at F for 100% FSD (you can do this using S3 and S4 once they are set up). Adjust RV7 for 1.06V at pin 3 of IC2. Adjust RV8 to get a suitable range of tones as S4 is switched through the FSD voltages. The range of tones is approximately 180Hz to 500Hz, dependent upon the setting of R8.

**COMPONENTS**

**Resistors**
R1 3.9k
R2 1k
R3 100R
R4 47k
RV1, RV2, RV3, RV4, RV5, RV6 5k pre-set
RV7 47k linear
RV8 100k log
RV9 5k linear

**Capacitors**
C1, C2, C3, C4, C5 100nF 25V
C6 470μF 25V

**Semiconductors**
IC1 741 op-amp
IC2 555 timer

**Other Items**
S1, S2 DP switch
S3 SP push-button switch
S4 SP rotary 5-way
L1, L2see text
Speaker 75Ω
Diecast box max 15 x 8 x 4.5cm
3-pin DIN plug and socket

## OPERATION

Connect up the modified reflectometer and plug in the audio SWR module. Switch on the module and set the FWD / REV switch S2 to forward power. A tone will come from the speaker telling the operator the unit is switched on.

With the transmitter on, increase power, the tone frequency should be increased to give a meter reading of 100% FSD; this is checked by pressing 'test' and a comparison made against a selected 100% FSD tone.

Switch S2 to reverse and adjust the ATU for the nearest tone to zero FSD.

# CHOOSING A MULTIMETER

If you're a newcomer to amateur radio and are looking to buy some kind of multimeter, you could well be asking yourself the question "Should I buy analogue or digital?". If so, this is a question to which there's no clear-cut answer, but to help you make a more informed decision as to which one to purchase, let's compare the two types of meter and their relative merits.

The words 'relative merits' were chosen carefully because, although analogue meters - the kind with a moving needle, are generally not as accurate at digital meters, accuracy isn't everything.

**ANALOGUE METERS**

This traditional type of 'moving coil' meter, shown in **Fig 1**, works by passing a current through a coil suspended in the magnetic field of a permanent magnet. A current flowing through the coil creates a magnetic field of its own, which interacts with the field of the permanent magnet, causing the coil to rotate.

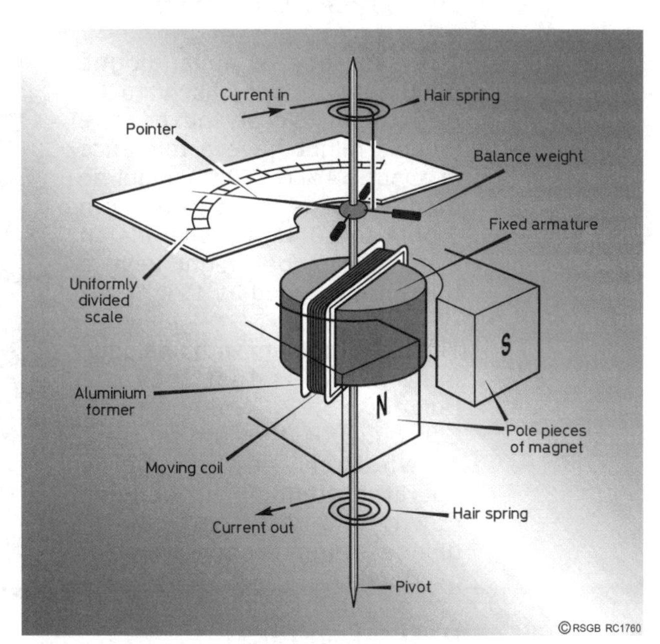


**Fig 1: Basic construction of a moving coil meter movement.**

The amount by which the coil rotates is directly proportional to the current which flows through it. Moving coil meter movements are available in a wide range of sensitivities, but, for example, if 10µA makes the coil rotate by 10°, 20µA would make it rotate by 20°, etc. Springs prevent the coil assembly from moving further than it should; and return the needle to the zero mark when the current is removed. The maximum angle of rotation of a meter of this type is typically around 90°, so if we wanted to measure more than 90µA, on the face of it we would have a problem.

This is where the 'shunt' and 'multiplier' resistors associated with multimeters of this type come in.

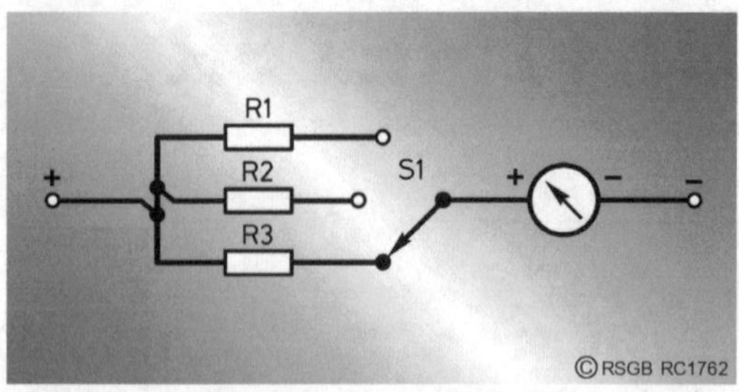


**Fig 2: Converting a moving coil meter movement into a multi-range DC voltmeter can be accomplished by adding a switch and 'multiplier' resistors. If you know what DC voltage ranges you want the meter to read, the DC resistance and the sensitivity of the meter movement, you can use Ohm's Law to calculate the values of R1, R2 and R3.**

To enable such a meter to measure a higher voltage than it naturally measures, 'multiplier' resistors are placed in series with the movement to reduce current flow through it. **Fig 2** shows the typical circuit of a multimeter with three DC voltage ranges.

To enable such a meter to measure more than 90μA, current must be 'shunted' away from the movement through a resistor placed across it. **Fig 3** shows how a moving coil meter can be made to read three different maximum values of DC current.

Practically all commercially-produced analogue multimeters measure AC and DC voltage, resistance and DC current, but measuring AC current is more difficult and only the more expensive models are equipped with this feature.

The accuracy of an analogue multimeter is very likely to be reflected in its cost. Inexpensive models might have a tolerance of about ±4% (ie 10mA might be displayed as anything from 9.6mA to 10.4mA), whereas an expensive model might have a tolerance as good as 1%.

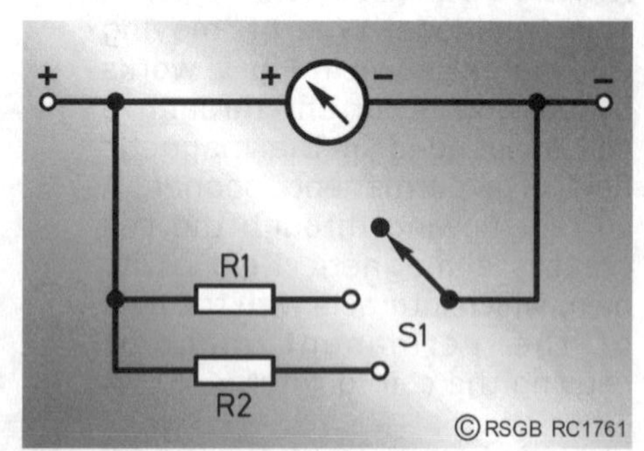


**Fig 3: Converting a moving coil meter movement into a multi-range DC ammeter can be accomplished by adding a switch and 'shunt' resistors. If you know what DC current ranges you want the meter to read, the DC resistance and the sensitivity of the meter movement, you can use Ohm's Law to calculate the values of R1 and R2.**

The main problem with analogue multimeters is that they are largely mechanical. Consequently, they are sensitive to whether they are being operated standing up or lying down, and they particularly dislike being dropped and/or overloaded.

## DIGITAL METERS

Modern, digital meters work by converting the voltage, current or resistance, from the analogue form which they interact with the outside world, into a digital form.

Typically this is done within an integrated circuit (IC) known as an Analogue to Digital converter (ADC). Once the analogue value being measured has been converted into digital form, ie it has been converted into a number (which initially takes the form of a large pattern of 'logic 0s' and 'logic 1s'), the number can be processed within the electronics and displayed, usually on a liquid crystal display. The basic layout of a digital multimeter is shown in **Fig 4**.

Pound for pound, digital meters are more accurate, with inexpensive models typically having a tolerance of about 1-2% and expensive models 1% or better. Because they employ a large amount of digital electronics, even moderately-priced digital meters tend to include features which analogue meters find it difficult to deal with, ie AC current, frequency, capacitance, etc.

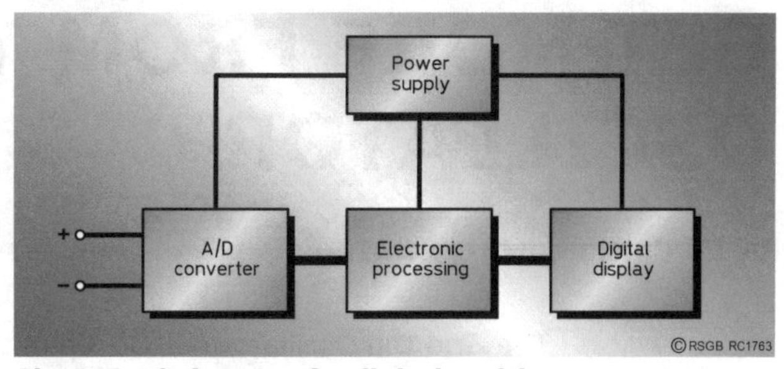


**Fig 4: Basic layout of a digital multimeter.**

**MAKING A DECISION**

So, coming back to the original question, despite the (often) higher price, the mechanical considerations, and the (generally) inferior accuracy of analogue multimeters, this older type of instrument still has two distinct advantages over the modern one.

1. Visual feedback is much better from an analogue meter. What is meant by this is that if you are adjusting a circuit and looking to see whether the value you are measuring is rising or falling, your brain will find the moving needle of an analogue meter easier to process than the changing numbers of a digital display. It's also worth pointing out that the display on a digital meter is delayed slightly while the electronics samples and processes the input, which increases the time required to take measurements.
2. Power (usually in the form of a battery) is required by digital meters. If the battery runs down, it doesn't work at all. This is unlike analogue multimeters, which only require a battery to measure resistance. At the end of the day, the option favoured by many engineers and service personnel is to equip themselves with one multimeter of each type; the digital model for measurements which require accuracy, and the analogue model for measurements which require speed and 'feel'.

# GET MORE FROM YOUR DIP OSCILLATOR

**by Ed Chicken, MBE, G3BIK**

The Grid Dip Oscillator (GDO) offers a quick and easy means of checking (to a degree of accuracy acceptable for experimental purposes) the inductance value of coils in the microhenry range and capacitors in the picofarad range, such as are commonly used in radio circuits.

This can be very useful, for example, when constructing an ATU, a crystal set, a short-wave receiver, a VfO or a band-pass filter for a direct conversion receiver.

**DETERMINATION Of L AND / OR C**

for this purpose I keep with my GDO two fixed-value Rf coils of known inductance - 4.7µH and 10µH - and one capacitor each of 47pf and 100pf (but the choice of values is yours). You may decide to keep one or more of each, to be selected from **Table 1** and **Table 2**.

| **for known L of:** | 1.0µH | 2.2µH | 4.7µH | 6.8µH | 10µH | 22µH |
|---|---|---|---|---|---|---|
| **unknown C(pF) =** | $25330 \div f^2$ | $11513 \div f^2$ | $5389 \div f^2$ | $3725 \div f^2$ | $2533 \div f^2$ | $1151 \div f^2$ |

**Table 1: To determine an unknown capacitor.**

| **for known C of:** | 10pF | 22pF | 33pF | 47pF | 68pF | 100pF |
|---|---|---|---|---|---|---|
| **unknown L(µH) =** | $2533 \div f^2$ | $1151 \div f^2$ | $768 \div f^2$ | $539 \div f^2$ | $373 \div f^2$ | $253 \div f^2$ |

**Table 2: To determine an unknown inductance.**

My personal choice of coil-type is the moulded Rf choke (Maplin) or Rf inductor (Mainline or RS). These have axial leads, are ferrite based, encapsulated, easy to handle, and readily available at low cost in a range of fixed microhenry values. The capacitors are 5% tolerance polystyrene, and also have axial leads.

To determine or verify the value of either an Rf coil or a capacitor, simply connect the unknown component in parallel with the appropriate known component to form a parallel LC tuned circuit, ie an unknown L in parallel with a known C (or vice versa), then use the GDO to determine the resonant frequency of the parallel LC circuit. The value of the unknown component can then be obtained easily to an acceptable approximation, by using the relevant formula from Tables 1 and 2 and a pocket calculator.

The formulae were derived from the accepted formula for the resonant frequency of a parallel tuned circuit:

$$f(\text{Hz}) = \frac{1}{2\pi\sqrt{LC}}.$$

But note that, in Tables 1 and 2, *f is the frequency in MHz as given by the GDO.*

**EXAMPLE 1**

An unknown capacitor in parallel with my known 10µH inductor, produces a dip at 6.1MHz, hence $f = 6.1$. from Table 1, the value of the unknown capacitor is given by:

C(pf) = 2533 ÷ $f^2$
= 2533 ÷ 6.1 ÷ 6.1
= 68pF.

**EXAMPLE 2**

An unknown coil in parallel with my known 47pf capacitor, produces a dip at 12.8MHz, hence $f = 12.8$. from Table 2, the value of the unknown inductance is given by:

L(µH) = 539 ÷ $f^2$
= 539 ÷ 12.8 ÷ 12.8
= 3.3µH.

Bear in mind that because the accuracy of results relies upon the frequency as derived from the GDO, it would be sensible to keep the coupling between the LC circuit and the GDO as loose as possible, consistent with an observable dip. This minimises pulling of the GDO frequency. Also, rather than relying upon the frequency calibration of the GDO itself, it might be useful to monitor the GDO frequency on an Hf receiver or a digital frequency meter.

A final point worth considering is that each fixed-value inductor of the type mentioned might have its own self-resonant frequency, but these would typically lie above the Hf range so should not be a problem. for example, the self-resonance of my own 10µH inductor is about 50MHz and that of 4.7µH is about 70MHz. You could quickly and simply find out the self-resonant frequency of an inductor, by taping it to each of the GDO coils in turn and tuning across the full frequency range.

It is best to make L and C measurements at frequencies much lower than the self-resonant frequency of your chosen test-inductor, but perhaps better be safe than sorry and stick with the lower µH values if your interest lies between 1.8 and 30MHz.

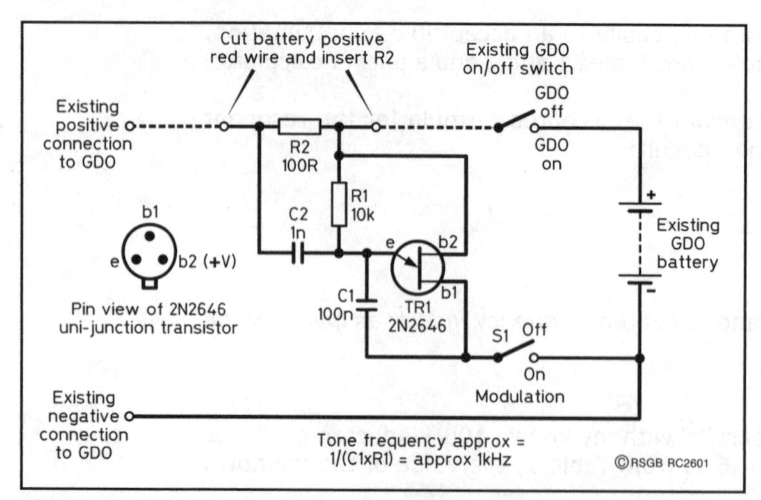


**Fig 1: This audio oscillator adds a 1kHz AM tone to a GDO.**

## TONE MODULATION

Sometimes it is useful to be able to hear an audio tone when using the GDO as an Rf signal source in association with a radio receiver.

If your GDO does not have tone modulation, you might like to construct the simple add-on 1kHz audio oscillator circuit shown in **Fig 1**. It uses a unijunction transistor, the frequency of oscillation being given approximately by 1/(R1 x C1). The 1kHz output tone connects via C2 to the positive supply line of the GDO, which it modulates. R2 acts as the modulator load and its value helps to determine the level of modulation.

This produces simple but effective tone modulation of the GDO's Rf signal, which can be heard on an AM or an fM receiver.

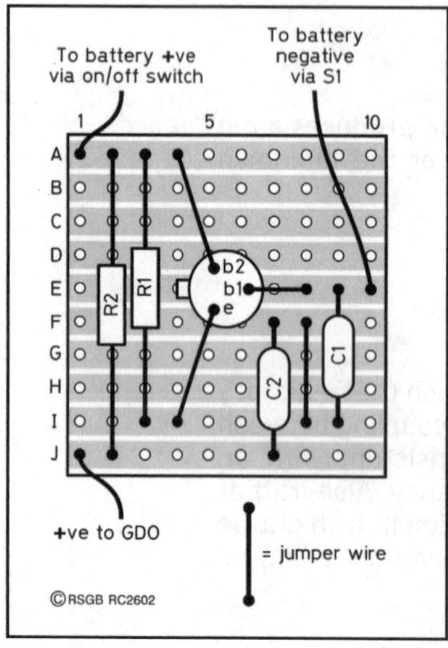


**Fig 2: Stripboard layout.**

**Fig 2** gives a suggested layout of the components on a small piece of stripboard, without the need for any track-cutting. The finished board might conveniently mount on one of the GDO meter terminals, provided care is taken to isolate the copper-tracks from the terminal.

**COMPONENTS LIST**

**Resistors**
R1 10k
R2 100R

**Capacitors**
C1 100nF
C2 1nF

**Semiconductor**
TR1 2N2646

# MAKE A SIMPLE 'MAGNETOMETER'

**by R G (Danny) Dancy, G3JRD**

As all hams know, radiation from the sun is responsible for creating the ionosphere. This, in turn, obligingly reflects some of our transmitted HF signals back to earth, so that they do not disappear for ever into space, thus enabling us to work DX. The same solar radiation also affects the earth's magnetic field and, during periods of high sunspot activity, the variations of both the ionosphere and the magnetic field can be quite considerable.

This article shows how to build a simple magnetometer and some electronic gear to go with it, and will hopefully encourage an interest in the fascinating field of geomagnetism. Perhaps 'magnetometer' should be in inverted commas, because this device does not measure the *value* of the magnetic field, but it does detect *variations* in the field direction. The design is not claimed to be refined, and there is plenty of scope for you to exercise your imagination and technical skills.

After studying an article by John Rowlands in the May 2002 issue of *Astronomy Now*, I have been conducting some very interesting experiments.

The first attempt was roughly as described by John, plus some ideas gleaned from the Internet; the latter were found by searching for 'Geomagnetism'. The information which trundled back down the line to my shack included the Danish Meteorological Institute's report on over 20 sophisticated magnetometers that they have operated since 1981 in Greenland.

Several simple designs for magnetometers are based on the idea of suspending a small magnet attached to a mirror on a fine thread, letting it swing freely in the earth's magnetic field, and using a beam of light reflected from the mirror onto a scale a metre or two away. The longer the distance from the mirror to the screen the more sensitive the detecting process is. The light beam is deflected by twice the angle of rotation of the mirror, which automatically doubles its sensitivity to changes in the field. By the time the light beam reaches the screen, moderate darkness is needed to see the narrow band of light reflected by the mirror.

### DEVELOPING THE IDEA

After some experiments (referred to by the XYL, predictably, as "Playing with the new toy"), the following can be -recommended as a working basis for experimentation. **Fig 1** shows the overall layout for the sensing

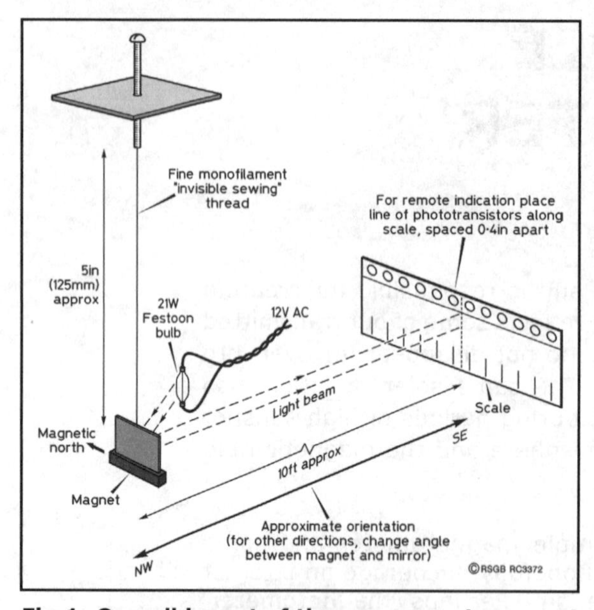


**Fig 1: Overall layout of the sensor system – not to scale.**

part of the magnetometer. It gives a mirror full-scale deflection sensitivity of approximately ±0.6°.

## The Light Source

Several light sources were tried, the best so far being as illustrated in **Fig 2** - a 21W festoon lamp (the type with a straight filament, -often used for car interior lights and readily available at your local friendly car accessory shop). It is mounted vertically, using 8mm copper pipe. The lower length of tubing was longer, flattened and bent at right angles for mounting to the base. Take care to handle the bulb -gently so that you do not have to go out and buy a replacement (as I did, being a clumsy oaf).

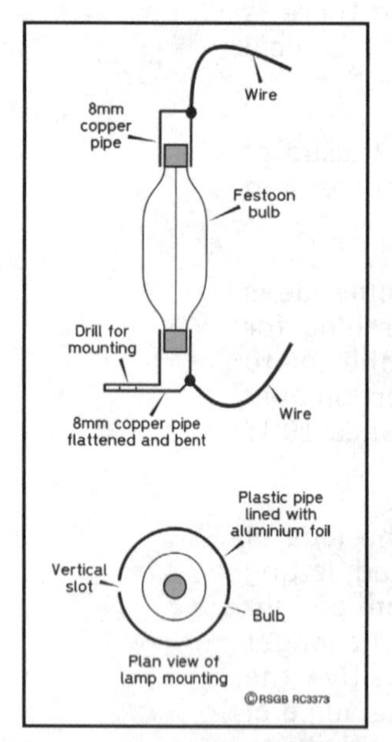


**Fig 2: Plan and elevation views of the lamp unit.**

The electrical supply to the bulb must be AC, via a step-down transformer, as DC would generate a steady magnetic field. For the same reason, any ferromagnetic material near the sensor should be avoided.

The lamp was enclosed in a length of $1^1/_2$in diameter plastic water pipe, with kitchen foil inside to reflect light. The pipe was split longitudinally with a small hacksaw, and two small pieces of wood about $^1/_8$in across wedged at top and bottom to form a slot down the length of the tube. Having obtained a light source, now for the 'compass'.

## Magnet and Mirror

Again, several variations were tried. In the end, the best results were with the smallest magnet available, to keep the inertia of the unit as low as possible. As suggested by John Rowlands, a discarded CD makes a good mirror, easily cut with a small hacksaw. Make the distance from the mirror to the screen at least two yards. I use a baseline of 10ft, but this is a point open to experiment.

To get some idea of the optimum curvature of the CD, a festoon bulb mounted without the surrounding tubing can be used as a light source, and by holding the CD behind it to project onto a wall the right distance away and flexing it gently into a slight curve [about an axis parallel to the filament – *Ed.*], a narrow line of light should be seen vertically on the wall. The lamp should be about four inches away from the mirror. Only a very small curvature is required.

Unfortunately, a CD does not stay like that without the application of some heat. Wait until the lady of the house is out of the way, and use a gas ring to heat the CD gently. [Younger experimenters are urged very strongly to have an adult present during this process, or to have the adult perform the heating and bending - *Ed*.] The temperature necessary is above the boiling point of water, but be careful not to heat it too much. Hold the CD in a piece of cloth to avoid burning your fingers. Once the correct heat is obtained, if you are lucky the CD will stay in a nice, very slightly curved shape after it cools.

Having achieved what looks like a consistent curve, check with the test rig, and find the area on the CD which gives the sharpest line of light on the wall, and cut out an area of about an inch square (25 x 25mm). Cement the small magnet underneath the mirror (Maplin have one which is suitable, coded FX71, (18 x 3 x 3mm). I used 'Plastic Metal', made by the Plastic Padding people to stick the magnet to the piece of CD, but any good permanent adhesive could be used.

## HOUSING THE SENSOR

To avoid draughts, which can be a big nuisance, the magnet and mirror must be mounted in some sort of container. Glass jars have been used by some people, mainly schools when demonstrating to children. I did not find this very satisfactory, and preferred to use a wooden box with a small sheet of glass on one side, with the bulb just outside it, as I was worried that any convection currents which would be set up by the heat of the lamp would disturb the detector unit. The glass is best mounted at an angle sufficient to avoid reflected light being transmitted onto the scale. **Fig 3** shows this. Painting the inside of the box matt black also reduces unwanted light reaching the scale.

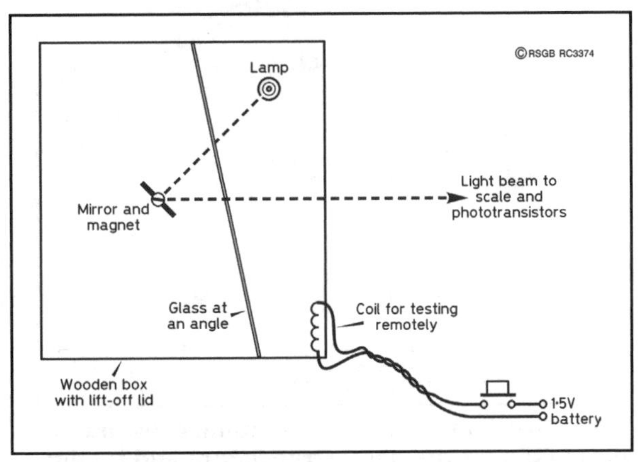


**Fig 3: Plan view of lamp, suspended magnet and mirror assembly.**

## THE THREAD

I tried suspending the sensor using several different threads, and found that a very thin monofilament was the best. This was obtained from the XYL's sewing box, and comes under the description of 'Invisible Sewing' thread. Monofilament is superior to any twisted thread for stability, as even the small weight of the magnet and mirror seems to make the multi-strand type unwind, causing a small, but annoying, rotation.

Gluing the thread with a small dollop of impact adhesive at the end of a brass woodscrew gave a useful method of adjusting the height of the sensor, and once that is correct, it allows a fine adjustment of the direction in which the mirror faces in order to get the beam onto the middle of the screen. Setting the whole contraption up so that the reflected light beam runs roughly from the NW to the SE gave the best results for me; choose a suitable orientation according to your setup. It is necessary to find the best spot for the light source before fixing it. Experiment!

Having made up a working unit in the shack, which is about 36ft away from our quiet road, I was surprised that most passing cars caused a significant disturbance of the field direction. Magnetic storms caused much larger swings, and the simple setup so far described would be enough to give some indication of geomagnetic variations. For those who want something a little more elaborate, read on.

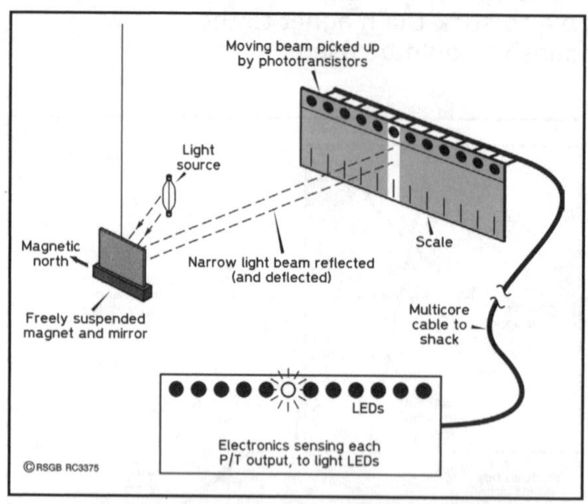


**Fig 4: Circuit diagram of the system, showing the disposition of the photo-transistors and testing coil (in the attic) and the indication circuitry (in the shack). The author has used several types of photo-transistor (PT1 – PT12) with equal success, the Siemens SFH309 (Maplin CY86) and the Siemens SFH 300-2 (Maplin NP64) being two of these. Each of the 12 channels has a 10kW sensitivity control associated with it. These controls can be individually adjusted in the shack to ensure that each LED will illuminate when the associated photo-transistor is active. The relay leads to the lamp and transformer in the attic are not shown.**

## A REMOTE INDICATOR

To reduce the interference from passing cars, to save space in the shack, and to install the sensor in a dark place, the attic was an obvious choice. This required some sort of remote indication of the light beam's deflection, and a series of photo--transistors was used in the circuit shown in **Fig 4**. These worked better than the cadmium sulphide photoresistors I had previously tried.

The sensor was positioned on a shelf in the attic, and it was easy to have a nice long base-line of 10 feet in a position as far away from the road as possible. A spot about 55ft away from the interference of passing cars was achieved though, surprisingly, even this distance was not enough to eliminate the unwanted deflections entirely. Multicore cables were run down to the shack, using two lengths of Maplin CW45, 4-pair cables, which happened to be the cheapest I could find, but any telephone cable would do.

For testing and setting-up purposes, an air-cored coil was wound on a piece of 1in plastic pipe. About 200 turns of 0.3mm diameter enamelled copper wire were used, and the coil placed about 6in away from the sensor. Via a pair of wires to the shack, a 1.5V battery connected for about one second generates a field for a time sufficient to make the sensor swing backwards and forwards over the full scale. The 10kΩ sensitivity potentiometers can then be adjusted so that each LED lights

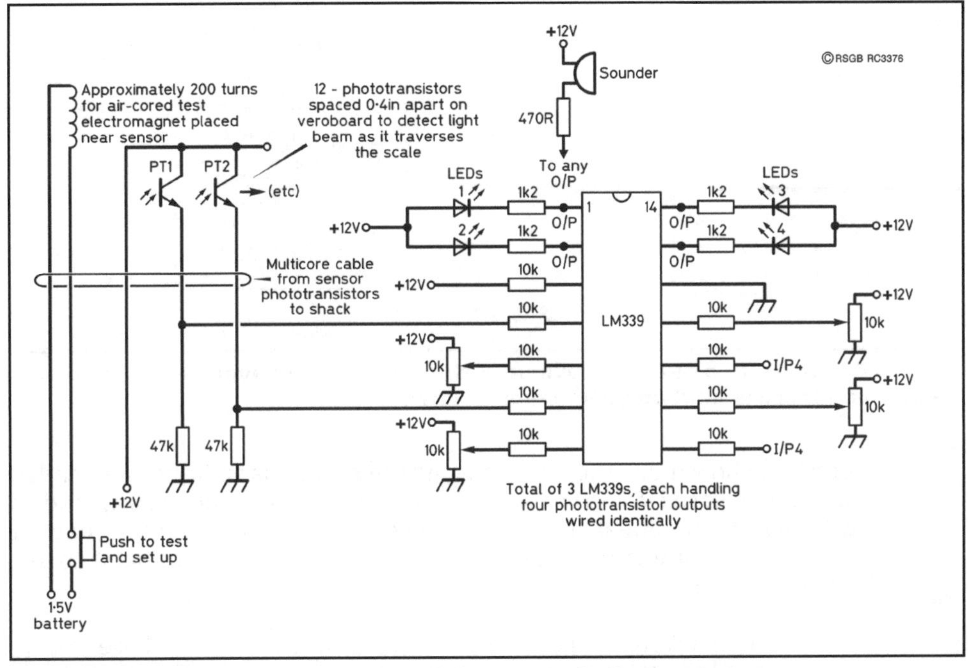


**Fig 5: Diagrammatic representation of the remote indicator system.**

momentarily as the light beam swings past each phototransistor. Small sounders connected to the output pins of the voltage comparators give audio alerts when there is some activity. Only two were found to be necessary, one indicating small swings and the other more extreme changes. Which pins to use is a matter of individual choice. On each IC, you have a choice of pins 1, 2, 13 and 14. It is possible to connect a counter to any of the outputs to count the number of times a particular level has been reached, but so far this refinement has not been attempted.

**Fig 5** shows the disposition of equipment in the attic, in the shack, and its interconnection.

## RESULTS

What were the results of all this pottering around? Firstly a lot of fun, and secondly, after several weeks of waiting when nothing very much happened, an exciting time when a magnetic storm caused the system to go crazy, swinging backwards and forwards over the whole length of the scale: **Fig 6**.

Most days are fairly quiet, as you would expect, though there does seem to be a shift between 0800 and 2200 each day, with a slow change of about 10 minutes of arc, sometimes in one direction and sometimes in the other. A pattern might be found after a time.

When there is a magnetic storm, the system goes mad, at times dashing backwards and forwards over the whole scale. It is not possible to

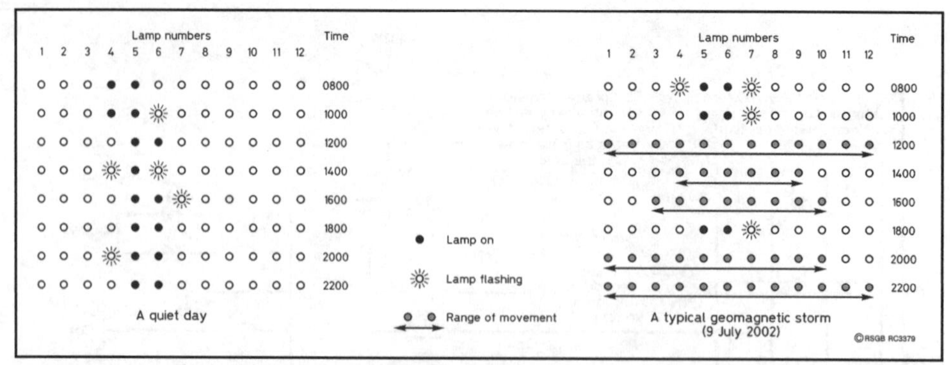


**Fig 6: An illustration of the behaviour of the system on an average day, and on a day when there is marked geomagnetic activity.**

determine how much the change is, as the system is so freely suspended that it oscillated for a minute or two before settling down, and is certainly not deadbeat. Damping the sensor by immersing the magnet in a small pot of water (what else!) was tried, but it was difficult to see if it reduced the sensitivity at all.

These experiments are continuing, and there could be many improvements to the setup described here. In theory, the perturbations should be an indication of radio conditions at the time they occur, but at this early stage it hasn't been possible to relate them positively. It is an open field for anybody who wants to dabble in some electronic experiments related to ham radio, and I would be very interested to hear of other people's experiences. Despite being a decrepit old-timer, I am up-to-date enough to be on email, so please let me know how you get on.

# SIMPLE DIODE-MATCHING CIRCUIT

**by Pat Hawker, G3VA, from 'Technical Topics'**

The changing nature of amateur radio with virtually 95% of equipment now factory-built, resulting in a decreasing interest in home design and construction, is regretted, if only passively, by many. A good example of its effects can be seen in the growing reluctance of our major technical libraries to take and display the overseas amateur radio publications that formerly graced their shelves. Another recent example is the merging of ARRL's *QEX* with *Communications Quarterly* published by *CQ* Communications Ltd as the successor to the still-missed *Ham Radio*. Clearly the interest in the technology of communications other than purely practical operating techniques is decreasing at an alarming rate. Much is being left to relatively small-circulation specialist journals such as the excellent *CQ-TV* and *Sprat*. Amateur radio is essentially a hobby, but for many years it has been recognised as a *technical* hobby providing "A radiocommunication service for the purpose of self-training, intercommunication and technical investigations", to quote from the long-standing international definition. Now, it sometimes seems that it is primarily for "intercommunication" and little else. That of course is unfair to the many who, while generally using factory designs, remain deeply interested in understanding and developing the technology, but it must be the impression given to newcomers.

J A Ewen, G3HGM, wonders if there are any constructors still about in these days of black boxes – with even simple wire aerials purchased – though he admits that may seem a rather curmudgeonly attitude brought about by increasing age. In the hope that there are still a few of what he calls "real hams" left, he offers the diode matching unit shown in **Fig 1**.

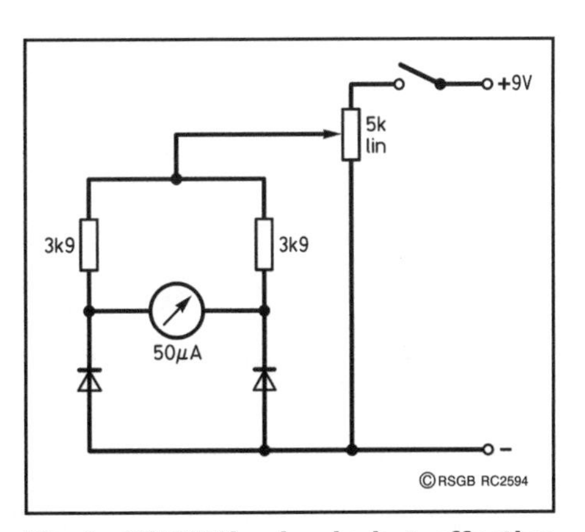

**Fig 1: G3HGM's simple but effective diode matching unit.**

He writes: "Users of this circuit who have wondered why their home-brewed double-balanced diode ring mixers [or indeed any balanced diode mixer] have underperformed will have it revealed that diodes selected for 'balance' by simply comparing their forward resistances at a single voltage (the voltage being applied by the ohmmeter) usually differ widely at different forward voltages.

"The circuit comprises a simple bridge with the two diodes under examination forming the lower arms. The two 3k9 resistors should be selected by measurement on a digital ohmmeter. Although their exact value is unimportant, they *must* be of identical resistance. The 50μA meter shows zero reading when the two diodes are passing identical currents (ie they are matched at that applied voltage). The 5k potentiometer allows a voltage varying from zero to approximately 9V to be applied. A current flows when the bridge is unbalanced by different forward resistances of the diodes. Ideally, the meter should be of the centre-zero type, but an end-zero meter is usable though not so convenient, since it requires the diodes to be interchanged when there is a negative reading.

"I believe that most users will be astounded by the spread of characteristics between diodes bearing the same type number, and will be driven to seek a pair of diodes where the needle virtually fails to move throughout a full sweep of the potentiometer. I have found that a meter reading of less than 1μA throughout the range indicates a match far better than that obtained by purchasing so-called 'matched diodes'. Due allowance should be made for the fact that germanium diodes do not start to conduct until approximately 0.2V is applied. The corresponding figure for silicon diodes is about 0.6V."

# THE LED BARGRAPH OSCILLOSCOPE

**by George Brown, M5ACN / G1VCY**

There is little doubt that the most flexible piece of test equipment available to any radio amateur is the Cathode-Ray Oscilloscope (CRO). What follows comprises a simple alternative, using a light-emitting diode (LED) display. It is by no means a replacement for the CRO, but it forms a very useful instrument which is not difficult to build, and which is a digital/analogue hybrid, giving those who lack practice in circuit-building a project which combines digital and analogue techniques.

## THE PRINCIPLE

The cathode-ray tube (CRT), at the heart of any oscilloscope or TV set, produces a display by scanning a spot of light from left to right across the screen; this is usually done so quickly that the eye cannot see the movement. In an oscilloscope, the signal being displayed is used to deflect the spot of light in a vertical direction, producing what is known as the waveform; in a TV set, the picture is built up by altering the intensity of the spot as it is scanned from the top left to the bottom right of the screen in a series of 575 parallel lines.

The technique to be used here is the simulation of the scanning of the spot in the CRT by repeatedly scanning along a horizontal row of LEDs, activating one LED at a time, from left to right. The signal to be viewed is fed to all LEDs, but the instantaneous value of the signal can be displayed by one LED only at any instant. This is an exact analogy of what happens in the CRT, except that the value of the signal in the LED system is indicated, not by the vertical deflection of a spot, but by the brightness of the activated LED.

A block diagram of the basic LEDScope system is shown in **Fig 1**. At its heart is a linear array of 30 LEDs. It is best for these LEDs to be in the form of bargraph modules which can be conveniently stacked to give a display of any desired length; the 'spot' is scanned across the line of LEDs by a series of shift registers. These shift registers activate only one LED in the whole array at any one time, but the lit LED appears to move from one LED to the next, from the left end to the right end, before reappearing at the left end and repeating the process. In the absence of any signal,

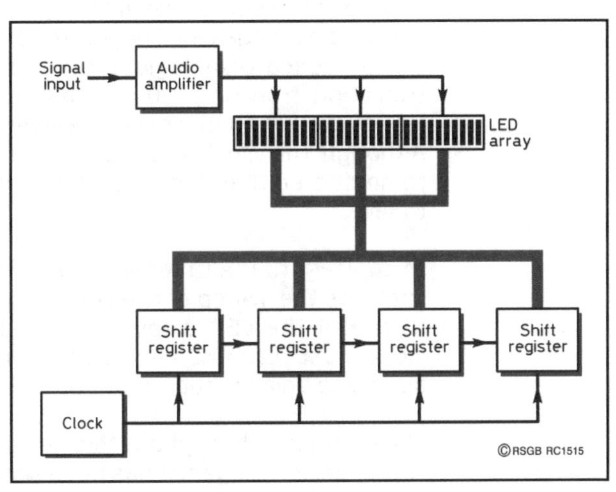


**Fig 1: The system block diagram.**

therefore, the spot of light simply scans from left to right along the array; the speed at which this happens can be changed by changing the frequency of the clock signal.

This is simply a rectangular pulse and, for every pulse from the clock, the light appears from the next LED to the right of the previously-lit LED. This process is equivalent to that of the timebase in a CRO. The only other part of the system is that which determines the size of the signal applied to each LED in order to change its brightness. What the signal circuitry must do is provide to the LEDs a signal having the correct maximum and minimum values to illuminate the LEDs correctly. Variable gain will be required to do this, because it cannot be guaranteed that all signals to be investigated will have the same amplitudes! Notice how the signal is applied to all the LEDs in parallel; this is quite a simple thing to do, and is possible because the only LED that will light as a result of the applied signal is that one which has been activated at that instant by the shift register.

The whole circuit is a low-voltage system, working on 13.8VDC, so it would be inappropriate to imagine that it would be suitable for monitoring high-voltage signals. Because of its display characteristics, it is suitable only from audio frequencies up to about 100kHz.

## THE DESIGN

This falls into three parts: the clock generator, which produces the incremental movement of the 'spot' along the LEDs; the shift registers which, together with the LED array, produce the scanning spot; and the audio amplifier and brightness control, which optimise the incoming signals for display on the LED array. The circuit is designed to run from any 12VDC (or the more 'normal' 13.8VDC) supply, since its operating current is quite low at 60mA.

### The clock generator

The clock generator is the means by which the 'spot' appears to travel along the row of LEDs. In the LEDScope, there are 30 LEDs in the row and, for the spot to travel from one end to the other, 30 rising edges are required from the clock generator. It should be noted that it is not primarily the presence of a logic 1 that causes the shifting operation, but the transition from logic 0 to logic 1. This shift register (in this case an HCF4015BE CMOS chip) is known as a rising-edge-triggered device. Although the clock produces a series of rectangular pulses, it must be remembered that it is on each rising edge that the who1e system springs to life.

In contrast to the timebase in a CRO, the clock does not run at the same speed as the sweep of the spot across the display, because of the discrete nature of the LED array. The 30 LEDs in the array determine that the clock must run 30 times faster than the sweep. In some ways, this may appear to be a limitation, but this has not proved to be the case in practice.

The emphasis throughout this design has been on simplicity and, in some cases, design elegance has been sacrificed for this aim without necessarily compromising the usefulness of the equipment. It is worth noting that, as a direct consequence of using the shift register as the 'timebase' for this design, the sweep is absolutely linear. Because of the limitations of the display, it is unlikely that waveforms much above 15kHz will need to be displayed. This means that, if one cycle of the wave is to be displayed on the LED array, the clock must run at 15,000 x 30 = 450,000Hz or 0.45MHz. This has been achieved with the simple clock circuit shown in **Fig 2**. The 100μF electrolytic capacitor (C5) shown between the power supply pins of the 555 is optional; all data sheets advise that this should be present and connected as close to the pins as possible. In this application, no change has been noticed if it is omitted. If irregularity in the output of the 555 is observed, this capacitor should be inserted.

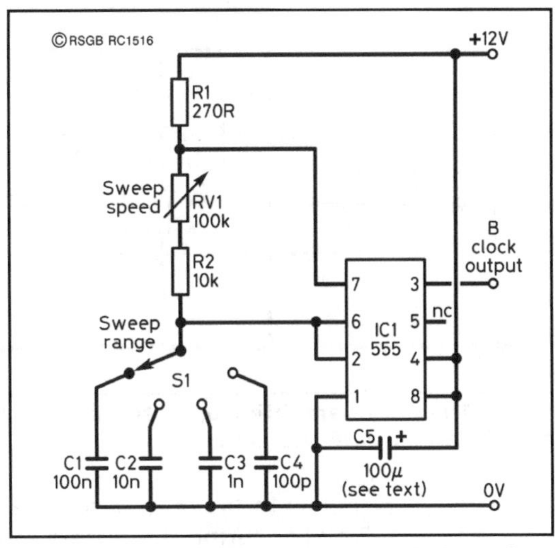


**Fig 2: The clock generator circuit based on 555 chip.**

### The shift registers

The circuit for the four shift registers and their associated three LED arrays is shown in **Fig 3**. Each shift register has eight stages, so four registers are needed, with two stages unused, to drive the 30-element LED array. All that need be said here is that its purpose is threefold: to reset the registers when the LEDScope is switched on; to flush the array of all zeroes except one; to scan the 'spot' sequentially along the array from left to right at a constant speed. It has two inputs, one (B) from the clock to provide the signal to increment the position of the active LED, the other (A) being from the audio amplifier, which determines that the

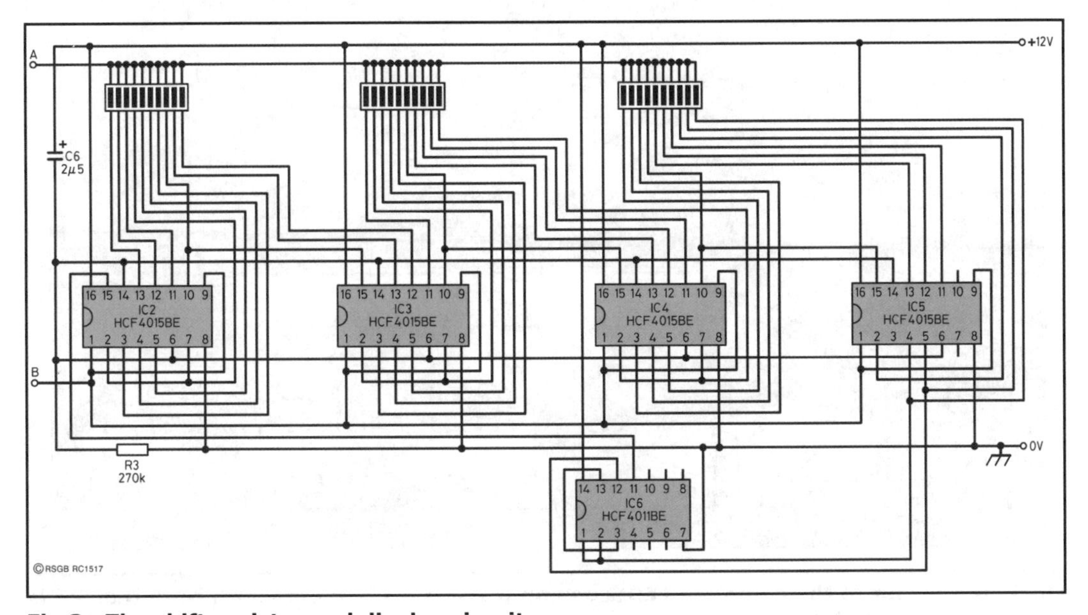


**Fig 3 : The shift register and display circuit.**

brightness of the active LED should be proportional to the signal voltage at that instant.

**The audio amplifier and brightness control**
The circuit for amplifying the incoming audio signal and for controlling the overall brightness of the LED array is shown in **Fig 4.**

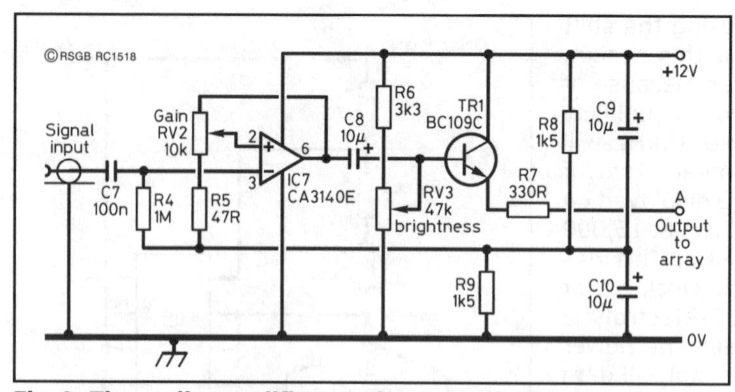


**Fig 4: The audio amplifier used to amplify the signal so that it can operate the LEDs.**

## CONSTRUCTION

Follow the instructions below in order to assemble and test the LEDScope in a structured way, thus ensuring the fewest errors in wiring. The layout of the prototype on pinboard is shown in **Fig 5**, and can be compared with the photo. *In all the tests, all points X, Y, Z, etc refer to the matrix-board layout of Fig 5.*

## THE CLOCK

Wire up the clock circuit as shown, leaving gaps for the connection of the variable sweep speed potentiometer (RV1) and the sweep range switch (S1). Do not insert the 555 chip into its socket until you are ready to perform the first test.

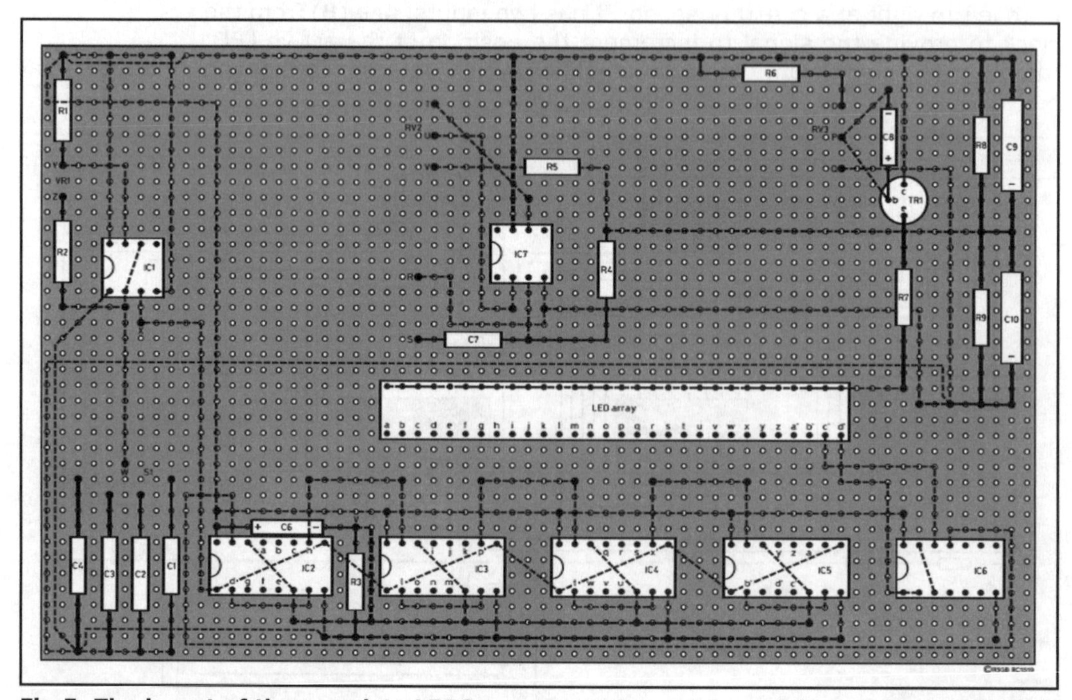


**Fig 5: The layout of the complete LEDScope on a 16cm x 10cm pinboard. Note that R6 is mounted underneath the board.**

The prototype layout, where all controls are mounted on the board itself. You may wish to put the circuit board into a box, in which case leads must be brought from the board at the positions shown in Fig 5.

### Intermediate test 1

Connect the 100kΩ potentiometer to the points Y and Z, and then *temporarily* connect a 2.5μF capacitor between point W and ground. The effect of putting in the large capacitor is to make the clock run very slowly (several seconds per cycle). Connect an 820Ω resistor from test point X, the output of the clock, to the anode of a discrete LED, with its cathode grounded. Insert the 555 chip and connect a battery or power supply (maximum 16VDC) to the circuit, ensuring that the polarity is correct. Switch on. The LED should flash very slowly, and its speed should be controllable by adjusting the potentiometer, RV1, connected between Y and Z. If the circuit does not work, switch off and check the wiring carefully.

The extra capacitor connected between point W and earth should be left alone at this point, as should the LED and resistor (unless they prevent the circuit board from being laid down flat); they will be needed later on to check the operation of the shift registers.

## SHIFT REGISTERS AND LED ARRAY

Wiring up the shift registers is the most complex part of the construction, and should be done with the aid of the layout diagram and the photograph. If the display is mounted separately from the board, it is recommended that coloured ribbon cable is used, thus minimising the potential for error. As each LED array is driven by more than one shift register chip, the wiring process is not a repetition of the same connections. The only thing that is common is the order in which the shift register pins are connected to adjacent LEDs in the array. Working from left to right along the LED array (as viewed from above), the order of the pins on each shift register is 13, 12, 11, 2, 5, 4, 3, 10. These connections are not shown on the layout diagram, as the constructor will no doubt want the LED array separate from the circuit board. In Fig 5, it is assumed that there are wires present which connect pin a on the array to pin a on the shift register, pin b to pin b, and so on.

### Intermediate test 2

Connect a wire between points Y and Z. *Do not insert the CMOS chips yet*. Switch on. As you have not changed the clock circuit at all, it should still work, and the LED should flash slowly. If not, switch off and look for dry joints where wires and components may have parted company as the board has been manipulated during the wiring of the shift registers.

When the clock is running, take a voltmeter on a range suitable for measuring up to 16V, and apply the probes directly to the supply pins of each CMOS socket. Looking at the upper face of the board, these are always the top left (positive) and the bottom right (negative) pins. Do not put the negative probe on the ground rail, because you will not detect discontinuities in the connections to the bottom right pins. Similarly, do not put the positive probe on the positive supply rail, or discontinuities in the wiring to the top right pins will go undiscovered.

While the equipment is switched off, remove the positive lead of the LED from test point X and, if necessary, attach to the LED anode a piece of wire long enough to reach all the shift register holders. Switch on again. Touch the wire briefly on test point X again, to ensure that the clock is still running. Then, go to each shift register socket and touch the wire sequentially onto pins 1 and 9. The LED should flash on each of the eight pins tested. Switch off.

Using the meter on its ohms range (or as a continuity tester) check that a connection (0Ω) exists between test point V (the junction of C6 and R3) and pins 6 and 14 of each 4015 socket; this confirms that the reset circuitry is properly connected.

Without having inserted any of the CMOS chips, the supply voltage, the signal, the clock and the reset connections have been checked and verified as correct.

From the common anode connection of the LED array, temporarily connect a single 820Ω resistor to the positive supply rail, and insert the arrays into their sockets. It is now time to fit the CMOS chips into their sockets.

Reconnect the LED and its series resistor to test point X. Make sure the circuit board is not resting on anything metallic, and switch on. The single LED should flash as usual; all 30 LEDs in the LED array should come on, with various intensities, as the registers are reset; then, from the left-hand end, the LEDs should go out sequentially each time the clock LED flashes; when the last LED extinguishes, the extreme left LED should light, and this should scan across the array in a re-circulatory fashion. The only fault which is possible at this stage is that the LEDs may come on in the wrong order; there is no alternative but to check the wiring of those LEDs which are at fault, after switching off.

Disconnect all temporary components which may impede further construction.

### THE AUDIO AMPLIFIER AND BRIGHTNESS CONTROL

This is a simple analogue circuit, and a typical layout is shown in Fig 5. No special constructional points here, except to notice that the amplifier derives its own balanced power supply by the resistor chain R8 and R9, decoupled by C9 and C10. Although IC7, a FET-input operational amplifier (CA3140E), is pin-compatible with the ubiquitous LM741, the use of an LM741 is not recommended because of its poor slew rate of $0.5Vms^{-1}$, compared with the $9Vms^{-1}$ of the CA3140E. Make sure that pin 7 of the 8-pin socket is connected to the positive supply rail, and that pin 4 goes to the negative supply rail (ground). The input load resistor (R4) and the gain-limiting resistor (R5) are the only components to be connected to the artificial ground (the junction of R8 and R9). If TR1 is being wired as shown in the layout diagram of Fig 5, ensure that the base leg of the BC109C is gently bent between the emitter and collector legs before being soldered to the board.

**Intermediate test 3**

Connect up the rest of the circuit almost as before – points Y and Z shorted together, but with point W connected to the top of C1, the 100nF capacitor. Do not insert the CA3140E yet. Set the brightness control, VR3, to about half-way around its travel. Switch on. After the short initialisation period, the scan across the LEDs will now be much faster than before, being visible as a flicker rather than as a directional scan. If all is not well, observe the fault and search for it after switching off. Check that the brightness of the display varies with the setting of RV3. With a voltmeter, check between pins 4 and 7 of the socket of IC7 that the full DC supply voltage is present. Then, check between ground and the junction of R8 and R9; approximately half the supply voltage should appear at this point.

## FINAL TEST

When the system is working, switch it off and insert the CA3140E carefully. It must be treated with the same care as the CMOS logic devices! If you have a signal generator, connect it to points R and S and set it for about 1V RMS sine-wave output at about 500Hz. If you don't have a signal generator use the output from an SSB or CW receiver tuned to a carrier to produce an audio tone.

Set the range switch to position 3, the sweep speed to minimum, and set the gain control, RV2, to about half-way round its travel. Switch on. Advance the sweep speed control (RV1) slowly. Initially, the display does not appear to change; it then begins to flicker, and then resolves itself into a moving pattern of two bright bands across the array. Further rotation of the control produces a single bright band.

The LEDScope is operational! When there is only a single band, the sweep frequency and the incoming signal frequency are the same. When there are several bands, the signal frequency, f, is given by

$f = N \times s$,

where *N is the number of bands, and*
*s is the sweep frequency in Hz.*

## PROTOTYPE PERFORMANCE

This device performs as it was designed. Despite the simplicity of the display, it is easy (with practice) to detect the differences between sine, square and triangular displays, although clipping is more difficult to pinpoint. It can be used to measure the frequencies of beats and of audio-frequency oscillators and has even been used as the display for an SSB two-tone test. Calibration is best performed with an accurate signal generator while watching the patterns produced on the display. The values of RV1 and R2 have been chosen to give approximate decade ranges with a degree of overlap between them. This can be seen in the Specification.

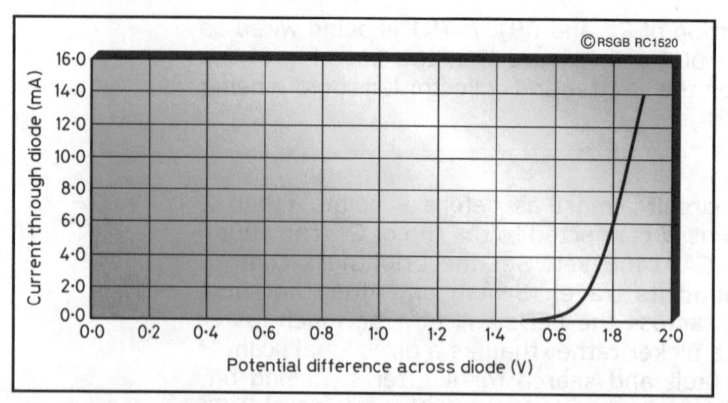


**Fig 6: The current/voltage characteristic of a single light emitting diode from a bargraph array.**

Range 4 is not a full decade range because the performance limit of the clock chip in a pin-board construction is fast being approached at these frequencies (390kHz).

## A NOTE ABOUT LEDS

The variation of current flowing through an LED as a function of the forward voltage across it is shown in **Fig 6**. This is a real characteristic, having been measured on a single LED from the arrays used in the LEDScope. Notice that no current flows through the LED until the voltage across it is about 1.5V. After that, it rises very steeply with little increase in voltage. This contrasts with the behaviour of normal silicon diodes, which begin to pass current at between 0.6V and 0.7V.

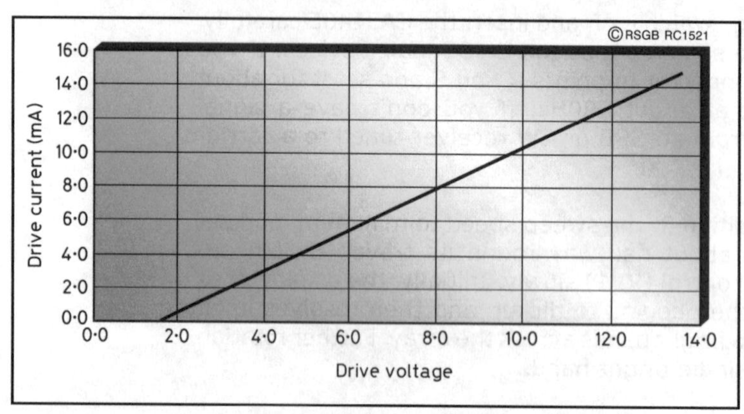


**Fig 7: The variation of diode current with drive voltage using a 820Ω series resistor.**

As soon as current begins to flow through the LED it begins to glow, and increases in brightness as the current increases. This behaviour would indicate that to intensity-modulate an LED, only a few millivolts would be needed to take the brightness from zero to maximum. This is true, but LEDs are never operated without a series resistor to control the current. **Fig 7** shows how the current varies through the LED when an 820Ω resistor is placed in series with it and the supply (or drive) voltage varied. What was a highly non-linear curve becomes almost a straight line, beginning at about 1.7V. This now means that any voltage between 1.7V and 14V

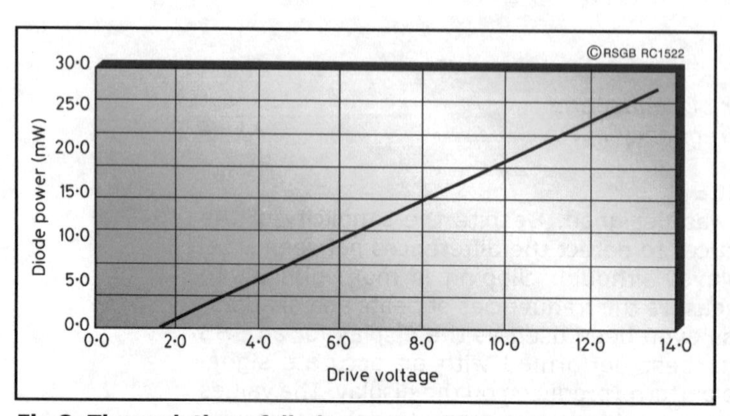


**Fig 8: The variation of diode power with drive voltage using a 820Ω series resistor.**

controls the current through the diode in a linear fashion.

This is an improvement on the situation without any series resistor and is shown in **Fig 8**. So, provided that the signal drive voltage supplied to the LED array is above about 1.7V, the light output will be directly proportional to the signal voltage. The audio amplifier of Fig 4 ensures that the voltage swing of the signal is large enough, and the brightness control provides the small offset that is necessary to overcome the 1.7V threshold.

**SPECIFICATION**

Note: All values measured on the prototype.

**Power requirements:**
10 - 16 VDC, 60mA at 13.8V

**Display frequency ranges:**
Range 1 2.0Hz - 24Hz
Range 1 20Hz - 240Hz
Range 3 190Hz - 2100Hz
Range 4 1700Hz - 13000Hz

**Signal input:**
Usable input range for full intensity modulation:
15mV RMS minimum - 3.28V RMS maximum

**Signal amplification:**
1.0 minimum, 218 maximum

**Useful range of frequency measurement:**
10Hz minimum, 100kHz maximum

**Components list on the next Page.**

## LED BARGRAPH OSCILLOSCOPE COMPONENTS LIST

### Resistors
All resistors 0.25W 5% (metal film or similar)

| | |
|---|---|
| R1 | 270R |
| R2 | 10k |
| R3 | 270k |
| R4 | 1M |
| R5 | 47R |
| R6 | 3k3 |
| R7 | 330R |
| R8, R9 | 1k5 |
| RV1 | 100k lin |
| RV2 | 10k lin |
| RV3 | 47k lin |

### Capacitors

| | |
|---|---|
| C1 | 100n, polyester layer, 5%, |
| C2 | 10n, polystyrene, 1%, 63VDC |
| C3 | 1n, polystyrene, 1%, 125VDC |
| C4 | 100p, polystyrene, 500VDC |
| C5 | 100µ, 35VDC |
| C6 | 2µ5, 64VDC |
| C7 | 100n, metallised polyester 250VDC |
| C8, C9, C10 | 10µ, 25VDC |

### Semiconductors

| | |
|---|---|
| IC1 | NE555N Timer |
| IC2 – IC5 | HCF4015BE 2 x 4bit shift register |
| IC6 | HCF4011BE Quad 2-input NAND gate |
| IC7 | CA3140E FET-op amp |
| TR1 | BC109C npn audio transistor |

### LED arrays
3 x 10-segment bargraph array (Maplin code RS code 588-027)
(The anodes of these arrays are identified as the same edge that carries the manufacturer's There is also a very small chamfer on the corner same long edge.)

### Other Items
Dual-in-line integrated circuit sockets
2 x 8-pin for IC1, IC7
3 x 20-pin for LED arrays
4 x 16-pin for IC2 – IC5
Switch S1 1 pole 4-way rotary

### Circuit board and pins
1 x plain matrix board 160mm x 100mm (Maplin code JP53H)
Pins as required (Maplin code FL23A)
4 knobs, as required

# Chapter 3 AERIALS
# FEED-LINE VERTICALS FOR 2m AND 6m

**by Rolf Brevig, LA1IC**

During decades of portable operation on the 2m and 6m bands, I have met amateurs who needed aerials that were efficient, simple to construct and also easy to hang up almost anywhere. The feed-line vertical aerials described here provide an excellent match to a transceiver, without a separate aerial tuner. They can be easily coiled-up and stored in your luggage or even stuck in your pocket, as they are made from only one piece of flexible coax cable. The basic design is extendable to any frequency segment between 50 and 150MHz.

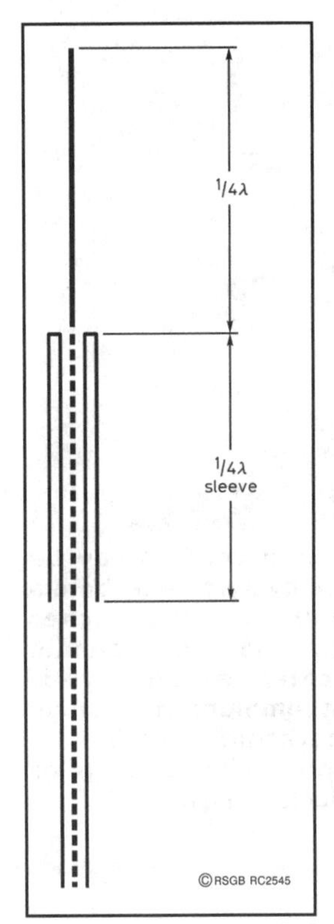


**Fig 1: The coaxial sleeve aerial. The best SWR that could be obtained was 2:1.**

The point of departure is the coaxial sleeve aerial (**Fig 1**), which was popular until the advent of modern SWR analysers. A true resonance will always be found, but I have never achieved a better SWR than 2:1 in such aerials, probably due to stray capacitance. The lesson learned, however, was that the RF current had no trouble in travelling up the inside of the coax and then making a 180° turn to travel back down the outer sleeve.

Because of this, perhaps we don't need the sleeve. Why not just use the braid of the coax itself? If we do this, however, how do we let the RF 'know' when it should stop flowing and reflect back towards the centre of the dipole, as it did when it came to end of the braid in the coaxial sleeve aerial?

After trying different wideband devices, I found that a coaxial cable choke resonating within the band segment in question was the best solution to meet my requirement. Very low SWR, broad-bandedness, and the possibility of working out reliable dimensions from the formulas in **Fig 2**.

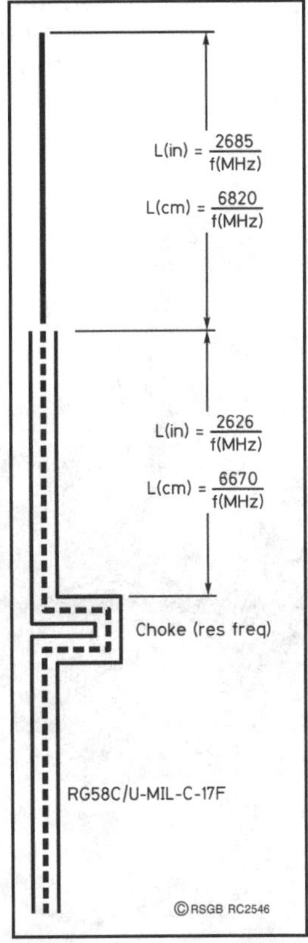


**Fig 2: The feed-line vertical. The best SWR that could be obtained was 1.1:1.**

## 2m AERIAL

In the photos you can see the 144MHz version of this aerial. It is made from a 387cm (152$^{1}/_{4}$in) long piece of RG-58CU coaxial cable, of which a quarter-wavelength (use the formula) of sheath and braid is stripped off, this forming the upper part of the dipole. Next, measure the lower part of the aerial (use the formula) and mark the starting-point of the choke. For the choke, wind 4.6 turns of the coaxial cable on to a piece of 32mm (1$^{1}/_{4}$in) diameter PVC tube. The caps on each end are not essential, but they are useful to centre the cable and lock the turns.

**The feed-line vertical can be suspended by nylon line from a tree limb or any other convenient support. It should dramatically extend your maximum communication range when operating portable, in comparison with a whip or 'rubber duck' aerial.**

A ring terminal or tag needs to be soldered to the tip of the dipole, bearing in mind that this will lower the resonant frequency a bit. Trimming, if necessary, should be done at the tip, outdoors, well away from objects that might affect the resonance.

Don't cut more than 6mm ($^{1}/_{4}$in) at a time. The SWR should be less than 1.3:1 and the impedance very close to 50Ω across the entire band. Observant readers will see that I have used exactly 5 electrical half-waves (340cm) of feeder. It is a good idea to make any additional feeder a multiple of 68cm (26$^{1}/_{2}$in).

**Rolled-up and ready to go, the 2m version of the feed-line vertical.**

## 6m AERIAL

For a 50MHz vertical dipole, you start with a 728cm (286$^{1}/_{2}$in) length of RG-58CU. Using the formulae, follow the same constructional procedure as previously. In this instance, the choke consists of 11.8 turns on 50mm (2in) diameter PVC tube.

Although not critical, you can centre the aerial to your favourite 6m frequency by cutting the tip little by little, and still enjoy a 1.3:1 SWR across the band. Adding feeder to this aerial should be in multiples of 198cm (78in).

# 'MAGIC' FRACTAL AERIALS

**by Pat Hawker, G3VA, from 'Technical Topics'**

Both Joe Kessel, G7RWH, and I found 'Aerial Magic' by Mike May (*New Scientist*, 31 January 1998, pp28-30) of more than passing interest. It claims: "Your mobile phone, the gas meter under the stairs and the vending machine down the hall may soon sport the world's smallest, weirdest aerials." Professor Nathan ('Chip') Cohen is presented as the inventor of a new family of 'fractal aerials' which may come to be used in everything from mobile phones to huge receiving arrays. The idea emerged from an experience in 1988 when Cohen wanted to use his ham radio equipment at a flat in the centre of Boston, where the landlords stipulated no outside aerials.

Instead of using a difficult-to-disguise conventional aerial, Chip Cohen cut a sheet of aluminium foil into the shape of a mathematical pattern known as an inverse Koch curve, and stuck it onto a sheet of A4- sized paper: "Cohen connected the foil to his radio receiver, to see if it might serve as a covert aerial if he mounted it outdoors. To his surprise, the fractal foil pattern worked well and Cohen was able to continue his hobby without arousing suspicion [no indication is given of the frequency band(s) concerned - *G3VA*].

"Today, Cohen's experiment has made him a pioneer in the new field of fractal aerial design. It turns out that fractal aerials have many advantages over their conventional counterparts. For a start, they are smaller - a fractal aerial for a mobile phone can be made the size and shape of a 35mm photographic slide and can be built into the casing... fractal aerials are also better at picking up signals and can receive them over a wider range of frequencies. But... physicists are being left behind. Nobody is exactly sure why fractals make good aerials. Now the race is on to find out." It seems, according to the article, that a fractal pattern aerial provides the 3 – 4dB gain of two half-wave elements in phase, but it is by no means clear how this relates to its usable bandwidth and what this amounts to.

Although the article describes in general terms what constitutes a fractal pattern, it does not attempt to provide precise details of the dimensions, impedances, etc, of a practical fractal aerial. Nor, in a quick library search, could I discover any further information on fractal aerials. My dictionary does not even list 'fractal'. I did, however, find out that fractal geometric structures are of particular interest to computer scientists, because of their use in computer graphics to depict landscape and vegetation in a very natural way. There is also apparently a strong connection between fractal geometry and the trendy chaos theory. Fractal geometric structures include the so-called snowflake curve, devised by Swedish mathematician Helge von Kohn in 1904 (**Fig 1**). Fractal patterns exhibit a 'self-similarity'

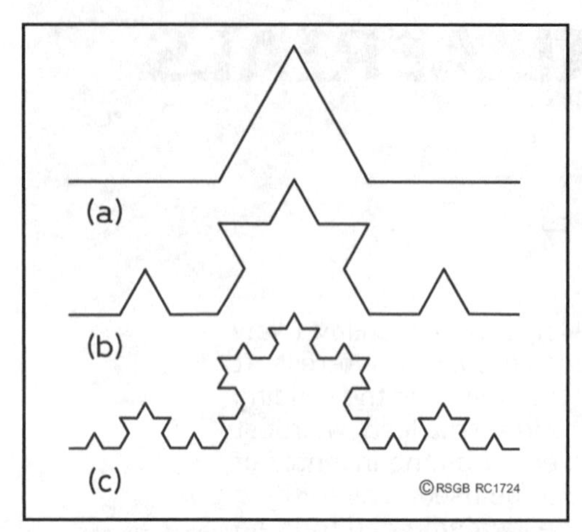


**Fig 1: Construction of a von Koch 'snowflake' fractal, formed from the basic curve of (a), with successive iterations shown in (b) and (c). As stages are added, the total length of the curve tends to infinity, although the curve is confined to a finite space.**

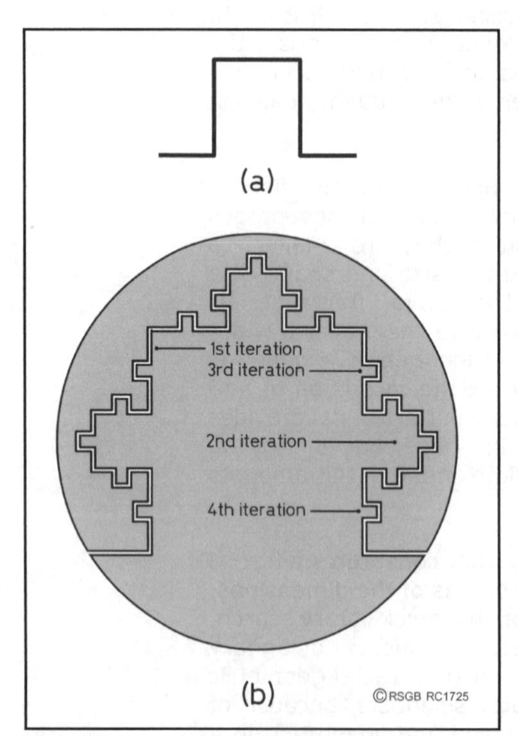


**Fig 2: Illustration of a fractal aerial that could be built into a military helmet, as described in the *New Scientist* article, built up from the iteration of the basic curve (a).**

when the distance at which they are viewed is changed. They consist of many repetitions (iterations) of a basic shape. The inverse Koch curve used by 'Chip' Cohen for his covert amateur radio aerial was apparently a fractal, comprising a series of triangles stacked up on top of each other like a pagoda, but his UHF micropatch fractal aerials seem to be variations of the Koch snowflake curve: **Fig 2**.

There is of course nothing new about using elements bent into regular shapes in order to reduce overall length, as in the 'Meander' and 'Zigzag' aerials (see for example *TT*, February 1992, from which **Fig 3** is taken).

## MORE ON FRACTAL AERIALS

Although apparently some readers thought the above item was an April Fool joke, I am grateful to Peter Chadwick, G3RZP, (and subsequently VP8AEM) for drawing my attention to the series of articles by (Professor) Nathan ('Chip') Cohen, N1IR, in *Communications Quarterly*. These articles, which I have located in the RSGB Library at Potters Bar, provide in considerable detail (totalling over 50 large pages) background

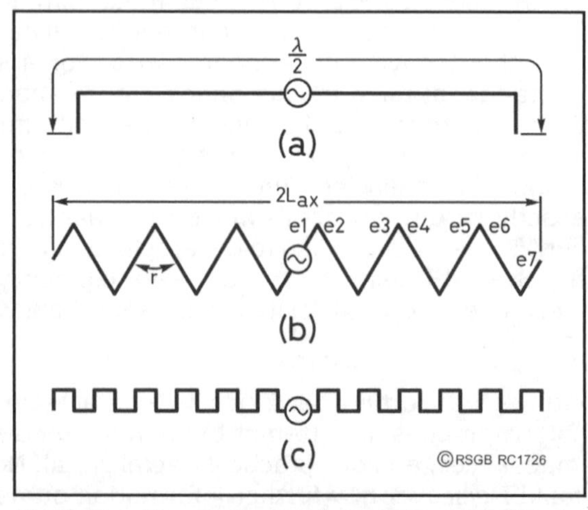


**Fig 3: Significant shortening of the overall span of a conventional half-wave dipole element can be achieved without significant loss of efficiency or directivity. (a) Ends dropped downwards or bent inwards. (b) Zigzag element, where e1 = e7 = 0.0208λ and e2 = e3... e6 = 0.0416λ. (c) Meander-line element as noted in 'TT' of February 1992.**

information on aerials having their elements shaped in fractal geometric forms as dipoles, folded dipoles and loop aerials. Many of the looped designs are based on various iterations of the Minkowski Island (MI) fractal, starting (MI0) from a quad (square) element. The basic Minkowski fractal, with four iterations, forms the basis of the military helmet aerial shown in Fig 2 of the April 'TT' (p59); see also **Fig 4** for M0, M1 and M2 geometry. Among the fractal aerials illustrated are an MI3 aerial for 144MHz, about 8in per side and a MI2 for 50MHz about 30in per side. **Fig 5(a)** shows the geometry of a first iteration Minkowski Island (MI1) and **Fig 5(b)** a second iteration, MI2.

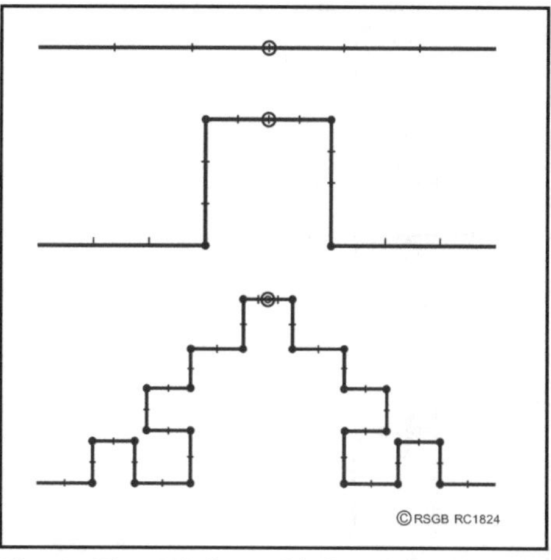


**Fig 4: The Minkowski (box) fractal for iterations 0 (straight line), 1 and (bottom) 2.**

In 'Fractal Antennas: Part 1 - Introduction and the Fractal Quad' (*Communications Quarterly*, Summer 1995, pp7 – 22), N1IR provides a good introduction to the geometry of aerial elements, pointing out that although aerial design has become a mature field, there is still room for a new approach that uses fractal geometry to produce very small aerials of high efficiency and with some other useful attributes. He presents a design for very small area single-element cubical quads, along with their comparative results. He makes it clear that although work on fractal aerials has blossomed recently, they are not entirely a new idea. One of the original forms of microwave log periodic aerials, as built over 40 years ago by DuHamel and Isbell, was a toothed spiral fractal.

Then, in a decade of work some 15 – 20 years ago, Prof F M Landstorfer and R R Sacher at the Technical University of Munich, investigated the use of bent aerial elements in order to develop an optimum shape. This work was noted several times in 'TT'. For example, 'Optimum- Shaped Antennas' December 1982, pp1054 – 55, from which **Fig 6** and **Fig 7** are reproduced. See also 'Directivity of Optimum Shaped Antennas' (February 1983, pp131 – 32). Although the Landstorfer element was not a fractal, nevertheless his work showed that the conventional straight element was not necessarily the optimum approach.

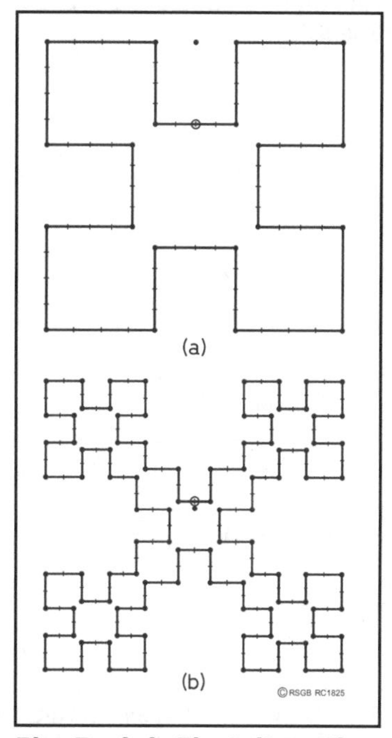


**Fig 5: (a) First iteration Minkowski Island (MI1) fractal developed from the quad square element and used in several N1IR designs. (b) Second Iteration Minkowski Island (MI2) has also been used by N1IR.**

'Fractal Antennas: Part 2 - A Discussion of Relevant, but Disparate Qualities' (*Communications Quarterly*, Summer 1996, pp53 – 71), came a year after Part 1. But, before then, a further article by N1IR on fractal aerials had appeared: 'Fractal and Shaped Dipoles - Some Simple Fractal Dipoles, their Benefits and Limitations' (*Communications Quarterly*, Spring 1996, pp25-36). Then, together with Robert G Holfeld, Cohen presented 'Fractal Loops and the Small Loop Approximation – Exploring

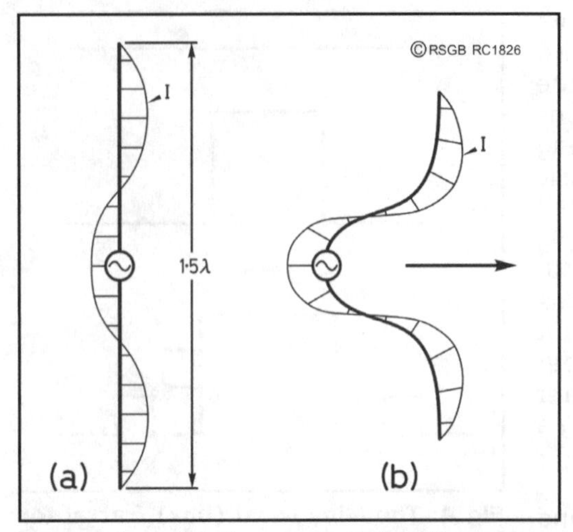


**Fig 6: Showing the current distribution on (a) a straight 1.5λ dipole where the phase reversal reduces radiation normal to the axis of the dipole; and (b) with a Landstorfer gain-optimised shape, which causes radiation to increase significantly in the forward direction as a single-element 'beam'.**

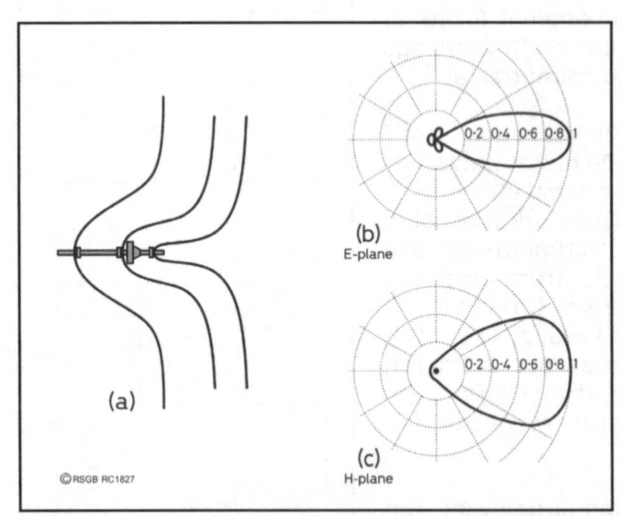


**Fig 7: A Yagi array using three gain-optimised elements each approximately 1.5λ long. A VHF array of this type, reported in 1982 by Dr Landstorfer, provided a forward gain of 11.5dBi, sidelobe attenuation better than 20dB and a front-to-back ratio of 26dB.**

Fractal Resonances' (*Communications Quarterly*, Winter 1996, pp77 – 81).

After attempting to digest, at least in outline, this mass of information, I remain rather uncertain as to its present practical value to amateur constructors. Clearly, as with the Landstorfer designs, it would not be easy to implement multi-iteration fractal elements on HF, where the space condensation would be most valuable.

On 144MHz, fractal designs could be implemented using aluminium foil pasted on a dielectric, as in N1IR's original 144MHz aerial (which he reveals was eventually spotted and cut down by his landlord). N1IR has also implemented VHF fractal loop-type aerials using heavy gauge wire for 50 and 144MHz, and these should not prove too difficult to implement and mount.

N1IR shows clearly that this approach appears to be quite practical at VHF or UHF, offering several resonances and capable of providing resonant feed-points with little reactance.

I suspect that a practical difficulty would arise for an experimenter in dimensioning the more complex fractal aerials so that the major resonance fell in the desired amateur band without a fair amount of trial and error. Nevertheless, it is clearly an important field for experimentation and may possibly come to represent the 'magic' breakthrough suggested in the *New Scientist* article on which the April 'TT' item was based.

After writing the above item, further references to published articles on fractal aerials have been sent in by readers, including the use of fractal radials for vertical aerials (*73 Amateur Radio Today*, September 1996, pp34 – 35) from PA3CQQ.

# MULTI-BAND WIRE AERIALS

**by Pat Hawker, G3VA, from 'Technical Topics'**

Over the years, 'TT' has provided details of many wire aerial designs for HF and VHF applications. Low-cost wire aerials can be home-brewed, can use fixed supports without the complexities of beam elements and rotators, are relatively unobtrusive and yet can provide respectable DX capabilities. On 28MHz, 50MHz and above, 'long-wire' aerials including rhombics and bi-directional Vs (both in classic and 'inverted' configurations), provide substantial power gain in specific directions and can be contained within average gardens. These points have been made a number of times in 'TT', yet there remains among many amateurs the belief that rotary beams are a *sine qua non* for reliable long-distance operation on HF and VHF. Admittedly, Yagi and quad arrays tend to dominate the HF pile-ups, but plenty of good DX can come to those prepared to listen and get in before the piles become too high.

Bob Wilmer, W3RW (*QST,* April 2000, pp46 – 48), describes a 28 and 50MHz long-wire aerial providing "better than dipole" performance on 28 and 50MHz without investing in a beam and rotator, including a matching section to deliver a near-50Ω match to coaxial feeder on both bands: **Fig 1**. The horizontal wire element (including the quarter- wavelengths formed from the 450Ω ladder line) represents 3λ (102ft 4in) on 28 MHz and 5λ (97ft 10.5in) on 50MHz. This results in a feed-point impedance of about 125Ω on 28.35MHz and 140Ω on 50.15MHz, transformed down to roughly 50Ω by the common 29ft of 73Ω (VF = 0.66) coaxial cable for connection to any required length of 50Ω cable. W3RW also provides information on an improved dual matching section.

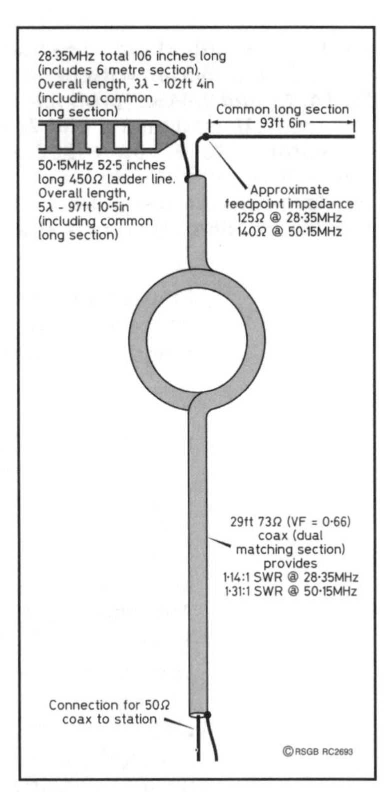


**Fig 1: Coax-fed long-wire aerial designed by W3RW for 50 and 28MHz. At 28.35MHz the total element length is 102ft 4in (3λ). At 50.15MHz the overall length is 5λ (97ft 105in).**

For many years, I tended to use a long-wire aerial approximately 132ft long, fed against quarter-wave counterpoises for the main HF bands. This had to be partly indoors, running through the roof space, then back outside and down to an ATU in an upstairs shack. This worked reasonably well, but eventually the wire broke where it passed over a branch of a tree.

It continued to operate as a random length fed from a π-coupler, but eventually my neighbour cut down the tree that gave the aerial some height and I had to be content with using a lower tree. I was never happy with the various counterpoises (which ran indoors) and

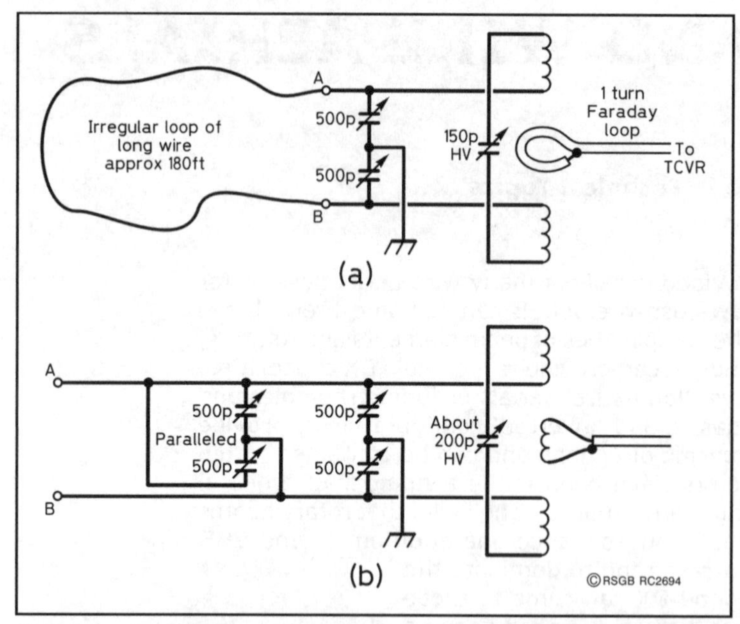


**Fig 2: An irregular long-wire horizontal 'loop' aerial, as used by G3VA in conjunction with balanced-output π-network ATUs (a) for 14, 21 and 28MHz (probably suitable also for 18 and 24MHz), and (b) ATU used on 3.5 and 7MHz, including a 3-gang loading capacitor (500 + 500 + 500pF) but a two-gang unit would probably be suitable. No great claims are made for this version, but large horizontal square or rectangular loop aerials can provide excellent results if at a good height above ground.**

converted the system into a form of a ragged - and much elongated - 'loop', by running back a return wire along a different route and feeding the result through a balanced π-network, using separate ATUs for 3.5/ 7MHz and 14/ 21/28MHz to avoid the need for tapped or roller-coaster coils, without having any clear idea of the total length of wire in the 'loop': **Fig 2**.

This aerial seems moderately directional towards the East, but has reasonable performance in most directions, despite being quite low (about 20 – 25ft above ground). Not a world-beater, but it happily provides contacts and can also be used on 1.8MHz with inductive loading. The large semi-horizontal loop aerial, whether square, rectangular or just made to fit whatever space and supports are available (the higher the better), whether random-length or resonant, seems a much neglected form of multi-band aerial.

Another useful aerial is an outdoor or roof-space 'doublet' essentially a non-resonant dipole feed with 450 or 300Ω ladder-line feeder acting as resonant open-wire line, preferably adjusted in length to provide a medium impedance to an ATU with balanced output. This can be the same as that shown for the long-wire loop.

Large loops and doublets were featured in a *QST* article 'HF Amplifiers vs Antennas' by Kirk A Kleinschmidt, NT0Z, reprinted in *Radio-ZS* (December 1999, pp10 – 11). To quote briefly from this article: "To save wear and tear on your neighbours, fellow hams, your wallet and even your house wiring, consider improving your aerial system before investigating in an amplifier... One almost universal way to get out more signal is to get your aerial(s) further up in the air... build a taller mast, find a taller tree, or put up a tower. If that dipole just isn't cutting it, put up a contest-winning and DX-catching secret weapon, a full-wave horizontal loop for 40 or 80m (up as high as possible, of course). Feed it with coax and use a tuner on bands above the fundamental frequency. That's a cheap way to snag an extra 2 to 10dB. depending on frequency.

"An alternative system is to disconnect the feed-line from your centre-fed single-band dipole and replace it with 450Ω ladder line: **Fig 3**. With a coax-fed dipole used on other bands, even if you may be presenting a happy impedance to the transmitter, the high SWR on the coax may slash your signal by 6, 10 or 25dB, depending on the band and the size of your dipole. By using 450Ω 'open-wire' line you will very likely retain most of that lost power. Now that's a 6 to 20dB shot in the arm that anyone can afford!

"You can increase the performance of a simple dipole by using low-loss open-wire 450Ω windowed feed-line. This is one of the easiest, most inexpensive aerials for the HF beginner. Just string up a dipole made of two equal lengths of copper wire. Don't worry about the overall length, but just make it as long as you can. Connect the feed-line to the centre-insulator and run it back to an aerial tuner with a balanced output. Attach coax between the tuner and the radio and you're in business on several bands!"

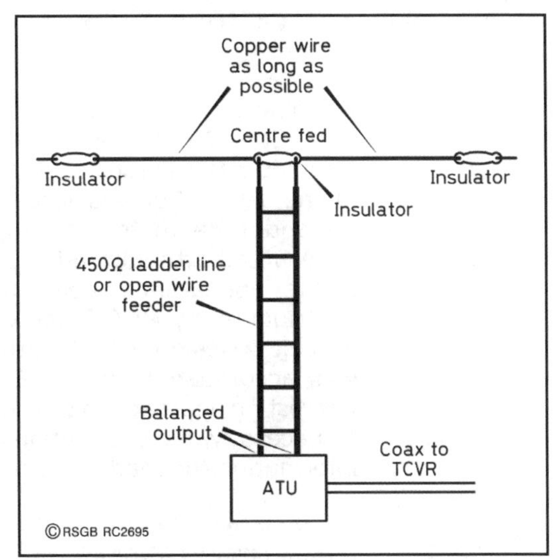


**Fig 3: The classic centre-fed 'doublet' (nonresonant dipole) using open-wire or ladder-line feeders and brought to resonance by the ATU can form an excellent multi-band aerial.**

A word of warning. The ATU needs to be adjusted (on low power) to present a low SWR on the coaxial feeder, particularly with a solid-state power amplifier. Similarly, beware of power losses in many ATUs, particularly on the lower bands, often denoted by heating up of the coils. Preferably, the length of feeder should be such that it presents reasonably low impedance (current-fed) to the ATU on all bands. This can be facilitated by providing an optional fairly short length of feeder that can be plugged into the chosen length on bands where this presents a high impedance (voltage-fed) situation. Balanced output π-networks, such as that shown in Fig 2, cope more readily with low rather than high impedance, but there is no need to worry about critical or resonant element lengths. A centre-fed doublet, no matter how long or short the span, should be capable of being resonated by the ATU to any band and provide a good match to the coaxial cable to the transceiver although, if less than a half-wave long at the lowest frequency band, there will be some small loss of radiation efficiency.

# VK2ABQ TRI-BANDERS, REINARTZ SQUARES & MOXON RECTANGLES

**by Pat Hawker, G3VA, from 'Technical Topics'**

By coincidence, two articles featuring the same antenna system appeared in the June 2000 issues of *QST* and *Practical Wireless.* Both traced briefly, if sometimes a shade misleadingly, the history of an antenna that originated in the 1970s but has since flowered in several versions: The VK2ABQ Tribander Beam for the 14/21/28MHz bands. This antenna received widespread attention after publication in 'TT' in January 1974, although its first appearance in print had been some two months earlier as a short contribution by Mr F Caton of West Merrylands, New South Wales to the 'Circuit & Design Ideas' feature of *Electronics Australia*, October 1973. It was accompanied by an Editorial Note: "This appears to be a very interesting approach to the never-ending search for the ideal aerial system. Of necessity, the description must be short, but there should be sufficient information for readers to duplicate the original."

Immediately after its publication in *EA,* the late Fred Caton, VK2ABQ, and former G3ONC, sent me an airmail letter containing the clipping and some extra information on construction and adjustment. From these I was able to concoct the item that appeared in 'TT' under the same heading previously used by *EA.* VK2ABQ frankly admitted that he did not know the forward gain, but claimed that it had a back-to-front ratio of 2 - 4 S-points, with substantial gain over a dipole and a low angle of radiation on long-haul contacts.

I admitted to having slight qualms about some aspects of the design, not reassured by the extremely scrappy nature of his notes, a characteristic that I came to recognise over the following years when he sent further ideas that never quite equalled his original tri-bander design. Nevertheless, I added: "the general concept seems interesting and a convenient way of constructing a compact tri-bander; so the information is presented as possibly experimental, but well worth investigating, certainly VK2ABQ/G3ONC has had good results." There can be do doubt that Fred was a gifted and determined experimenter in the days before NEC computer design began to dominate the antenna field.

It needs to be stressed that VK2ABQ designed a wire tri-bander that could be fixed or rotated. He made no claim to having developed the basic square-shaped compact element/reflector. As noted below, this compact form of directional antenna can be traced back to 1937, shortly before the recognition of the greater gain that can be achieved by close-spaced arrays – a major development based on the work on MF antennas by Dr George Brown of RCA.

VK2ABQ himself wrote: "After a lot of experimenting, I am now using what I consider to be the simplest and best home brew tri-bander yet. It has no traps or coils and so no losses related to such devices. Also, no mysterious blobs of electronics hanging on the array. Also, mechanically the system is very simple, it has no boom and a 14, 21 and 28MHz version has a turning radius of only 12 feet."

Subsequent reports from several UK amateurs who checked out the VK2ABQ soon confirmed that it fulfilled a useful role for amateurs still seeking a simple tri-bander as a low-cost alternative to commercial rotary beams using tubular elements. The design is simpler to construct and to erect than a typical condensed-quad array, although the forward gain is likely to be rather less.

Care is needed in adjustment for optimum performance. Variations in dimension were suggested for example by G3FRB and these are shown in **Fig 1**, although G3FRB stressed that a GDO should always be used to check for resonance. This is a relatively high-Q design, making for critical dimensions. Tuning for resonance and low SWR is facilitated if it is easy to raise and lower the antenna. Although reports were received of satisfactory performance when a VK2ABQ-type antenna was erected as a fixed beam in a roof space, this is not really advisable for any high-Q antenna - as it will be affected by nearby metal objects, etc.

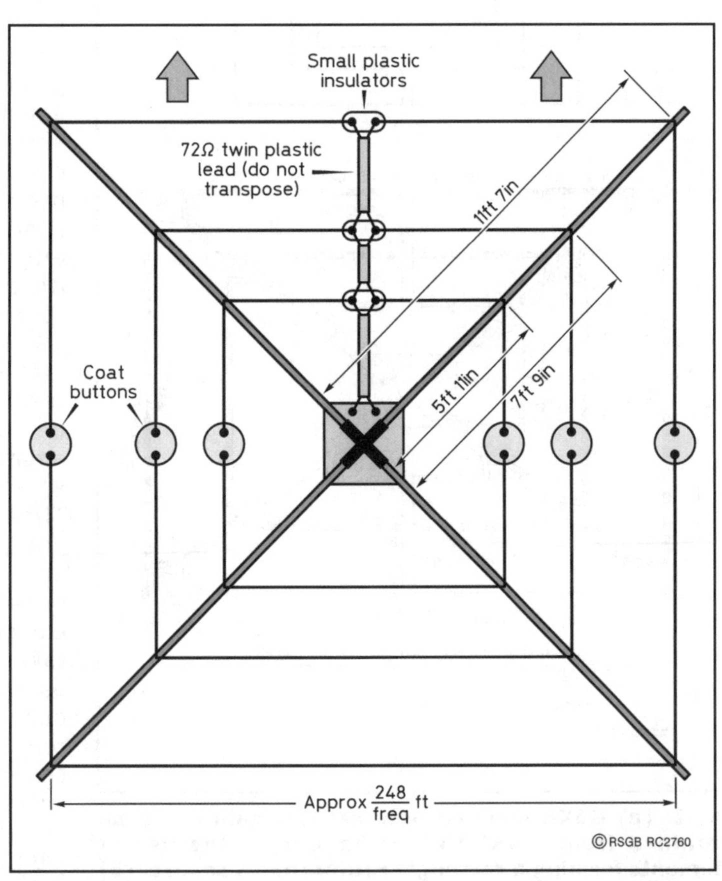


**Fig 1: The VK2ABQ tri-band beam for 14/21/28MHz, seen looking down on the array.**

The power gain of a VK2ABQ will be less than can be achieved with a well-designed close-spaced 2-element Yagi array or the Moxon Rectangle discussed below, but it remains valid to claim that the VK2ABQ deserves attention as one of the few mechanically simple tri-band designs suitable for home construction with a minimum of tools.

The first publication in book form was in my *Amateur Radio Techniques* (5th edition 1974, subsequently in the 6th and 7th editions, all now out-of-print). It has also appeared in several other RSGB books on antennas, most notably *HF Antennas for all Locations* (see below).

The same January 1974 'TT' in which the VK2ABQ appeared included a short report on an RSGB London lecture by Les Moxon, G6XN. I wrote: "A full house, but why [even then] so few younger members? A right royal evening of myth destruction and constructive hints. Looking through my notes I find such nuggets as: What's so important about front/back ratio if the beam has side lobes: tune for maximum gain... 14MHz folded dipoles work very nicely on 21MHz (the trick is to use resonant feeders)... There is no optimum height for horizontally-polarised aerials – get as much height as you possibly can (even if this means using a two-element rather than a three-element Yagi)... Unless a low SWR has been achieved after great care, this usually indicates *poor* performance... Interaction between different aerials is important (and for the amateur who has everything, including two different beams, he can get an extra 3dB gain without difficulty)... Loops are not single-band resonators... Trees have a great effect on vertically-polarised aerials...There is no limit to the possible errors when aerials are compared on the basis of ground-wave measurements." G6XN is a long-time professional scientist/ engineer and his book *HF Antennas for All Locations* (published by RSGB, first edition 1982, second edition 1993) although not always easy reading, remains one of the very best antenna books for the radio amateur!

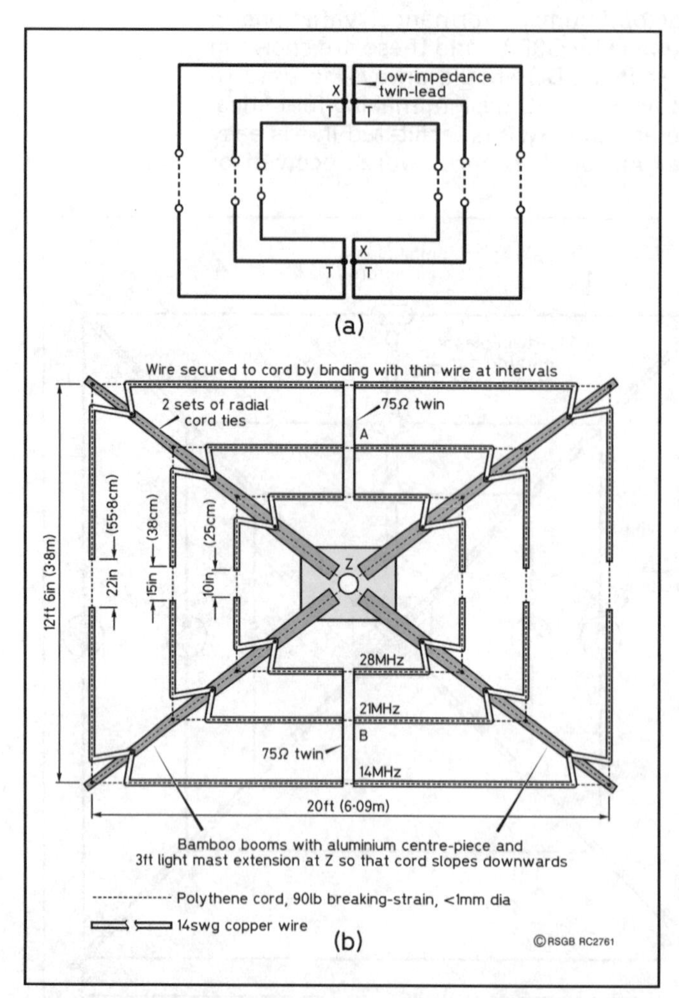


**Fig 2: (a) G6XN showed how greater gain could be obtained from a VK2ABQ tri-bander by the use of elements forming a rectangle rather than a square. (b) Implementation of G6XN's rectangular tri-bander without increasing the turning radius.**

These memories have been provoked by the two June 2000 articles concerned with the later version of the VK2ABQ elongated into a rectangle. This takes advantage of the extra gain possible with closer spacing of the driven element and the reflector. It was developed by Les Moxon, G6XN, and described in his book: **Fig 2**. It has later been modified and simplified by L B Cebik, W4RNL, who calls his version the 'Moxon Rectangle'. W4RNL's design appeared first in the article 'Modelling and Understanding Small Beams: Part 2, VK2ABQ Squares and the Modified Moxon Rectangle', *Communications Quarterly* (Spring 1995, pp55 - 70) and has been followed by a number of articles in the UK, USA and Australia. Recent examples include: 'Antenna Workshop: The Moxon Rectangle Revisited', by Peter Dodd, G3LDO (*Practical Wireless,* June 2000, pp 42-43), and 'Having a Field Day with the Moxon Rectangle' by L B Cebik, W4RNL, in *QST,* June 2000, pp38 - 42.

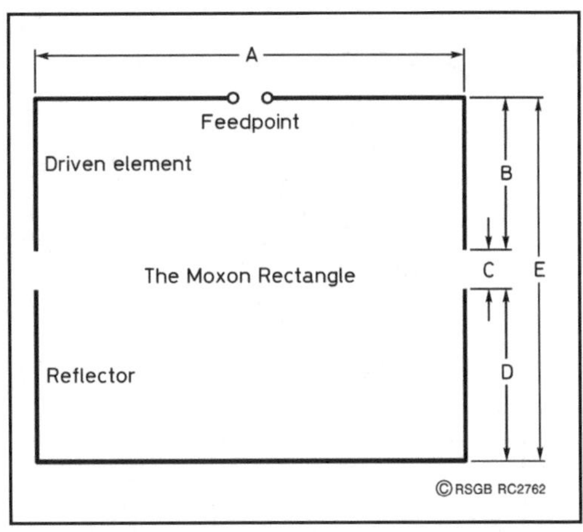


**Fig 3: W4RNI's outline of a Moxon Rectangle with various dimensions labelled as in Table 1.**

**Fig 3** and **Table 1**, from W4RNL's *QST* article, provide an outline of a Moxon Rectangle with the dimensions in feet applying to 14AWG [16SWG - *Ed*] bare-wire antennas for the HF bands. However, it should be noted that C, the spacing between the two sections, is fairly critical for optimum performance and

| Freq (MHz) | A | B | C | D | E |
|---|---|---|---|---|---|
| 3.6 | 99.98 | 15.47 | 2.16 | 18.33 | 36.96 |
| 7.09 | 50.69 | 7.82 | 1.15 | 9.35 | 18.32 |
| 14.175 | 25.30 | 3.87 | 0.62 | 4.70 | 9.19 |
| 21.225 | 16.88 | 2.56 | 0.44 | 3.14 | 6.14 |
| 28.3 | 12.65 | 1.90 | 0.35 | 2.36 | 4.61 |

Source: W4RNL in *QST.*

**Table 1: Dimensions of Wire Moxon Rectangles for 3.5 - 28MHz with reference to Fig 3. Dimensions are in feet and apply to 14AWG [16SWG - Ed] bare-wire antennas.**

it would be advisable to refer to his article for detailed information, including radiation patterns, etc. His article includes expected forward gain figures for various heights above ground, and also includes the outline of a direction-switching Moxon Rectangle by changing the feed-point from one section to the other and using transmission-line stub loading to lengthen the reflector element electrically.

It seems a useful addition to this growing family of directional wire antennas. No claim was ever made in 'TT' that the basic idea of folding a driven element and a reflector element into a square with sides equal to a one-quarter wavelength was novel to the VK2ABQ design. Indeed, such a mono-band configuration had been illustrated in G6CJ's 'Aerials' chapter in the first two editions of the RSGB *Amateur Radio Handbook*, published in 1938 and 1939, and rightly attributed to John L Reinartz, W1QP. It stemmed from his classic article: 'Concentrated Directional Antennas for Transmission and Reception - Rotatable Loops and Antenna-Reflector Systems of Reduced Dimensions' (*QST,* October 1937, pp27 - 29). **Fig 4,** reproduced from my much-battered copy of a pre-war *Amateur Radio*

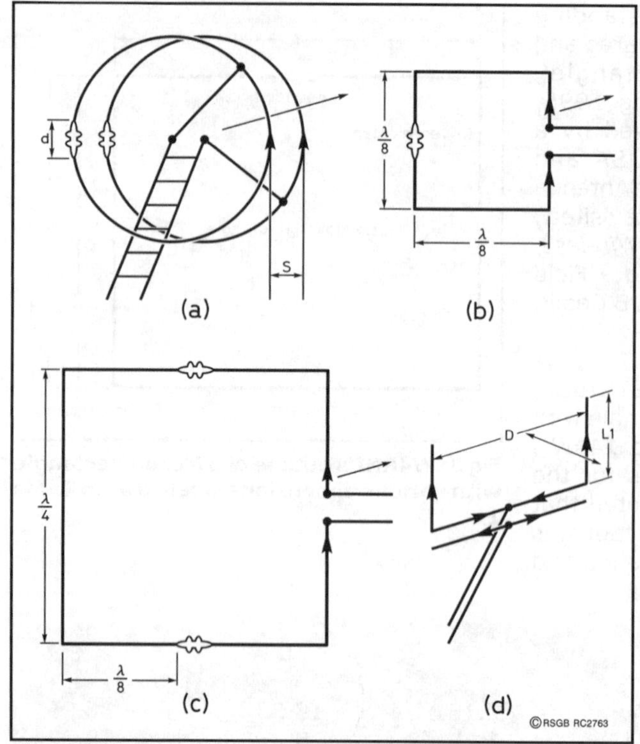


**Fig 4: Part of a collection of compact directional antennas as presented in the pre-war editions of the RSGB's Amateur Radio Handbook. (a), (b) and (c) were derived from the classic 1937 *QST* article by John Reinartz.**

*Handbook* shows the first four of nine illustrations of compact directional antennas, three of them directly attributable to John Reinartz's 1937 article.

John Reinartz, who held the professional experimental call W1XAM as well as his amateur call, W1QP, had earlier attracted world attention when, in the April 1925 issue of *QST*, he presented his theory of ionised layers to account for the skip distances experienced on 'short-waves'. Previously, in November 1923, he was one of the main participants in the first trans-Atlantic contacts between ARRL and Leon Deloy, F8AB, on about 100 metres. In 1924, his pioneering 14MHz signals were heard in California, the first time this feat had ever been accomplished.

The Reinartz antenna loops and squares attracted worldwide interest, particularly the double loop system for VHF operation. The basic elements were included in *Amateur Radio Handbook* as three items in a nine-part (A) to (K) diagram as the (a), (b) and (c) items of Fig 4. I have no idea whether VK2ABQ / G3ONC ever saw this material, but certainly he never claimed that he originated the basic square element configuration.

It is for his simple tri-bander design that, rightly, he should be remembered. Presented alongside Reinartz's article was a description by Burton Simpson, W8CPO, of 'A Square 'Signal Squirter' for 14MHz', based on a suggestion from W1QP. This was a hefty mono-band affair using *copper* tubing with quarter-wave sides to the square in the form shown in **Fig 4(c)**, with many wooden and brass struts. This elaborate approach was soon to be overtaken by the flood of two-element, close-spaced Yagi designs that followed the presentation by Walter Van Roberts, W3CHO, of 'The Compact Uni-Directional Array' (*Radio,* January 1938 pp19 – 23, 173). W3CHO's is believed to be the first amateur radio Yagi design to take advantage of a portion of Dr Brown's classic paper in *Proc IRE,* January 1937.

Until then, reflectors had always been placed at least a quarter-wave behind the driven element. Another pioneer of close-spaced arrays was W8XK although, in his antenna, both elements are driven.

# Chapter 4 OPERATING A BEGINNER'S GUIDE TO RTTY CONTESTS

**by John Barber, GW4SKA, Phil Cooper, GU0SUP, and Dick Whittering, G3URA**

Since the advent of the computer soundcard to decode digital signals, there has been more activity on RTTY than ever before. One now hears more stations during a single day on RTTY than were normally heard in a week a few years ago. Much of this increase is due to the very effective, and free, *MMTTY* [1] programme, by Mako Mori. This, in itself, has brought thousands of people world-wide on to the mode to enjoy the casual, friendly world of RTTY.

**Co-author Phil Cooper, GU0SUP, operating RTTY from Guernsey.**

Many RTTY operators use RTTY contests simply as a way of increasing their DXCC totals, working towards a 'Worked All States' (WAS) or 'Worked All Zones' (WAZ) award, or picking up new prefixes and many of them continue to use *MMTTY* to this end.

**RTTY CONTEST PROGRAMMES**

While this is all very well, and can also give the 'novice' contester a flavour of what can be worked during an RTTY contest, *MMTTY* is *not* designed as a contesting programme. It therefore does not have the capabilities of a dedicated contest package such as, say, *WriteLog* (2) or *RCKRtty* (3). These programmes can recognise incoming callsigns preceded by "DE", track your score, show needed calls, multipliers and duplicate contacts and do much more to help you maximise your score.

While *MMTTY* will not always produce a contest log in the format specified in the rules (particularly if a 'Cabrillo' log is required), all specialised RTTY contesting programmes, like the two mentioned, will produce the correct format needed when sending in your log. A 'Cabrillo' log, by the way, is a standardised log format in which all the information is in one file that includes the contest log, your name and address, callsign and comments. An alternative for producing logs is *Cabrillo Tools* by WT4I (4) which will convert various files into the correct Cabrillo format for you. This type of log is now required in all RSGB HF contests and most RTTY contests.

Most logs can be e-mailed to the contest address and will be acknowledged. Any comments you have about the contest can be included in the 'soapbox' section of the Cabrillo log, but not in your e-mail text, as this may only be seen by a 'robot' mail handler!

*Always* send in your log, no matter how many or how few contacts you made. Your log, yes *yours*, with only 20 or so QSOs in it, is used to verify the points claimed by the other contestants whom *you* worked. "But with only 20 QSOs, I'll come last," I hear you say. I doubt it! While trying out some new software, Dick, G3URA, ended up working just 16 stations in one contest and when the results came out, there were at least five stations below him. Despite a 'low' score, there is also the possibility that you may be the *only* G, M or 2E station to submit a log in any particular class, meaning you could end up with a certificate for being the top G in that class. Before submitting your log, do read through it and make sure it looks OK. Ensure there are no obviously wrong calls or daft exchanges. Printing the log out and then looking it may help you spot anything untoward.

**BEFORE THE CONTEST**

Entering a contest can be a bit daunting the first time. The following few paragraphs are aimed to help the beginner take the plunge and offer a few tips from both avid contesters and a contest manager.

The first rule in any contest must be: 'Read The Rules'! The rules for all major RTTY contests can be found on the web [5], and are also published in the British Amateur Radio Teledata Group (BARTG) [6] monthly magazine, *Datacom*, together with detailed operating tips for all the major contests. The rules will state date and time, the exchange required, where to send logs, and in which format they must be. The rules will say if there are different classes of entry, and a single band entry may suit you if you are restricted by time or aerial considerations. Always keep a copy of the rules handy.

Before the contest, set up some simple memory 'buffers' containing only the minimum information required for calling another station and exchanging the relevant information. It is also a good idea to add a couple of 'return' characters to the beginning and end of the exchange, as this can make your exchange stand out a little more.

For a contest where the time also has to be exchanged, the buffers might look like this:

**DE GW4SKA GW4SKA K**

(Use this to answer a CQ. Never send his call and always send the 'DE', see below). **RGR UR 599 001 001 1254 1254 DE GW4SKA K**

(meaning I have your message, this is mine for you).

These will be fine in most conditions but be prepared to repeat the serial / time etc several times if copy is poor, like this:

**001 001 001 TIME 1254 1254 1254 QSL? BK**

Set up a separate buffer for this. There is no need to repeat the RST as it is always 599 no matter what the conditions!

### DURING THE CONTEST

Never send any unnecessary information such as names, your rig, power or aerial details. Also, even if the station worked is a new country for you, never ask for his QSL information, as you can find this out after the contest. Remember that the serious contester will be aiming to make about two contacts every minute, so stick to the essential information only. Remember too, to call exactly on the other station's frequency and keep the AFC and NET controls turned *off* when answering a CQ.

Think about your exchanges and watch what others are doing. Most of us know our own callsign so seeing it three times before we see your callsign just once, is a real 'no-no'. For example, sending "M5ACN M5ACN M5ACN DE G3URA PSE K" will probably not get you well-liked!

Know what the 'multipliers' are. Are they countries? Prefixes? Zones? This will be explained in the rules, which you will have read before the contest, right? Are there bonus points for working different continents? If you run a 'little pistol' station where anything outside Europe is a bonus, don't forget that the Canary Islands, EA8, counts as Africa, and Cyprus, 5B4 or ZC4, counts as Asia. Both are fairly 'local' and easily worked.

Having a 'little pistol' station can have some advantages, as most of the time you will be in 'S&P' ('search and pounce') mode while the 'big guns' will sitting on one frequency calling CQ. You can pick and choose whom you work; they can't. Never forget that they *want* your call and will do their best to get you in their log. This is especially true if you have a regional locator in your callsign, such as GM or GI.

If they are rare DX and have a huge pile-up, worry not. If it is a 48-hour contest, wait 24 hours and call them on the second day, when they will be crying out for contacts and will want you in their log. If you do have to wait patiently for your turn to work the DX, again, watch what is going on: there should be no need to ask for a repeat of his serial number, for example.

In some contests, such as the Australian ANARTS, points are based on distance worked and in these types of test it is far better to trawl the bands looking for DX rather than just work mainly European stations.

Watch out for time limits on band changes or off times. These will be in the rules. If, for example, you are limited to two band changes in 10 minutes don't work that one multiplier on 10m if you can't hear any other stations, or you will then have to sit on a quiet band until the 10-minute time period has elapsed (however, if you just can't resist working that VP6 before returning to 20m, you can always use up the rest of the time with a 'comfort break'!)

If you are keen to try CQing, even with your 'little pistol' station, think about doing so in the dying hours of the contest. Then, many of the big boys will start to search and pounce for those extra contacts that escaped them during the main part of the contest.

**INTERESTED?**

More information about RTTY contesting and the datamodes in general can be found by joining BARTG [6], from which several RTTY awards are available. There is also a popular RTTY reflector [7] where you can ask questions, find QSL routes, and compare contest scores. Other helpful information can be found on the 'RTTY Info' website [8], where there is an excellent RTTY tutorial for those wishing to learn more about the mode.

New RTTY operators will find contesting a very easy way to make a start on the mode without the need to type at furious speeds. Those with more experience will know that in any of the major contests held each year, they can find well over 1000 stations to work. Look at the contest calendar [5], read the rules, join in, but most of all, have *fun*!

We look forward to seeing you on our screens and seeing your calls listed in the results. Oh, and one last bit of advice: Read the rules - *again*!

**WHAT'S ON THE NET?**

[1] http://mmhamsoft.ham-radio.ch/
[2] www.writelog.com
[3] www.rckrtty.de
[4] www.wt4i.com
[5] www.rttyjournal.com/contests
[6] www.bartg.demon.co.uk
[7] http://lists.contesting.com/mailman/listinfo/rtty
[8] www.rttyinfo.net

# BUGAMBIC – SON OF SUPERBUG

**by Chas Fletcher, G3DXZ**

Superbug [1] was a project of 1990 intended to cure the serious shortcomings of my dearly-loved Lionel Bug key. When I was asked for a copy of the original circuit by a long-time user who had lost his copy, I thought an update might well be in order.

PIC-based designs of radio peripheral equipment are becoming quite commonplace in recent times, especially those based on the very useful Microchip 16F84. Although there are cheaper and faster PICs, the 16F84's EEPROM, (electrically erasable and programmable read-only memory), which allows one to re-write the code without cost until it is right, keeps it high in the 'most-useful' table. Thus, when thinking of updating my 1990 design, I decided to try to reduce the number of parts and the circuit complexity of the project by substituting machine instructions for components. The 'machine', of course, is the Microchip 16F84.

A micro-controller solution combines purely logical programming with physical construction of the electronic circuitry, and when it works, it's magic. For anyone thinking of taking up the challenge of learning assembly-language programming, a look at the articles (see the list at the end) proves easy reading. Those readers wishing to go straight to the finished article are catered for in the final paragraph.

**WHAT IT DOES**

A Morse key with mechanical contacts is always prone either to dirty contacts or to contact bounce. Listening on 40m reveals that the venerable Bug is still alive and kicking, in fact it seems more prevalent than ever. Not all, however, have a clean keying characteristic, and exhibit a scratchy sound due to one or other of the problems mentioned above. The Bugambic keyer will clean up contact action whatever mode is used and should be a worthwhile addition to any mechanical key.

This circuit will permit a single- or twin-paddle keyer to act as a Bug key and let the user experience semi-automatic operation with very little outlay. In fact, the Bug key simulation is more elegant than the standard mechanical Bug, as it not only produces clean dots, but it also forces a one-dot space between dashes and between dots and dashes.

Simple software designs that claimed iambic operation failed, when tested, in one respect. When holding the dash paddle, a brief touch on the dot paddle during the dash period did not insert a dot into the dash stream. The reason was that the paddles were scanned only at the ends of dot or dash periods. The problem was solved using the processor's interrupt facility to set 'paddle-pressed' flags on a virtually-continuous basis, and by raising the clock speed. The performance was then found to be indistinguishable from other hardware-based iambic keyers.

A few older amateurs will remember the Side-Swiper key. Since all the necessary functions were available in the software, I have included the option - just in case.

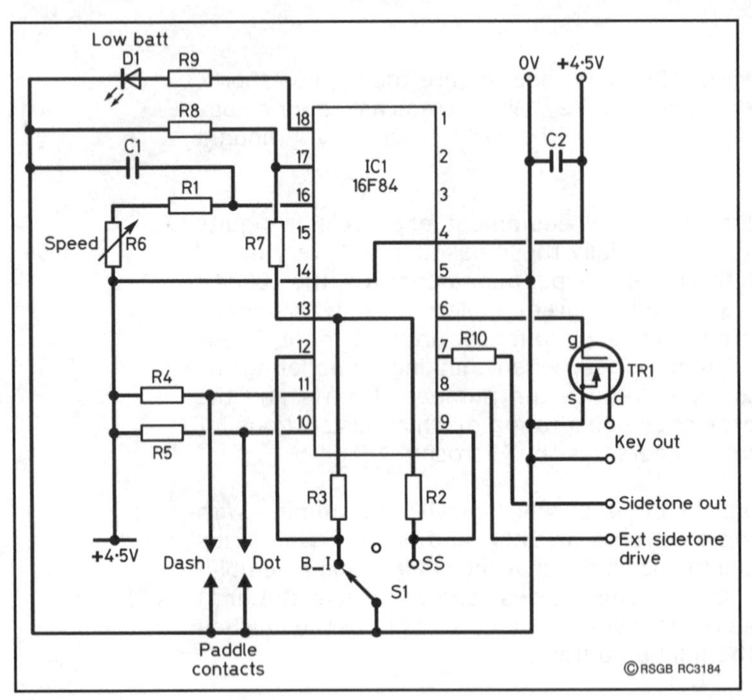


**Fig 1: Circuit diagram of the keyer.**

## CIRCUITRY

The complete circuit of the keyer is shown in **Fig 1**. Inputs consist of a 'mode' switch, S1, with two halves, labelled B_I and SS. Only one of these switches can be closed at any time. The other inputs are the 'dot' and 'dash' contacts. There are three outputs. The keyed circuit is made by MOSFET TR1 which should be operated with the positive keying lead to the drain connection. Two sidetone outputs are provided. Pin 7 is a square-wave output of 4.5V p-p and intended to drive an efficient transducer typified by the telephone earpiece insert. Driven through R10, an earpiece of this type will provide adequate volume for normal, quiet, operating conditions. However, a further output on pin 8 is identical to the MOSFET drive and is intended to feed an external or more powerful sidetone generator. The output is at +4.5V while the key is pressed.

Speed control is via the PIC internal clock which is of the RC type, because absolute stability is of little importance. With the object of achieving long battery life, the clock rate is kept down to around 250kHz. This is a compromise between speed and power consumption, in that the keyer is fast enough to respond instantly, apparently, to the paddle, but the running power drain, which rises with the clock rate, is minimised. When the keyer is activated, pin 13 of the PIC is set to 4.5V, which supplies the mode switches and the voltage monitor divider resistors. This is done because when the PIC is in the SLEEP mode, the output on pin 13 is set to 0V and no power is wasted on these features. For the same reason, a high-brightness red LED is used for the low-battery alarm to give a good light at low current.

The internally-generated sidetone has the unusual feature of changing pitch with the speed setting. With a dot speed of 25WPM, the sidetone frequency is around 800Hz. For any user wishing to operate consistently at some other speed, the sidetone frequency may be adjusted by the software.

## POWER SUPPLY AND CONSUMPTION

The circuit was intended for battery operation, although any steady supply between 4.5 and 6V would do. Because the actual key circuit is made by the action of a small MOSFET, which is the PIC's output device and will happily 'make' 200mA and withstand 50V, the supply voltage to the PIC needs to be not less than 4V in order to switch the MOSFET efficiently. The PIC itself will function happily down below 3V, so 4.5V was chosen, because three 1.5V AAA dry cells in series will do the trick.

During standby, ie between paddle contact closures, the keyer consumes 250μA. In iambic mode, with both paddles squeezed, the drain rises to 1.4mA. If left inactive for about three minutes, the PIC will go to sleep and, in this mode, the current drain is below 5μA. Hence, no power switch is needed and the battery life is as good as its shelf-life. These figures apply to the simple Bugambic keyer without sidetone output. If internal sidetone is used, the current consumption rises to around 2.5mA during keying. This latter figure translates to around 400 hours of continuous key-down operation from good AAA cells, which should not cramp the style of all but the most hardened contesters.

One adverse aspect of long battery life devices is that one either forgets when the battery was fitted or feels the need to check when it is still OK. To offset this nagging doubt, the PIC monitors its own battery supply and, if the voltage falls below 3.9V, it will blink an LED at the end of each dash sent, giving reasonable notice for replacement.

Pins 1, 2, 3 and 15 are shown as not connected in the circuit diagram and they should be left unconnected rather than be tied to ground. Pins 1, 2 and 3 are configured as 'output' bits, are set to 0V by software and hence are not at risk from static. Pin 15 is the internal clock output of the PIC and carries a square wave of 4.5V p-p when the PIC is running.

## SOFTWARE

A full listing of the software used for this project is available from the '*RadCom* Plus' area of the RSGB members-only website. It is written in the assembly-language dialect, MPASM, that is used by Microchip Technology Inc for its *MPLAB* software. For those interested, its website offers free downloads of the software. Compilation into machine code is nicely done by the *MPLAB* suite in the *Windows* environment, but programming the chip needs a programmer and compatible software.

After a stumbling start (the cure for which remains unknown, but was accomplished by re-installing the program), I have found Microchip's *PICPROG* shareware software effective and very easy to use. Beware, however, of trying to programme the chip at too fast a rate. I found a setting of 200 for both long and short timing intervals consistently successful after failing at faster rates. (See Options > Advanced in *PICPROG*.)

Basically, the programme surveys the state of the keyer paddles, the mode selection switches, the battery state and the length of time since the last paddle was pressed, in cyclic order. Action is taken (virtually) immediately a switch is made to service whatever the contact state demands and, afterwards, the loop cycle is resumed. Separate routines are provided for Bug, iambic or Side-Swipe/straight key operation, and the processor will move between them according to the setting of the B_I and SS switches.

During iambic operation, any change of state at the dot or dash paddle inputs is recorded by setting bits in the appropriate byte in memory. This byte is tested regularly by the processor and no fleeting contact of the paddles is missed.

Battery-checking is accomplished by comparing a fraction of the battery voltage supply to the input threshold voltage of one of the PIC's inputs, RA0. If the voltage at RA0 falls below the input threshold, a routine is invoked to flash an LED indicator for about 30ms after each dash is sent.

The published software is reasonably well commented, in order that the sharp-eyed or inventive can correct my minor blunders and untidiness, or add their own ideas to improve the program. It is also invaluable, when reviewing the software after the passage of time, to remember what on earth one had intended it to do!

## CONSTRUCTION AND TESTING

The circuit is low-power and low-frequency. The RC oscillator used by the processor has a 1ms period (mid-speed range), resulting in the system clock occurring at 4ms intervals. Hence, almost any method of construction is admissible. My prototype was built on Veroboard and worked identically to subsequent versions that were mounted on PCBs.

A PCB layout is offered for those who enjoy the artwork and subsequent neat result. It is advisable to use an 18-pin socket for the 16F84 to ease subsequent reprogramming or re-use. Although the PIC cannot be described as delicate, it can be damaged by applying unlimited power to the wrong pins and, hence, is best fitted last after connections are verified.

Having assembled the circuit, it should work without fuss, assuming all is as it should be. An oscilloscope will show if the PIC is active - a sawtooth waveform of 5ms period will appear at pin 16 and a square-wave of 4.5V amplitude, 20ms period, at pin 15. If no oscilloscope is available, pin 15 should show about 2.25VDC on an ordinary meter. The voltages on all other pins of the PIC are DC levels, either 0V or +4.5V, with the exception of pin 17.

I would recommend that, for primary testing of the basic keyer, only the paddles, speed control pot and the supply are connected. In this state, the circuit will emulate a Bug key and should be active immediately the supply is connected. A high-resistance voltmeter or oscilloscope should detect almost the full supply voltage at pins 14, 13, 12, 11, 10, 9, 4 and also across the paddles and at the B_I and SS tags. Pin 17 will show around 1.2V. Pressing and holding the dah paddle will cause pins 6 and 8 to go high. Pressing the dit paddle should produce alternating high and low at these pins according to the dit speed setting. If these actions occur, you are in business and the other connections to the mode switches can be made. Remember, only one of the B_I or SS mode switches may be made at any time, otherwise operation will be abnormal. I used a centre-off, three-position slide switch as a combined mode switch which guarantees that only one can be made at a time.

To test the low-voltage alarm feature, the supply voltage should be slowly reduced to below 3.9V while sending repeated dahs. When the supply falls below the warning threshold, the LED blinks after each dah. If not, check that the voltage at pin 18 goes high briefly after each dah, that pin 17 voltage is 0.9V or less, and that the LED polarity is correct.

In operation, either a single paddle key with separate dot and dash contacts or a twin-paddle key may be used. For straight key clean-up operation, use the dash input and select either Bug or Side-Swipe action.

Seldom does one achieve anything entirely without help and, in this case, my thanks go to GM3HBT and GM3KCY for their efforts in constructing prototypes and bringing me down to earth!

Finally, for anyone wishing to have a go, but lacking a PC and/or an Internet connection, a pre-programmed PIC, PCB layout and components (less box, switches and sidetone transducer) is available from the author at £11 inclusive, as is the software on disc, at £2.

**REFERENCE**

[1] 'Superbug Simulator', by Chas Fletcher, G3DXZ, *RadCom* August 1990, pp48 / 49.

**COMPONENTS LIST**

**Resistors**

R1 ................ 10k
R2 - R5 ......... 47k
R6 ................ 22k lin pot
R7 ................ 82k
R8 ................ 33k
R9 ................ 68R
R10 ............... 1k

**Capacitors**

C1 ................ 220p
C2 ................ 100n

**Semiconductors**

D1 ................ Red LED (see text)
TR1 ............... BS170
IC1 ............... 16F84 (4MHz)

# ECHOES FROM THE LEONIDS

**by Trevor Sanderson, PA3BOH / G4OEY**

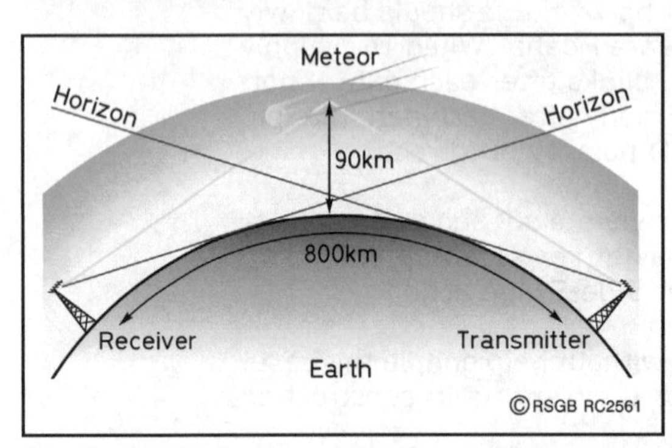


**Fig 1: Typical set-up for detecting meteor echoes. The receiver is tuned in to a VHF station beyond the horizon. Echoes are received each time a meteor leaves behind a trail of ionisation that scatters the radio signal.**

Using a radio to listen to meteor echoes is nothing new. Many amateurs and professionals alike have used the reflections from meteor trails either to detect the presence of the meteor, or as a means of communicating over great distances (see **Fig 1**).

Now a new technique is available. Pioneered by Peter Martinez, G3PLX, the technique uses modern DSP software to look for the Doppler shift of the frequency of the received signal, which is due to the motion of the meteor trail in the upper atmosphere winds. Using these DSP techniques it is possible to translate these signals into visible spectrograms, or into audible tones, and so look and listen to the sound of the echo from the meteor trail. There is now a wide range of PC-based Digital Signal Processing software that allows one to measure the effect of the upper atmosphere winds on the meteor trail. Now it is possible for anyone with a good receiver and a PC to look and listen to the echoes.

**INTRODUCTION**

In November 1999, the Leonids returned again. This time it was expected that the display would be particularly spectacular, the main reason being that the earth would sweep through the dust cloud left behind by Comet P55/Tempel-Tuttle when it passed close to the earth's orbit back in 1966. Not only that, but when the earth would pass through the cloud, Europe would be on the dawn side of the earth, and therefore moving into the cloud (had Europe been on the dusk side of the earth at the time of maximum, as was the case in 1998, then the earth, acting as a shield, would have prevented us from seeing most of the meteors).

Already in 1998 it was expected that the high numbers of meteors might pose a threat to the multitude of satellites which encircle our globe. Measures were taken within the major space agencies to try to predict, and if possible, minimise the risk to their satellites.

In 1998 the progress of the dust particles could be followed, almost in real time, as several professional and amateur groups were using the forward meteor scatter techniques to observe and count the number of meteors observed and were posting their results on the Internet. At the times when Europe was shadowed by the earth, groups in the Far East were making measurements, and vice-versa.

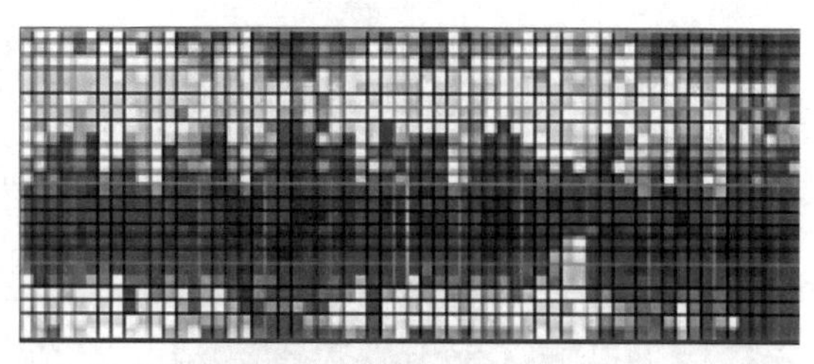

**Fig 2: Number of meteors per hour recorded during November and December 1999 by the University of Ghent Automated Meteor Observation Station, taken from their website.**

A well-known way of doing this is to tune to an FM broadcast station about 1000km away. Not quite as easy as it sounds because, usually, several much closer station are transmitting on the same frequency. However, East European stations such as the Polish broadcast stations located in Krakow (66.89MHz) and Wroclaw (72.11MHz) are popular for this sort of study, since they are outside of the band of Western European FM stations. For a description of this method, see the website at **www.dmsweb.org/radio/ radio.html**, or the website of the radio section of the International Meteor Organisation at www.imo.net/radio/

As was said earlier, there is nothing new in the principle of these techniques. What is new however is the availability of the results to everyone via the Internet, and the beautiful presentation software that they use.

A good example of the sort of results obtained is shown in **Fig 2**. This shows data for November and December 1999 taken from the Internet site of the University of Ghent Automated Meteor Observation Station in Belgium (**http://allserv.rug.ac.be/ ~hdejongh/astro/meteor/ meteor.html**). Here, the days are plotted from left to right, each box being one day wide, while the hour of the day is plotted from top to bottom, each box being one hour wide. The colour code shows the number of meteors per hour. Two peaks can be seen, the Leonids on 17 November, and the Geminids on 12 and 13 December. Notice also how the maximum every day is always around 5 or 6am.

These observations are all part of a network called the Global Meteor-Scatter Network which is coordinated by NASA's Research in Planetary Astronomy and Planetary Atmospheres programme. For more information, see **wwwspace. arc.nasa.gov/~leonid/GlobalMSNet.html**

## DOPPLER TECHNIQUE

The Doppler technique differs considerably from the technique mentioned above. The method enables us to get far more information about the meteor itself. Again, there is nothing new in the Doppler technique. Doppler radar has been around for decades, and Doppler direction finding has also been around for a decade or two. However, the use of the Doppler technique is now within the grasp of anyone with a good receiver and a PC. This has only become possible within the last year or two, thanks to the availability of DSP hardware and software.

My interest in this technique was first aroused by the feature in *RadCom* by Peter Martinez, G3PLX, 'Using Doppler DSP to Study HF Propagation', in May 1998. As usual with such an article, I did not really understand it all. I was able to get hold of the DSP module needed. G3PLX kindly sent

**The great Leonids shower storm of 1833. Reproduction of a wood-cut engraving by Adolf Vollmy, based upon an original painting by the Swiss artist Karl Jauslin.**

me by e-mail a copy of his software. In his feature he showed how this software could be used to detect meteor trails, but at the time this had not registered within me. I had actually intended to use the software to study some aspects of propagation on 80m, but that's another story.

It's very simple really, but the G3PLX software does requires a separate DSP evaluation module (later on in the feature I'll tell you about other software which is available that does not require such a module, and which can work on any PC with a sound card). With the G3PLX software, a dedicated DSP chip and a microcomputer on the evaluation board are used to sample the signal and then perform a Fast Fourier Transform, before sending the data stream via the COM port to the PC for display.

The additional hardware, an Analog Devices DSP56002 evaluation module, is clearly out of the reach of anyone wanting just to try out this software.

The technique requires some experimentation to get it right. First, find a short-wave AM broadcast station in the skip zone, ie high enough in frequency and close enough to you that the signal goes off overhead and is not reflected back to you. I used the BBC World Service around 17MHz which, I think, originates from a transmitter about 500km away from me. Using the SSB mode, tune what little of the signal you can hear (due to scatter, etc) so that the carrier is heard at around 1kHz. Then feed the signal to the DSP module and with luck, you will get a trace like the one in **Fig 3**, reproduced from the G3PLX article.

This is a spectrogram with time going from left to right, and positive Doppler shifts above the carrier and negative below. The range of Doppler shifts is 100Hz full scale. This shows several meteor trails, labelled A to E. G3PLX calls this picture a Dopplergram. The traces show the meteor trail (called a 'train' in meteor circles). The trail of ionisation reflects the signal. As the trail passes through different layers of the upper atmosphere, it is blown around by the upper atmosphere winds. Different layers move in different directions and with different speeds. The part of the signal scattered by each of these different layers

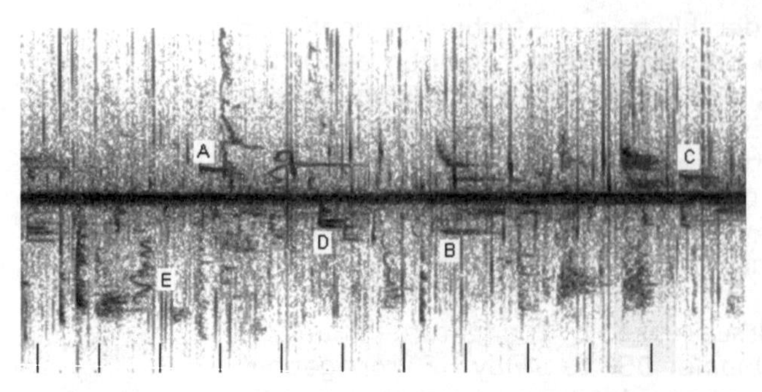


**Fig 3: Spectrogram of meteor echoes obtained by G3PLX.**

experiences different Doppler shifts, hence the squiggly nature of the traces in the spectrogram.

## THE PC VERSION

On most modern PCs, there is now enough DSP hardware in the sound card to make it possible to perform these tasks without resorting to a dedicated DSP module, and suitable software using just the PC hardware can now be found on various sites on the Internet.

One of the best is the *r_meteor* software. This software was specifically developed for meteor analysis by Bev Ewen-Smith, CT1EGC/ G3URZ/ VK5AES, who runs the Centro de Observação Astronómica no Algarve (COAA), an observatory in the Algarve, Portugal. Check it out on **www.ip.pt/coaa/index.htm**

This software was inspired by G3PLX's *RadCom* feature. A free trial version of this software can be downloaded from **http://sapp.telepac.pt/coaa/r_meteor.htm** **Fig 4** shows a typical spectrogram obtained with this software. Again, time is plotted along the x-axis, and frequency (which corresponds to the drift velocity of the meteor trail) is plotted along the y-axis. The straight line shows the signal from the carrier which, in this case, was a short-wave broadcast station in Lisbon. This spectrogram shows the reflections from meteor trails. Also seen is the signature of an aeroplane as it slowly crosses the sky.

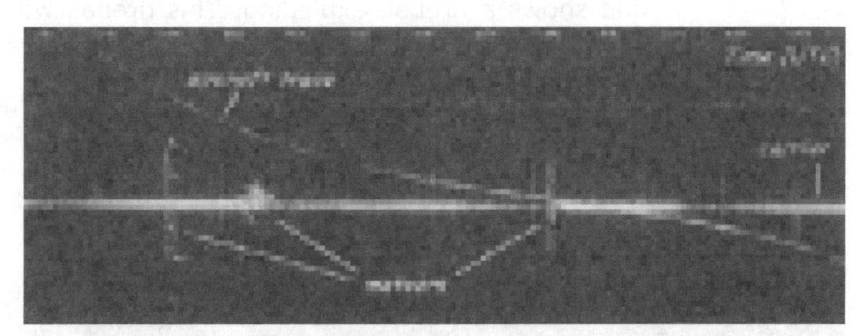

**Fig 4: Typical spectrogram obtained with the *r_meteor* software. This spectrogram clearly shows the reflections from meteor trails. Also seen is the signature of an aeroplane as it slowly crosses the sky.**

With the *r_meteor* software, the output of the receiver is connected to the input of the sound card (if the PC has a microphone, it can also be used to pick up the audio signal directly from the receiver's loudspeaker, but with a small loss in quality). The software then uses the analogue-to-digital converter of the PC's sound card to digitise the incoming signal. Then the PC itself performs a Fast Fourier Transform, which is then plotted on the screen. Using the PC this way means that a dedicated DSP module is not needed and everything can be done on the PC. Other programs equally suitable are the audio analysis packages that are now available from a multitude of commercial companies. Much audio software can be found on sites such as **www.hitsquad.com/smm/** One which I found very useful is *Spectrogram*, available from **www.monumental.com/rshorne.gram.html**

## LOOKING AT THE ECHO

As part of its 1999 Leonid activities, the European Space Agency (ESA) established a website pointing to a range of activities which included satellite operations, Meteosat observations, meteor observations in Spain, and a joint ESA/NASA aircraft campaign (**www.estec.esa.nl/spdwww/leonids/ index.html**). It also included a chat line with scientists and engineers on line. One activity included as part of ESA's Science Directorate Outreach activities was to try out G3PLX's ideas and listen to the sound of the reflection.

At the time we proposed this we were uncertain about the results. A website describing the method was posted and members of the public were invited to try out the method for themselves (**www.estec.esa.nl/spdwww/ leonids/leolisten.html**). Included in these activities were Udo (PA3EZI) and myself (G4OEY) from ESA's Space Science Department in Noordwijk in the Netherlands. Also in touch with us was Bev, (CT1EGC/G3URZ/ VK5AES) in Portugal in the Centro de Observação Astronómica no Algarve (COAA), who had written some of the software described above.

Already early on during the meteor shower, good results were being obtained by Bev in Portugal, as shown in the top of **Fig 5**. The top part shows a spectrogram taken in Portugal on the night of 15/16 November, and shows a fireball exploding. This fireball was seen visually as well, and lit up the whole of the countryside around. As before, time is plotted from left to right and frequency from top to bottom. The range of frequencies is only ±50Hz. The line across the middle is the carrier. When the fireball in the top part exploded a cloud of ionisation was released which expanded and drifted off in all directions, hence the bright patch which shows both positive and negative Doppler shifts. The patch of ionisation lasts for several seconds.

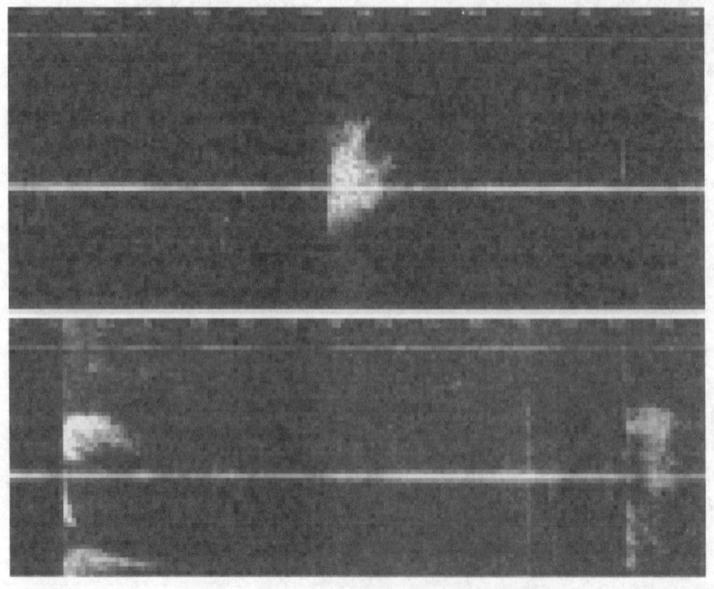

**Fig 5: Spectrograms obtained in November 1999. The top part, taken on the night of 15/16 November, shows a fireball exploding; the lower part, taken on the night of 17/18 November, shows two fireballs exploding.**

The bottom part of Fig 5, also taken in the Algarve by COAA on the night of 16/17 November, shows the signatures of two more fireballs disintegrating over the sky in Portugal. These fireballs show more structure. The three yellow patches on the left are probably due to patches of ionisation in different layers of the atmosphere which drift off in different directions, and therefore impart three different Doppler shifts to the reflected signal. The fireball on the right is much more complicated and shows evidence of some stratified motion prior to the final disintegration.

## LISTENING TO THE ECHO

As well as being able to display this data in a spectrogram, it is possible to listen to the signal reflected by the meteor. As mentioned earlier, the receiver has to be operated in the SSB mode so that the carrier is audible. Then, if the received signal is fed to a loudspeaker, a more or less constant tone, corresponding to the carrier, is heard. Each time a meteor enters the atmosphere it releases a trail of ionisation between transmitter and receiver, and the signal strength increases due to the presence of the additional scattering material. The amplitude of the tone will increase for as long as the meteor trail persists. The signal consists of the steady tone due to the carrier, and the additional signal due the meteor trail. On an expensive receiver, most of the carrier can be removed by using a notch filter, leaving only the sound of the reflection of the meteor. The remaining signal is more or less centred at the frequency of the tone corresponding to the carrier, and drifts around in frequency a few hertz.

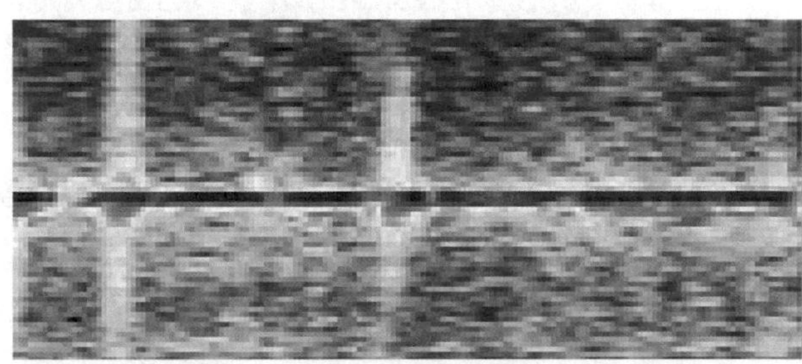

**Fig 6: Spectrogram of a 1.25-minute period of data taken around the time of maximum of the Leonids shower on the morning of 18 November.**

**Fig 6** shows a spectrogram of a short 1.25-minute period of data taken around the time of maximum of the Leonids shower on the morning of 18 November. The carrier has been notched out, so all that is left is the sound of the Leonids. These are the sounds of the reflection from meteors. This sound is available online as a 2MB .wav file on **http://helio.estec.esa.nl/ wave5c.wav**

At first, we expected the sound would be a short 'ping', corresponding to the meteor entering the atmosphere. However, most trails persist for a few seconds and so the echoes should also persist for a few seconds. In fact the sound of the echo is more like a short whistle that lasts for a second or two. The pitch of the tone is actually decided by the operator of the radio. Tuning up or down in frequency changes the pitch of the carrier, so any pitch can be chosen. Modulating this tone are the very small changes in frequency due to the different drifts experienced by the meteor trail. We had hoped to be able to hear these small variations, but so far we have not really succeeded as these variations are only of a few hertz. All we need to do now is expand the frequency range of the signal.

## CONCLUSIONS

The radio technique of listening to meteors is now within the reach of everyone with a good receiver and a PC. Much software is available on the Internet, and it is now just a matter of downloading and trying it. As well as the Leonids, there are many other meteor showers all year round. Look at the IMO website at their calendar for the year 2000 for a list of forthcoming showers (**www.imo.net/calendar/cal00.html**) and look at the website of the Ghent group (**http://allserv.rug.ac.be/~pdegroot/meteor/mc1_99.html**) to see at what time of day to expect them. Try it!

**ACKNOWLEGEMENTS**
I WOULD LIKE to thank Peter Martinez, G3PLX, for permission to use his software, Herwig DeJonghe for permission to use Fig 2, and Bev Ewen-Smith, CT1EGC, for permission to use his software and Figs 4 and 5. Also, his colleagues from the European Space Agency who participated in this exercise – in particular, Peter Faulkner; Detlev Koschny; Jean-Pierre leBreton; Udo Telljohan, PA3EZI; and Andrea Toni.

**WHAT ARE THE LEONIDS?**

***Each year, the night sky is illuminated by dozens of meteor showers. During these showers, fragments of cosmic debris leave glowing trails as they are incinerated during entry to the Earth's upper atmosphere. We see them as short-lived trails of light streaking across the sky. Most meteors are caused by cosmic dust burning up as it enters the Earth's upper atmosphere. The dust comes from giant dirty snowballs called comets. For most of their elliptical orbits, comets remain in deep freeze, far from the Sun. When they approach the Sun their icy surfaces are warmed and start to vaporise, generating powerful jets of gas and dust which spurt into space. The ejected dust lingers around the comet for a while, but eventually spreads out around its orbit.***

***One of the most famous meteor showers is known as the Leonids, so-called because their light trails all seem to originate from the constellation of Leo. The Leonids meteors are associated with dust particles ejected from Comet P/55 Tempel-Tuttle, which pays periodic visits to the inner solar system once every 33.25 years. The meteors appear every year between November 15-20, when the Earth passes very close to the comet's orbit. However, the numbers on view vary tremendously. In most years, observers may see a peak of perhaps 5-10 meteors per hour around 17 November, but roughly once every 33 years the Leonids generate a magnificent storm when thousands of them illuminate the night sky. The most memorable of these storms in recent times occurred in 1833, when tens of thousands lit up the heavens over North America. Unfortunately, not all the comet's appearances are marked by such wondrous sights. While meteors appeared in large numbers during the last perihelion passage of Tempel-Tuttle in 1966, the previous viewing opportunity of 1933 proved to be a damp squib. The reason for this spasmodic and unpredictable behaviour is that, although the main stream of debris trails for millions of kilometres behind the comet, it is not very wide, perhaps 35,000km across. Within this narrow stream, the dust ejected during each of the comet's close approaches to the Sun forms a series of separate ribbons. Their characteristics vary considerably. Generally, the most recent dust streamers are thin and dense, while the older material, which has had time to spread out, forms wider, less densely populated bands. The location of the stream also changes with time, as the gravity of the planets - especially Jupiter - exerts an influence. Sometimes the Earth ploughs right***

*into a dense stream of debris, causing a storm of bright meteors. Sometimes it misses almost all of the tightly confined dust trail, so very few meteors are seen.*

*The Leonids are renowned for producing bright fireballs which outshine every star and planet. Their long trails are often tinged with blue and green, while their vapour trains may linger in the sky like enormous smoke rings for five minutes or more.*

*Although the incoming particles are small, ranging from specks of dust to objects the size of small pebbles, the Leonids glow brightly because they are the fastest of all the meteors. A typical Leonids meteor, arriving at a speed of 71kms$^{-1}$ (more than 200 times faster than a rifle bullet), will start to glow at an altitude of about 155km and leave a long trail before it is extinguished. The reason for this high-speed encounter is that, like their parent comet, the particles travel around the Sun in a direction which is almost directly opposite to the orbital motion of the Earth. The result is a head on collision between the planet and the dust.*

*Detlef Koschny, ESA Space Science Dept*
*/ESA Science Communications:*
*http://sci.esa.int/leonids99/*

# HOW TO HANDLE A SPECIAL-EVENT PILE-UP

**by Don Lamb, G0ACK, and Roy Clayton, G4SSH**

***You are operating a special event station on 40m SSB and you are overwhelmed with callers, all wanting to make a contact with you. How do you handle the situation? The experience of handling a pile-up can be very satisfying, if done well and efficiently. But the special event pile-up can also be very daunting to newcomers. And yet newcomers are often invited by their radio clubs to help out with the operating when the club puts on a special event station. There are few things more frustrating to the experienced operator than hearing an inexperienced one struggling to handle a pile-up - and making a real hash of it!***

***Don Lamb, G0ACK, has admitted to having a dilemma when trying to handle the sort of pile-up a special event station can generate. However, Don has developed his own methods of dealing with this, which he outlines here. To obtain further tips, we asked Roy Clayton, G4SSH, of the Scarborough Special Events Group, for his advice.***

***It should be emphasised that in this article we are addressing the style of operating required for domestic special event stations. The techniques needed for the DXpedition operator are quite different.***

**QSL from German special event station DL0EM, one of the stations operated by Don Lamb, G0ACK.**

Don Lamb, G0ACK, writes: "I am actively involved in managing a permanent special event station, GB2DHH. I also visit friends in Germany and, as a native English-speaker, operate special event stations DL0EM and DA0IMD over there. Naturally, such calls are very popular and I am often overwhelmed by the response. I have a dilemma. I am never sure how best to work the pile-ups.

"Because I only run modest power at home, I am conscious of the difficulties many operators are having when trying to make contact. I want to be fair and work as many stations as possible. Of course the loudest stations are easily heard, but what about those that lie beneath the strong signals, those that do not run high power?

"My preferred way is to ask for the last two letters of the callsign and make a list of, say, 10 or so. This method can quickly remove the strongest callers from the cacophony and enable me to winkle out some of the weaker stations trying to make contact. On occasions I have been able to reach a readability threshold. Although I find this process fairly slow, it has the advantage of there being two letters already in the log and, by removing the powerful signals, it gives those amateurs with weaker signals a chance. As far as my response is concerned, I often feel guilty that after a station has tried for some considerable time trying to contact me, I can really only give a signal report and thanks for being patient.

**Another of Don's German special event stations: the International Marconi Day activity from Borkum Island in the North Sea.**

"Another option, which seems to be favoured with DX stations, is just to call and then to try to pick out what I can. This seems unfair, as the strongest stations will be dominant. Perhaps another way, which I have heard but not used, is to use the number in the callsign and run through from zero to nine.

"These are just some personal thoughts with an explanation of how I prefer to manage any pileup I create. However, I would be interested to hear from others how they think a special event station should respond to many people calling at once."

**EXPERT ADVICE**

Roy Clayton, G4SSH, responds to Don's dilemma. Roy is a founder member of the Scarborough Special Events Group (SSEG), which celebrated its 15th anniversary in June 2003 and has activated numerous special event stations during this time. This is what Roy had to say:

**Roy Clayton, G4SSH (foreground, second from left), with other members of the Scarborough Special Events Group, takes delivery of a Kenwood TS-570D transceiver from David Wilkins, G5HY, of Kenwood (UK). The transceiver is used on many of the Scarborough group's special event station operations.**

"The overriding priority of a special event station is to showcase amateur radio to the general public, usually to commemorate some unique occasion. While running a special event station, the operator is an ambassador for the United Kingdom and amateur radio as a whole. The person at the microphone has a unique opportunity and a responsibility to ensure that certain basic operating standards and aims are met. It is his job to organise efficiently, manage and control the calling stations to ensure that as many stations as possible make a contact, while giving all callers an equal chance of getting through. One method is to call CQ, pick the loudest station from the ensuing melee, exchange details, then repeat the process with the next powerful station. However, this is absolutely soul-destroying for many callers, such as the hesitant newcomer, the QRP station, 2E and M3 stations with limited power, and the person with an indoor aerial. These stations will often call for hours without success.

"The operator of a special event station is in an ideal position to assist and encourage such modest stations to enjoy the hobby. These people are the future of amateur radio and should be considered to be the target audience.

**LEVEL PLAYING FIELD**

"The only way to get these target stations into the log is to create a level playing field. The most efficient way is to call CQ, ask for full calls only, then make a list of stations calling by acknowledging one at a time, giving each callsign a number in the list and asking the station concerned to stand by. Continue this process right down to the noise level and when the frequency is quiet; only then can you be sure that you have the weakest stations in the log. At this point make a brief announcement explaining the reason for the event, the QSL route for direct cards and the name of the operator. Then run smoothly down the list, exchanging reports, names and QTH, until it is time to repeat the exercise.

*Take control!* Any station calling as you run down the list should be politely asked to stand by whilst you complete the current list. This method will slow the rate, but you are now reaching those elusive stations who do not usually manage to make a contact with a GB station. However, it is important to be flexible, this method will work well for most of the time, but the operator should recognise abnormal conditions such as rapid or deep fading, and revert to single calls until propagation improves.

The Scarborough Special Events Group has used this procedure when operating every GB station over the past 14 years, making in excess of 300 contacts per day. The most common comments on the incoming QSL cards are thanks for the orderly operating procedure which allowed them to make contact. The majority of calling stations do not mind the short wait, they are just delighted to be in the log. This method is often frowned upon by some 'big guns' whose pleasure is to crack the pile up with a single call and move on in triumph, but you cannot please everyone.

Running a special event station should not be confused with running a DX pile-up, where the objectives are completely different. To a DX station, maximising the number of contacts is of paramount importance, often resulting in a feeding-frenzy of undisciplined callers shouting

**Don Lamb, G0ACK, with his wife beneath the massive tower at the DL0EM special event station.**

over the exchange of signals in an effort to make contact. This must never be allowed to happen when you are demonstrating amateur radio to the general public.

**M3 LICENSEES**

With January 2003 marking the first anniversary of the Foundation Licence, it is interesting to note that M3s in particular are interested in 'collecting' special event stations. In 2002, M3s accounted for more than 25% of our total QSOs for each event. The system of making a list appears to both encourage newcomers to call in, perhaps for the first time, and also it allows them to get through without the discouragement of a pile-up. Most newcomers write direct for a QSL card and the following comments from M3 stations have been received following recent SSEG events: 'Many thanks for my first QSO on HF. I got my call in June'. 'Many thanks for letting me work my first GB station. I am disabled and sat my Foundation Licence at home'. 'I am limited to simple aerials in a small garden, so I really appreciate the procedure used by your group to encourage low-power stations".

"'Many thanks for your patience in digging around in the mud at the end of a list to get me in the log. I shall cherish my first-ever GB card'. 'Your system of taking callsigns then calling them in strict rotation works 100% and should be adopted by other special event stations. More power to your elbow'. 'I have been using my M3 call for the last four weeks on the HF bands, but have been unable to have a contact with any special event station until you answered me at the end of a list, so I am absolutely delighted. I tried for three hours for a contact with a GB station yesterday and was drowned out by high-power stations. Two friends of mine have suffered the same way... With more and more M3 calls coming on the air I feel that it is very important for other GB stations to make a list and give us the incentive to make a contact'."

"It is clear that collecting commemorative QSL cards from special event stations has become one of the fastest-growing specialist interests amongst newcomers to amateur radio. To these stations, a contact with a GB station is a moment to treasure that allows them take part in a piece of history brought to life by amateur radio. Your choice of operating procedure can make the difference between delight or disappointment."

# IMPROVING LANGUAGE SKILLS USING AMATEUR RADIO

**by Norman Shackley, G0OSX**

***Amateur radio is an absolutely ideal medium for improving your school French or German, or for brushing up on your Spanish before your next holiday to the Costas or the Canaries. Yet very few UK amateurs, at least, take advantage of the wonderful opportunities provided by our hobby, relying instead on the overseas amateurs' ability to speak English. Norman Shackley, G0OSX, says "Try something new in amateur radio!" and provides a number of ideas based around using your foreign language skills.***

The pages of *RadCom* regularly draw readers' attention to both new and different aspects of our multi-faceted hobby. One aspect which has not featured prominently for some time, however, is the use of amateur radio for developing foreign language skills. Recent changes in the licensing rules and the welcome influx of younger M3 Foundation Licensees, many of whom are still pursuing academic studies, make this a good time to reflect on how we might give greater emphasis to this important aspect of the hobby.

As someone who has been a long-standing student of the French language – for more years than I care to remember – I have to say that whenever conventional language students outside amateur radio see the facilities we have for live, on-air language practice, the reaction is often one of disbelief. The opportunity for voice and text contact with foreign nationals who share similar interests to our own is something we frequently undervalue.

Like many other UK Francophile operators, I use 80 and 40m and the lower portion of the 20m SSB segment to make contact with French radio amateurs. The contacts can be mutually advantageous in the sense that the language used can easily be switched according to need. English is used if French amateurs wish to practice their hard-learned vocabulary, and French is used if British operators wish to test their language skills. In practice, the operator with the weaker vocabulary tends to defer quickly and lapse into his or her own language, but this need not always be so. Experience has shown that the simple request to be allowed to continue in the other language will almost certainly be met with patience, courtesy and a genuine willingness to assist.

In fairness to overseas operators, however, this can only work in the longer term if language students do their homework *before* going on air. In this respect, foreign language development in amateur radio is really no different from the long-standing approach to good radio operating practice in general. Putting the microphone away and *listening* to experienced foreign amateur operators will frequently ensure better progress than bulldozing one's way into a QSO, ill-prepared and inadequately-equipped linguistically for a meaningful two-way exchange of information.

Perhaps a closer parallel to good learning practice is the building-block approach to learning the Morse code - which is itself a language, after all. Many of the good linguists in amateur radio first started foreign language QSOs using the simple, rubber-stamp exchanges. Typically, they then progressed to more detailed contacts by steadily expanding their QSO vocabulary using those increasingly-common expressions 'lifted' from the valuable listening sessions mentioned above. If the commitment is there, before too long a facility is developed for dealing with standard radio exchanges and the operator begins to cope with increasing confidence when pushed to the limits of a restricted vocabulary.

**Saint-Jean - a watermill in the hamlet of Polincove in the Pas de Calais, the first mill activated on the day described in the text.**

Speak to any radio amateurs who have scaled the language barrier at whatever levels and in whatever languages, and all will agree on one thing. The satisfaction and pleasure they now gain from the hobby makes all the months and years of effort worthwhile.

At the present time, several UK operators are regularly active working stations across the channel. Harry, G0CSS, from Nottingham can be heard most days, as can Ray, G0RKQ, in Scunthorpe. Ray has for many years enjoyed regular QSOs with Patrick, F5RXE, from Castillonia-Bataille in the Bordeaux region of south west France. On the French side, Henri Lohier, F5MDM, a retired air force officer and dedicated Anglophile who has been studying English for six years, actively seeks and is always delighted to make contact with UK stations to supplement his conventional language studies. If any reader would like to register an interest in developing language skills through amateur radio I would be happy to co-ordinate replies and put like-minded members in touch witheach other under the various language groupings.

**FRENCH AWARDS PROGRAMMES**

For any English-speaking operator wishing to develop language skills and radio links with French operators directly, the awards programme organised by REF-Union (the French equivalent of the RSGB) offers real opportunities for regular contacts. Awards such as the regional departments award (DDFM - *Le Diplôme des Départements Français de la Metropole*); the French castles award (DFCF - *Le Diplôme des Forts et Chateaux de France*) and the French windmill and watermill award (DMF - *Le Diplôme des Moulins de France*) all provide real operating challenges, combined with the opportunity to improve language skills for those wishing to do so.

In all of these award programmes, the level of French language required for a successful contact can be as simple or as complex as desired. They therefore make an ideal starter-pack for the aspiring linguist.

Les Allwood, G3VQO, setting up the 40m mobile antenna on the 4 x 4 by the windmill in the village of Watten.

**TILTING AT WINDMILLS**

My own interest in the DMF award was triggered by Les Allwood, G3VQO, following a 40m contact in May 2002, when he was operating as F/G3VQO/P from a windmill in the Pas de Calais region of northern France. The ensuing QSL and e-mail correspondence spoke of the detailed preparations involved in identifying and locating mills suitable for activation by invading Brits. I expressed an interest in learning more about the DMF and was quickly invited to join the next trip planned for November 2002. And so it was that I found myself one dark, Saturday morning in Horsham, Sussex, at 0420, shaking hands with G3VQO.

After brief introductions we climbed into a 4 x 4 and set off for Folkestone and the 0630 cross-channel tunnel departure for Calais. We quickly cleared Calais port controls by 0830 French time, and were rolling through the French countryside just 15 minutes later.

The day's activities called for a number of skills, all of which were amply demonstrated by G3VQO: map reading, antenna assembly and adjustment, language skills for operating as well as for seeking route directions as a last resort (some mills were so remote that not even all local residents were aware of their existence). Even natural history knowledge was tested on occasions as we tried to identify the wealth of wildlife observed during the course of the day.

Under the rules of the DMF award programme, the portable station can be located up to 500 metres from the mill. In practice, we were able to set up the station immediately adjacent to the mill in most cases. Les was adept at springing into action and converting the 4 x 4 into a working station within minutes. The equipment used for the project was a Yaesu FT-847, powered by a 110Ah 12V leisure battery with a variety of HF whip antennas available for connection to a magmount on the roof of the vehicle.

Les Allwood, G3VQO, at the mill at Inglinghem.

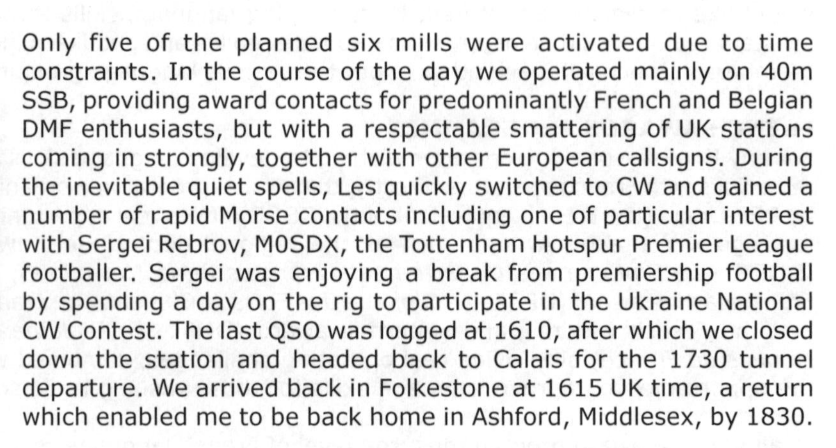

Only five of the planned six mills were activated due to time constraints. In the course of the day we operated mainly on 40m SSB, providing award contacts for predominantly French and Belgian DMF enthusiasts, but with a respectable smattering of UK stations coming in strongly, together with other European callsigns. During the inevitable quiet spells, Les quickly switched to CW and gained a number of rapid Morse contacts including one of particular interest with Sergei Rebrov, M0SDX, the Tottenham Hotspur Premier League footballer. Sergei was enjoying a break from premiership football by spending a day on the rig to participate in the Ukraine National CW Contest. The last QSO was logged at 1610, after which we closed down the station and headed back to Calais for the 1730 tunnel departure. We arrived back in Folkestone at 1615 UK time, a return which enabled me to be back home in Ashford, Middlesex, by 1830.

### ONE THING LEADS TO ANOTHER...

Would I consider doing it again? I certainly would. The benefits of overseas operating in this unusual but very popular award programme are many and varied. I particularly enjoyed the strengthening of links with French operators on their home soil in the true spirit of amateur radio. Being able to operate in the French countryside was most enjoyable, as our surroundings were always pleasant. Perhaps most of all I enjoyed the getting up and about and trying something *different* in amateur radio for a change, a step I had not taken for many years.

**Haute Escalles - a rural windmill in the Pas de Calais countryside.**

The reward here was being able to watch a skilled CW operator at close quarters. Les Allwood, G3VQO, belongs to the French UFT *(L'Union Française des Télégraphistes)*, membership of which requires a set number of successful French-language CW contacts with, and nomination by, other UFT operators.

Not unexpectedly, the experience has fired my ambition to follow suit and since my return I have been busy brushing up my CW French-language vocabulary.

Would I want to change anything the next time I go over? I would certainly want to reward myself just a little more for all of those many DMF award contacts issued. I would also like to benefit from the location, once the radios have been safely stowed away, by tucking in to a large meal with lots of good wine, ideally in the company of a few French operators to do our bit for l'*Entente Cordiale*, followed by an overnight stay in a comfortable hotel!

### WHAT'S ON THE WEB?

DMF (*Diplôme des Moulins de France*) http://f5pez.free.fr/Diplome.htm
DDFM (*Diplôme des Départements Français de la Metropole*) www.dxawards.com/DXAwardDir/france-ref.htm
DFCF (*Diplôme des Forts et Chateaux de France*) http://perso.club-internet.fr/f6fna/dfcfa.html
REF-Union (*Réseau des Emetteurs Français*) www.ref-union.org
UFT (*Union Française des Télégraphistes*) www.uft.net

### WHERE TO LOOK

Where can you find foreign-speaking radio amateurs to practise your language skills? The answer is, of course, all over all of the HF bands. But certain nationalities tend to congregate around specific frequencies so specific languages can often be heard in particular parts of the bands, particularly on 20m. Here are a few.

| | |
|---|---|
| ***Français* French** | 80m (all), 40m (all), 20m 14,120 - 14,130kHz |
| ***Deutsch* German** | 80m (all), 40m (all), 20m (all) |
| **Ελληνικα Greek** | 20m ±14,285kHz |
| ***Svenska* Swedish** | 20m 14,300 - 14,310kHz |

# MAKING THE MOST OF SPORADIC-E AT VHF

**by Tim Kirby, G4VXE**

To the VHF enthusiast, Sporadic-E is one of the most exciting forms of propagation that exists.

When Sporadic-E is about, even an amateur with a simple VHF station - running just a few watts - can suddenly and unexpectedly make contacts of over 2000km. In this article, we'll try to cover some basics about how it works, when it happens, and discuss some tricks that you can use to help you make some exciting DX contacts.

### WHAT IS SPORADIC-E?

Sporadic-E, or Es, as it is often known, is the reflection of VHF signals back towards earth from the E-layer of the ionosphere. The E-layer is located about 100km above the earth's surface, although this may vary slightly as we will see later. Normally, the VHF signals will pass straight through the layer but, during periods of intense ionization, the signals may be reflected back towards the earth.

The Maximum Usable Frequency (MUF) is the highest frequency that will be reflected back towards the earth, rather than passing through the ionosphere and going on into outer space. As the ionisation increases in intensity, so the MUF rises. [*The Sporadic-E Maximum Usable Frequency must not be confused with the F-layer Maximum Usable Frequency. It is the F-layer MUF that is usually meant when referring to 'MUF'. The Sporadic-E MUF is usually much higher in frequency than the F-layer MUF* - Ed.]

What causes Sporadic-E? Many fascinating theories have been proposed over the years and there is generally something compelling in each one. Wind-shear over mountain ranges, meteoric dust and thunderstorms have all been discussed as possible causes.

Geoff Grayer, G3NAQ, has published some fascinating work in *The VHF / UHF DX Book* [1] which is well worth reading, and also in the American publication *Ham Radio* [2]. Jim Bacon, G3YLA, has also published some most interesting thoughts on why Es occurs [3]. The conclusions of Geoff and Jim are not identical but, if you are interested in a more scientific approach than will be provided here, I commend these two authors to you.

However Sporadic-E is caused, it is clear that signals are reflected back to earth by the E-layer. Reflective areas move around and we generally refer to those areas as 'clouds'. Clouds will vary in size but, on occasions, areas capable of reflecting the MUF may be only a few tens of metres across. This goes some way to explaining the very selective nature of Es at the higher frequencies - where you can be listening to an amateur across town working someone at S9 and you can't hear a thing! Wait for a moment, though, because in seconds, as the reflective cloud moves, the tables can be turned!

## WHICH BANDS DOES IT AFFECT?

The 28, 50, 70 and 144MHz amateur bands are all commonly affected by Sporadic-E. In the USA, where the 220MHz band is allocated, a small number of contacts have been made via Es on that band, but it is considered extremely rare. It's fair to say that the higher the MUF rises, the less likely it is to continue to rise.

For the amateur, 50MHz contacts via Es will be plentiful. Many observations of Es propagation are noted at 70MHz, although the lack of other countries using that band hinders detailed analysis. By the time we get to the 144MHz band, there are far fewer openings - perhaps a handful of good ones each year - whereas 50MHz may be open almost every day during the Es 'season'.

At HF, 28MHz is a prime band for Sporadic-E and many DXers use the Es season as a method of working countries relatively close to home that tend to be difficult to work on 28MHz by 'normal' HF (F-layer) propagation. When the Es MUF is particularly high and the ionisation particularly intense, paths of 300 or 400km are possible on 28MHz. Normally, of course, a signal approaching the ionosphere at such an angle of near-vertical incidence would pass straight through. So, as we'll discuss later on, this is a really useful indicator of exciting things about to happen on VHF.

## WHEN DOES IT OCCUR?

As the name suggests, Sporadic-E is just that – sporadic. However, for us in the northern hemisphere, the main season is from early May until the end of August. Certainly, openings will occur outside these times, but they will be less frequent. An urban myth suggests that there will always be a 144MHz opening during the first week of June, but regrettably, life is not that simple!

There is also a secondary peak in Es activity from mid-December to early January. 144MHz openings during this 'mini-season' are very infrequent, but not impossible – but it is unusual for there not to be some good 50MHz Es openings at this time.

During the 'seasons', it is usual to find that the highest MUFs will occur in the late morning, fall away during the afternoon and reach another peak during the early to middle evening. Multi-hop openings, discussed shortly, appear more likely to occur later in the day. Openings on 50MHz from Europe to the eastern part of North America
generally occur during the mid to late evening in Western Europe, although the author has noted openings at almost all other times of day!

Anecdotal evidence suggests that there are fewer Sporadic-E openings towards sunspot maxima, suggesting that a quiet sun is a contributor to good Es conditions. To my knowledge, there is no accepted scientific proof that this is the case, although my own observations do seem to bear it out.

**WHERE DO THE OPENINGS OCCUR?**

The bad news for those of you in Scotland or the north of England, I'm afraid, is that the higher you are in latitude, the less likely you are to see a high MUF by Es. That is not to say it won't happen. It will, but you will have to be more patient. The further south you are, the greater the likelihood of finding an opening.

Operators from the Channel Islands for example, have noticed many more openings than those in Southern England. A difference of 60 - 100km can make a marked difference.

Geoff Grayer, G3NAQ, notes in the *VHF/UHF DX Book* [1] that some paths seem to open repeatedly over a short period of time. I have noticed this myself. You may hear nothing from a particular area for several years only to find openings on consecutive days. Is there a residual level of ionisation that causes this to happen?

**WHAT DISTANCES CAN BE COVERED?**

Assuming that the E-layer resides at a height of between 90 and 120km, a theoretical maximum of between 2130 to 2450km exists. However, many contacts over greater distances have been made. For example, some years ago, I was able to work an EA8 station on 144MHz Es at a distance of 2880km. During that opening, I could also hear EA1 stations around half way from me to the Canaries, which suggested two different reflecting areas. One was perhaps half way from me to EA1 and the second, half way from there to EA8. Multi-hop paths do seem to exist and contacts have been made out to around 3500km.

On 50MHz, during the summer, there is often the opportunity to work into the US. Once again, opinions vary on the exact mechanism providing this propagation. The most often-suggested method is multiple-hop Es. Some quite remarkable contacts have been made from the Mediterranean and the Middle-East into W5 and W7. It is considered that there must have been a multi-hop mechanism in action here.

Perhaps because of the lower latitude, the incidence of ionisation is greater and there is a possibility of more reflecting areas being present to sustain the longer path.

Perhaps too, this is straying into the realm of 'Equatorial Zone Es', which provides much 50MHz propagation around the magnetic equator.

**PUTTING IT ALL INTO PRACTICE**

How, then, can you make the most of the openings that occur? In the 'old' days, monitoring various key frequencies was all you could do – perhaps Band I television, Eastern European radio around 70MHz, Band II radio between 88 and 108MHz and perhaps aircraft navigation beacons around 120MHz. You'd monitor all those and when interesting things started to happen, the DX Telephone Net would spring into action.

Nowadays, the same frequencies are still useful to monitor, but we have a wealth of other useful data available to use. The first is using the *DXCluster* to see what is happening elsewhere.

Using the *Cluster*, we can see openings taking place in other parts of Europe, or even the world. These may simply alert us to the fact that a band is open in a particular direction and that we should get on *post haste*! 29MHz FM can be a very useful indicator of good VHF conditions. When the skip distance comes down to 200 - 400km, you can be certain something interesting is going on.

However, there is more interpretation of the information that we can do. What, for example, if an amateur in Slovakia (OM) spots an amateur in the EA3 area of Spain on 50MHz? Assuming that the reflecting area is half way along the path (not necessarily true, of course!) this puts it over the very north-western part of Italy. So, located here in Windsor (IO91 square), I should be able to 'see' that reflecting point. If I point my antenna there, I may find a 50MHz Es opening to the very far south-east of Italy. How do I know this? Well, there is some handy software that I can use to plot the paths for me. There are a couple of packages that you can use. I found *SEProp* by K9SE ideal. **Fig 1** shows what *SE-Prop* displays in this case.

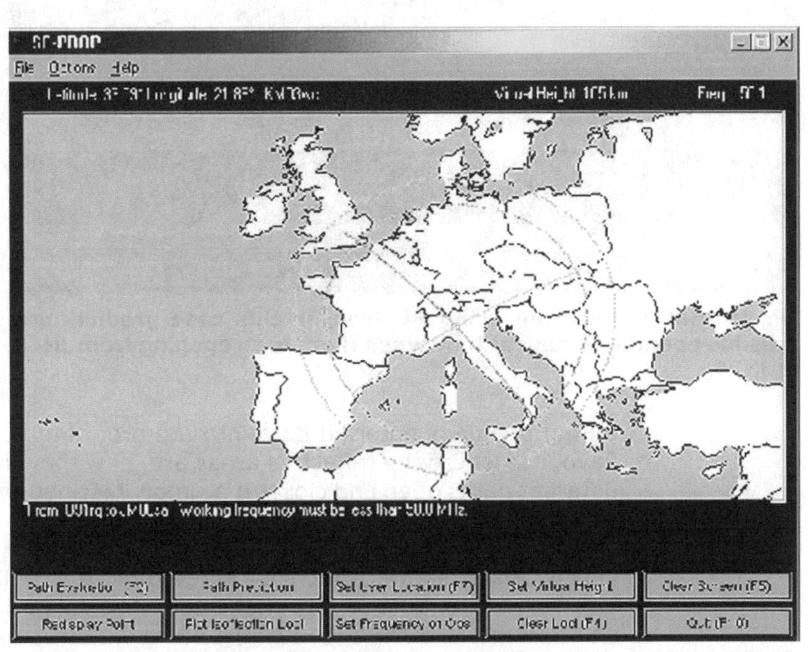

**Fig 1: *SE-Prop* display of predicted 50MHz Es propagation from the author's location to south-east Italy when there is known Es from Slovakia to Spain.**

Using the 'Path Evaluation' feature, you can select the two end points of a path. The mid (scatter) point will automatically be displayed for you. You can then use the 'Path Prediction' feature, select the scatter point that you just discovered from the OM / EA3 path, click on your own location and see where the reflected signal goes. Sometimes, it will drop into the sea somewhere useless! Sometimes, you will see that the MUF for that path falls below 50MHz, and so won't be particularly useful to you. But other times, you will find that the program predicts a path being available. When that happens, get on the band and look in that direction! You will be surprised how often it works.

Let's take another example. You see a spot on the cluster from a DL in JN48 square of a French station in JN26 on 50MHz. You can tell instantly that that's a very short distance, so the MUF must be pretty high to support that contact. Put the details into *SE-Prop* and it will show you the

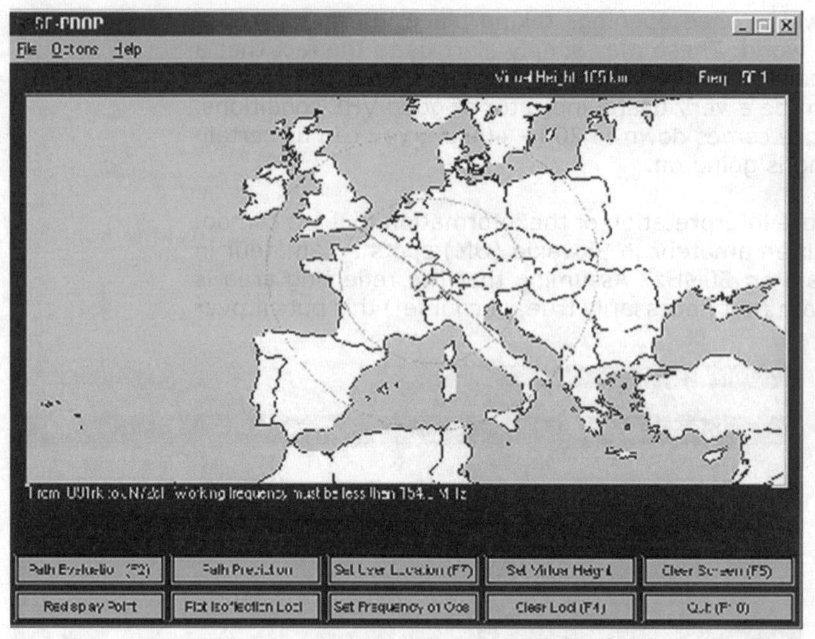


**Fig 2: Another example using *SE-Prop*, in this case predicting a possible opening to central Italy when there is an opening from JN48 to JN26.**

scatter point and the MUF – in this case around 169MHz! Then use the 'Path Prediction' feature, enter the scatter point, your own location as one end-point and see where the other one goes. In this case, a 144MHz contact looks possible into the I6 area (see **Fig 2**).

That example was, actually, a case of working back to get an interesting result and is probably a little unlikely in practice. However, as you spend more time during openings entering data into the program, you will begin to get a flavour of where the reflective areas are, how they move around and how different paths open and close. It's simply fascinating and becomes rather compulsive!

Finally, **Fig 3** shows what your screen might look like after an opening lasting a while! This is data that I collected on 28 July 2001 - the paths shown were on 50MHz. You can see two distinct reflecting areas, one up the east coast of England and the second over the Bay of Biscay. As time went on, the cloud over England moved from almost directly overhead of JO01 square in a northeasterly direction.

## OTHER POINTS OF INTEREST

Sometimes, you may enter the two end points of a path, and the program may report that the distance is beyond the theoretical maximum. In some cases, the program may suggest that you try repeating the calculation with a different height for the E-layer (perhaps altering the height of the layer from 105 to 110km). In other cases, it may simply suggest that the path is multi-hop.

## SOFTWARE

Although I have described the use of *SE-Prop*, there are other programs available that work in the same way. If you have a *DOS*-only machine, you may like to try the *SE-Prop for DOS* program. There is also a very useful command-line utility called *MUF.EXE*, which was written by Mark, G4PCS, which calculates the MUF based on the frequency and end-point locators of a particular path.

It is outside the scope of this article, because the mode of propagation is not commonly noted in the UK, but *SE-Prop* also has features to allow the study of Field-Aligned Irregularity (FAI) paths on 144MHz. The serious enthusiast may find this of interest.

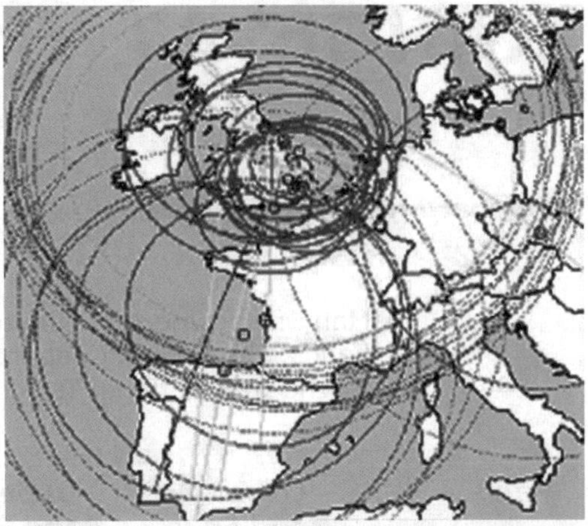

**Fig 3: The 50MHz Es opening on 28 July 2001, showing two distinct reflecting areas, one over the east coast of England and the other over the Bay of Biscay.**

## CONCLUSION

I hope that the reader will have enjoyed this introduction to the mechanisms of Sporadic-E at VHF. Certainly, for me, it is one of the most fascinating aspects of propagation I know.

Sporadic-E is a great 'leveller'. Because of the nature of the propagation, an amateur with a very simple station who is in the right place may make some marvellous contacts. During the RSGB 144MHz Backpackers contests, in the 3W class, we have several times made Es contacts with stations in central and eastern Europe. So don't assume that just because you have a small station that you can't enjoy this propagation. You can.

Developing your own ideas and theories about the mechanisms of Sporadic-E is at least as interesting as operating in an opening! If I have been able to impart some of my interest and enthusiasm for the subject, I shall be delighted.

Finally, in a research program dedicated to the memory of Serge Canivenc, F8SH, Jim Bacon, G3YLA, is collecting data of 144MHz Es contacts. If you make (or have made) some contacts, please consider forwarding details to Jim, in order that they may be recorded and analysed, in the hope of finding out more about this fascinating mode of propagation.

## FURTHER READING

[1] *The VHF / UHF DX Book*, edited by Ian White, G3SEK. (Chapter 2: 'VHF / UHF Propagation' by Geoff Grayer, G3NAQ).

[2] 'Sporadic-E and 50MHz Trans-Atlantic Propagation During 1987', by Dr Geoff Grayer, G3NAQ, *Ham Radio*, July 1988.

[3] 'An Introduction to Sporadic-E', by Jim Bacon, G3YLA, (Parts 1 - 4, *RadCom* May 1989 – August 1989).

## WHERE TO LOOK ON THE NET

*SE-Prop* software http://pwl.netcom.com/~wb9qiu

*MUF.EXE* www.dh0ghu.de/download.html

F8SH challenge http://challengef8sh.ifrance.com/challengef8sh/indexen.htm

# 'RANGE' AT VHF AND UHF

by Richard Newstead, G3CWI

Many newcomers to amateur radio begin operating on the VHF or UHF bands. The frequency spans of these bands, shown in **Table 1**, are defined by the International Telecommunications Union.

| Band | Lower Band Edge | Upper Band Edge |
|---|---|---|
| VHF | 30MHz | 300MHz |
| UHF | 300MHz | 3000MHz |

**Table 1: The internationally-agreed limits of the VHF (Very-High Frequency) and UHF (Ultra-High Frequency) bands.**

We all know that during a 'lift', distances of 1000km or more are possible in the higher VHF and UHF bands, but how much range can be expected under 'flat' conditions? Even before starting to transmit, we all have some idea of the range we might achieve based on our experience of listening within the bands. For example, the FM broadcast band is in the region of 100MHz and anyone with an old style FM radio in the car knows that it generally needs to have the channel changed every 30 to 40km. Newer radios with RDS do the re-tuning automatically.

The television band lies in the 500-800MHz region and the maximum range is usually about 50km. So what is it that determines the range that can be achieved? An expression that is sometimes used in these bands (particularly at UHF) is that propagation is 'line of sight'. However, this tends not to be very useful - even as a rule of thumb - and it's easy to see why with a few simple calculations.

## INDUSTRIAL PRACTICE

When professional engineers design radio systems, they are usually interested in the range that can be achieved. It can be critical to get this right, as an underestimate of only a few percent would mean huge additional costs with cell-phone systems which have many thousands of sites across the country.

The first stage in all these calculations is to work out how much signal you can afford to lose along the path between the transmitter and receiver, so that there is enough signal left to be readable. This is actually quite easy to work out. Let's assume your 50Ω receiver input needs 0.5μV for a readable signal. Using the familiar equation $P = E^2/R$, we can calculate that 0.5μV ≡ 5 x $10^{-15}$W (a very small amount of power). To make this easier to handle, we can turn it into decibels relative to 1W,

ie $10 \times \log_{10}(5 \times 10^{-15}) = -143$dBW.

If we assume that the aerial system has no gain (or loss), then this is the signal level that we need for acceptable reception. Now, looking at the transmitter end, let's assume that it runs 10W (10dBW) and, again, that the aerial system has no gain (so that the Effective Radiated Power is 10dBW). In this case we can tolerate a loss of 153dB (the difference between 10dBW and -143dBW). This loss is sometimes called the *Basic Transmission Loss*.

If we imagine our path is actually line of sight (ie from the transmitting aerial we can see the receiving aerial), we can work out how far we can communicate using the Free Space Path Loss Equation which is:

Path Loss (dB) = $32.4 + 10\log_{10}(d) + 20\log_{10}(f)$,

where d is the distance in km,
and f is the frequency in MHz.

With a bit of rearranging and calculating, we can construct the range graph shown in **Fig 1**. Judging by this graph, on 6m (50MHz) you might expect a range of 22,000km and on 2m (145MHz) a range of around 7,000km. These distances would be accurate in space, but the actual ranges you might achieve here on Earth under flat conditions are much less, so the conclusion is that there are some other losses involved that we have not accounted for.

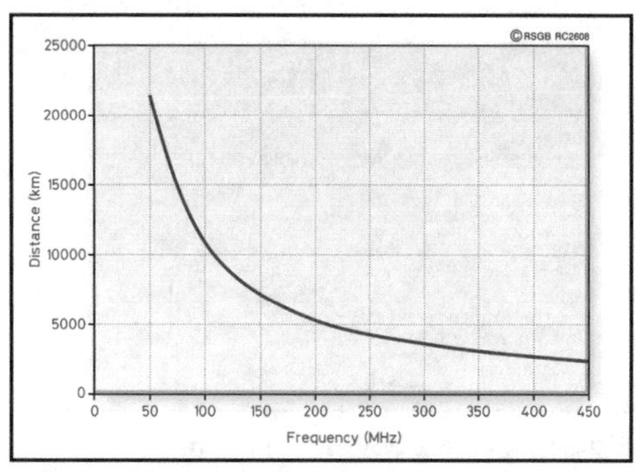


**Fig 1: How far you can expect VHF and UHF signals to reach in free space.**

These losses come from a variety of sources, but the two major ones are blocking of the signals due to terrain (the signals don't bend around hills very well), and blocking of the signal due to trees and buildings. The other major factor is that the earth is not flat. After the horizon is reached, the curvature of the earth itself gets in the way. Now we are into some much more complex calculations to work out the range.

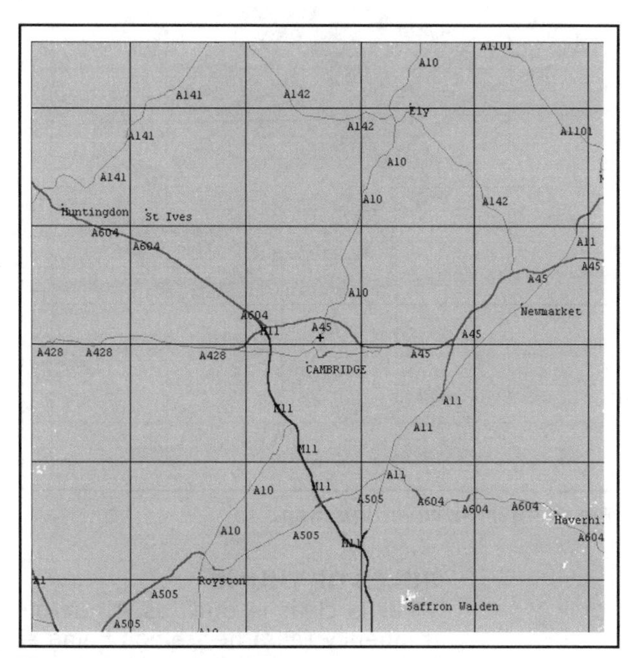


**Fig 2: 50MHz coverage map.**

## ENGINEER'S APPROACH

The professional engineer uses a variety of techniques to estimate range, taking into account the various obstructions we have identified. The most commonly used techniques calculate the diffraction losses over the obstructions. This can be very laborious, because to calculate the range in any one direction the calculations need to be done for all the points along the path. Fortunately, computer programs take the hard work out of this.

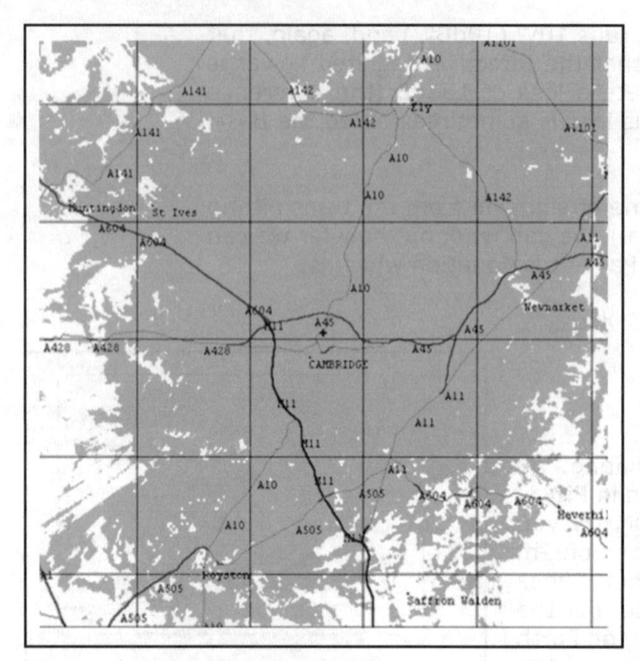


Fig 3: 144MHz coverage map.

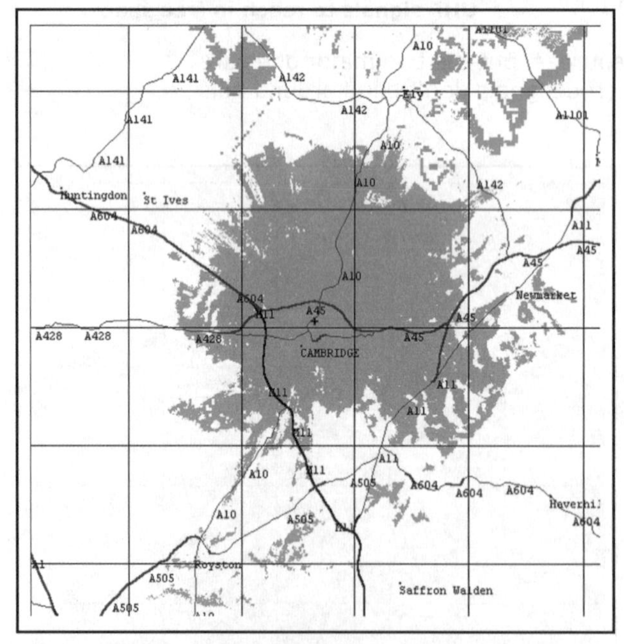


Fig 4: 432MHz coverage map.

Three coverage predictions are shown in **Figs 2 – 4** from a sophisticated software propagation-modelling package. It uses terrain heights accurate to 1m, every 50m along each path. It also has details of other obstructions in the area, such as houses and trees. To generate these predictions, I have selected a site somewhere in Cambridge and simulated a transmitter there with a vertical dipole aerial 10m above the ground. Using the figures above, the shaded areas show where we would expect to get coverage with a probability of 90%. The grid on the plots is 10km. On 6m (50 MHz) Fig 2 shows that virtually the whole area is covered.

Moving to 2m (145MHz), Fig 3 shows that there are some white patches, indicating areas that are not covered to the standard we specified.

On 70cm (435MHz), Fig 4 indicates that the range has dropped dramatically, with coverage only reaching about 10 – 15km in each direction. Considering the modelling problems presented by the complexities of the environment, it is perhaps surprising that professional engineers have developed some fairly simple formulae for predicting average ranges that don't need any terrain information. The best know of these was proposed by Hata. His formula allows a range to be calculated from frequencies between 150 and 1000MHz in a variety of situations. Unfortunately, it is not much use to radio amateurs, as the minimum aerial height that can be modelled is 30m.

### RULES OF THUMB

What is clear is that, as a rule-of-thumb, range drops with increasing frequency (all other things being equal). But that's not the whole story. In built-up areas, the higher frequencies tend to reflect better and can penetrate buildings better, as the shorter wavelengths can enter through windows. This effect is often quite marked in road tunnels, where 2m fades quickly, and 70cm goes on much further. So, in built-up areas, higher frequencies may prove more effective.

Another factor to consider is that at the higher frequencies, due to the shorter wavelength, aerials become smaller. Consequently, for a given physical size of aerial, more gain can be achieved, countering the increased losses.

**GETTING THE BEST**

So, how can we get the best range? Getting your aerial as high as possible is the first thing to do. For the best range, it certainly needs to be clear of the rooftops and ideally clear of any other obstructions too. Aerial gain helps too. In general, gain can be achieved more easily with a Yagi than with a vertical such as a colinear or ground plane. However, a Yagi must be rotated, making the whole aerial system more complex. Choosing the best coax you can afford to connect your aerial to your radio is the second thing to do. Losses in the coax feeder reduce your transmit power, and you can lose those weak signals on receive too. For the ultimate in performance, use a masthead preamplifier – but choose one that won't get overloaded by strong signals. Finally, if all else fails, look for a house on top of a big hill – the extra height gained this way will make a *big* difference to your range on the VHF and UHF bands!

# SPURIOUS SIGNALS

**by Ian White, GM3SEK, from 'In Practice'**

***Q: If using a modern transceiver with digital readout, is it still necessary to have and to use frequency-checking equipment such as a crystal calibrator or wavemeter?***

***A:*** This specific requirement was dropped from BR68 some years ago, and replaced with two much more general requirements in clause 4:

- *The Licensee shall ensure that the apparatus comprised in the Station is designed and constructed, and maintained and used, so that its use does not cause any undue interference to any wireless telegraphy.*
- *The emitted frequency of the apparatus comprised in the Station is as stable and as free from [spurious emissions and out-of-band emissions] as the state of technical development for amateur radio apparatus reasonably permits.*

In other words, you are required to transmit good-quality signals, and not stray outside the amateur bands, but the RA [now Ofcom – Ed.] no longer specifies exactly how you must achieve this. A frequency-synthesised transceiver with a digital readout is a fairly good assurance that you are transmitting on the frequency displayed. However, as I've stated before in this column, your digital readout does not actually count the real frequency - it is only based on the instructions sent to the synthesiser. In practice, this is hardly ever a problem, because it would take a very unusual kind of synthesiser fault to mislead you seriously. Even so, you are required to *conduct tests from time to time to ensure that the requirements of clause 4 are met*, so an occasional check with a separate frequency counter would be in order.

Similarly, there is little point in checking for harmonics with a wavemeter. All commercial solid-state transceivers have built-in low-pass filters and a simple wavemeter is usually incapable of detecting the residual levels of harmonics. Home-built rigs are a different story, but again the sensible solution is to build in a low-pass filter as a matter of course. These days it's irresponsible to put any rig on the air without a low-pass filter, because either your harmonics will fall inside our own bands where they may interfere with other radio amateurs, or else they fall outside the amateur bands and you risk interference to another service. For example, the second harmonic of 144 – 146MHz falls in the Aircraft band, the third harmonic is in 'our' 430 – 440MHz band (where we aren't the primary user), the fourth to sixth harmonics are in the TV band... need I go on?

Many of us have built transistor power amplifiers using the simple kind of output network you see in the amateur handbooks and in the transistor datasheets. A typical home-built transistor PA can produce quite significant levels of harmonics, often no better than 30dB below the fundamental. Similarly, PA modules can have second and third harmonic levels as high as -30dB, with higher harmonics not much better suppressed. Minus 30dB means that for every watt you transmit on the 144 - 146MHz band, you're transmitting a milliwatt of harmonic – and in today's EMC conditions, that just won't do! If you could see the output from your home-built transmitter on a spectrum analyser, you'd know that you need a low-pass filter; but since most of us don't have access to that kind of test gear, the only responsible policy is to use a low-pass filter anyway.

Dealing with VHF first, **Fig 1** gives details of very simple 'no-tune' low-pass filters for 50, 70 and 144MHz [1]. The only tuning you may need to do is re-peak the PA a little for maximum output. Also in Fig 1 are details of a modification to provide an extra notch to reduce the second harmonic of 50MHz in the FM broadcast band. These filters are small enough to be tucked inside existing equipment on a scrap of PC board. You can use small ceramic plate capacitors for powers up to a few watts, and ordinary silver-mica capacitors with short leads for powers up to a few tens of watts. Above these power levels you'll need to use metal-clad silver mica capacitors, or large porcelain chip capacitors, and coils made from larger-diameter bare copper or silver-plated wire [2, 3]. For higher power levels still, or for better attenuation of specific harmonics, use one of the notch / low-pass filter designs published by G4SWX [4].

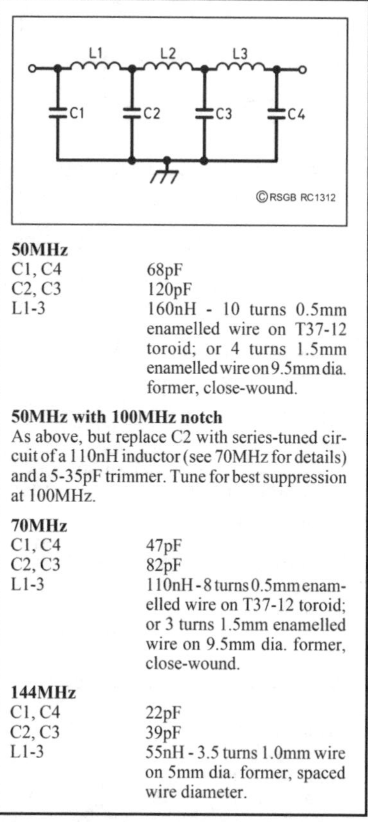


**50MHz**

| | |
|---|---|
| C1, C4 | 68pF |
| C2, C3 | 120pF |
| L1-3 | 160nH - 10 turns 0.5mm enamelled wire on T37-12 toroid; or 4 turns 1.5mm enamelled wire on 9.5mm dia. former, close-wound. |

**50MHz with 100MHz notch**
As above, but replace C2 with series-tuned circuit of a 110nH inductor (see 70MHz for details) and a 5-35pF trimmer. Tune for best suppression at 100MHz.

**70MHz**

| | |
|---|---|
| C1, C4 | 47pF |
| C2, C3 | 82pF |
| L1-3 | 110nH - 8 turns 0.5mm enamelled wire on T37-12 toroid; or 3 turns 1.5mm enamelled wire on 9.5mm dia. former, close-wound. |

**144MHz**

| | |
|---|---|
| C1, C4 | 22pF |
| C2, C3 | 39pF |
| L1-3 | 55nH - 3.5 turns 1.0mm wire on 5mm dia. former, spaced wire diameter. |

**Fig 1: Simple 'no-tune' low-pass filters for 50, 70 and 144MHz transmitters [1].**

Filters with specific notches in the pass-band are especially useful for amateur VHF operation because we tend to make only small relative changes in frequency, so the G4SWX designs can be highly recommended. **Fig 2** shows two even simpler stub filters, again from G4SWX, which will notch-out the 2nd, 3rd and 4th harmonics, and also odd-numbered multiples of these. The inductor in the VHF design provides some general low-pass response to help with the other harmonics where none of the coaxial stubs is resonant. However, extensive testing and computer modelling has shown that the response of LC low-pass filters at frequencies well above the fundamental can be very dependent on parasitic reactances in the components used: series inductance in the capacitors, and distributed capacitance in the inductors. These effects may cause completely unexpected notches and peaks to appear in what should be a smoothly falling stop-band response. As an example, in the 144MHz version of the filter in Fig 2(a), the distributed capacitance of L1 may well provide a notch response at around 500 – 700MHz, which could be very convenient to reduce the 5th harmonic at 720MHz, but depends critically upon the construction of L1. In contrast, the notch responses of the coaxial stubs are much more reliable and predictable, so Fig 2 shows the option to add

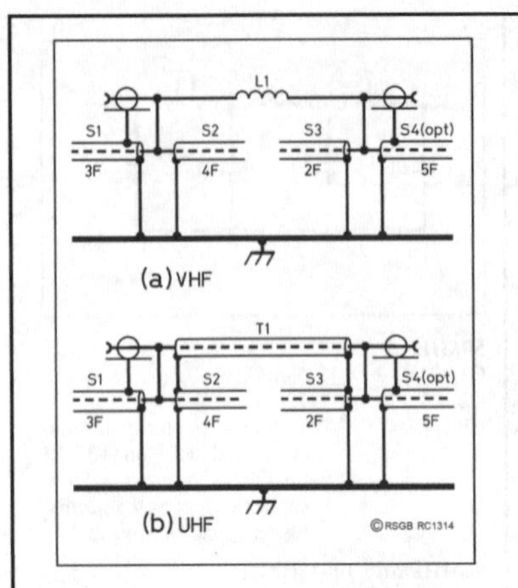


(a)

| Open-circuit coaxial stubs | Lengths in UR43/RG58 (mm) 51MHz | 70MHz | 145MHz |
|---|---|---|---|
| S1: 0.25λ at 3f | 327 | 238 | 115 |
| S2: 0.25λ at 4f | 247 | 179 | 86 |
| S3: 0.25λ at 2f | 489 | 355 | 172 |
| S4: 0.25λ at 5f (optional) | 196 | 142 | 68 |

For L1 see /Fig 1.

(b)

| Coaxial lines & stubs | Lengths in UR43/RG58 (mm) for 432 MHz |
|---|---|
| T1:0.265λ at f | 121 |
| S1:0.265λ at 3f | 38 |
| S2:0.265λ at 4f | 29 |
| S3:0.265λ at 2f | 57 |
| S4: 0.265λ at 5f (optional) | 23 |

**Fig 2: Two new harmonic notch filters from G4SWX - even simpler than before. (a) For 50, 70 and 144MHz; use inductor values from Fig 1. (b) For 432MHz or above. Construction requires very short leads, so lay-out in an 'H' shape.**

a specific stub for the 5th harmonic. The penalty is a slight increase in VSWR at the fundamental operating frequency. For 144MHz, 70MHz or 50MHz transmitters, use Fig 2(a) with the inductor details given for the same band in Fig 1. All you need to do to adjust this filter is squeeze or stretch the inductor for the best VSWR at the fundamental frequency. **Fig 3** shows the frequency response up to 1GHz - harmonic suppression really can't get much easier than this! With stubs cut from RG-213 and a physically larger coil of the same inductance, these filters will handle 500W.

For 432MHz and above, L1 would be physically too small to be usable at power levels above a few watts, so Fig 2(b) shows a coaxial alternative which is easier to build at UHF but has no general low-pass properties. The May 1995 'In Practice' column gives general constructional details for these stub filters. If you carefully measure the required lengths of coax for the open-circuit stubs for the 50MHz, 70MHz and 144MHz filters and use short connections, the harmonic rejection notches will usually be pretty deep. However, the stub lengths become quite critical at the higher frequencies, and especially in the 432MHz harmonic filter, so be prepared to do some cut-and-try. If you have any kind of receivers for the harmonic frequencies (UHF broadcast TV, 1.3GHz equipment, scanner, satellite TV receiver etc), Fig 6 of the May 1995 column shows a suitable test set-up for trimming the stub lengths.

**WARNING** - do not use the old-fashioned type of high-*Q* bandpass filter to suppress VHF harmonics! Their harmonic suppression and insertion loss are both poorer than a well-designed low-pass filter, and vary greatly with the state of tuning and loading. If mistuned, they can also present a very high VSWR to the transmitter, which may cause intermodulation distortion or even physical damage. If you need any further persuasion, try this mock RAE question:

*The correct type of filter to suppress harmonics is: (A) Band-pass (B) High-pass (C) Low-pass (D) Fibreglass*

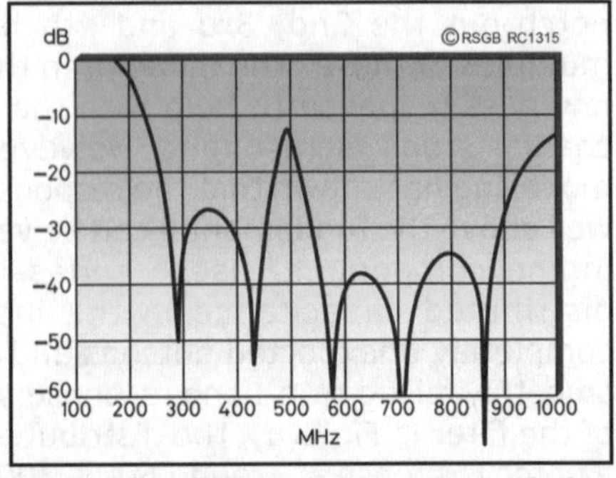


**Fig 3: Frequency response of the 144MHz filter in Fig 2a. Note the deep notches at all harmonic frequencies up to the 6th.**

Except in very special circumstances, the best place for high-*Q* single-tuned filters is in a museum!

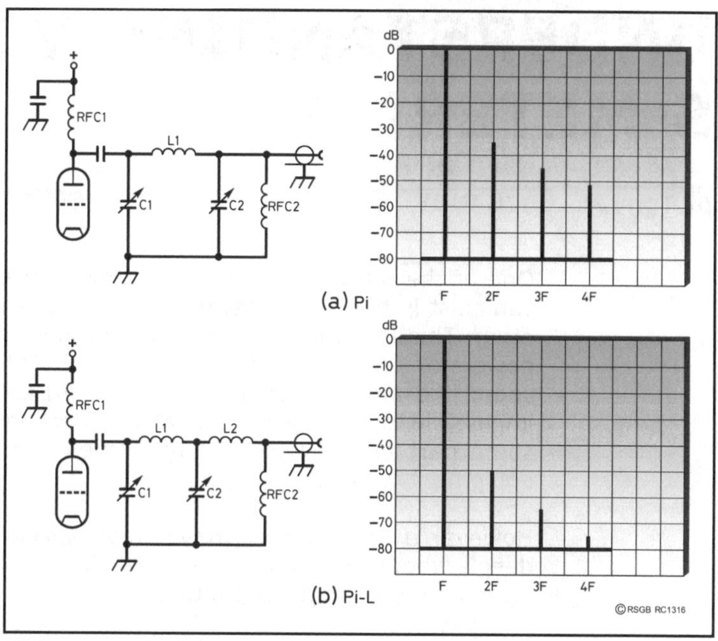


**Fig 4: π and π-L transmitter output networks, showing typical harmonic suppression performance (*ARRL Handbook*). RFC1 is the anode supply choke; RFC2 is a DC safety choke.**

Harmonics from HF power amplifiers need extra care, because add-on low-pass filters intended for TVI suppression will only deal with harmonics above 30MHz. On all amateur bands up to 14MHz, your harmonics will fall somewhere *inside* the general 3 - 30MHz HF band. Solid-state HF power amplifiers usually include a bank of band-switched low-pass filters, but valve PAs rely on their output π-networks. The harmonic suppression of a typical Class-AB PA using a simple π-network (**Fig 4(a)**) will depend greatly on the type of valve, its DC operating conditions, RF drive and the state of tuning and loading. Even so, some general trends emerge and the *ARRL Handbook* quotes typical values of 35dB for the second harmonic, 45dB for the third and about 52dB for the fourth. These figures are marginal by modern standards: in particular, 35dB suppression of the second harmonic would not comply with the FCC regulations in the USA, which require a minimum of 40dB. Looking at it another way, when transmitting 400W PEP on 3.5MHz into a multi-band antenna such as a trap dipole, 35dB of second-harmonic suppression still leaves 127mW, which could be quite a competitive signal in the QRP segment of the 7MHz band! Modern good practice is to use a π-L network (**Fig 4(b)**) which gives much better suppression of the second, third and fourth harmonics, typically 50, 65 and 75dB - see the *ARRL Handbook* for design details.

To sum up, the modern way to deal with harmonics is not to keep a wavemeter in the cupboard as some kind of lucky charm. Instead, build adequate low-pass filtering into every transmitter as a matter of policy.

## REFERENCES


[1] Chapter 12, *The VHF/UHF DX Book* (RSGB).

[2] 'Solid State 600W 6 metre Linear Amplifier', John Matthews, G3WZT, *RadCom*, November 1996 – February 1997.

[3] 'Guidelines for the Design of Semiconductor VHF Power Amplifiers', John Matthews, G3WZT, *RadCom*, September and December 1988.

[4] 'Stub Filters Revisited', John Regnault, G4SWX, *RadCom*, November 1994.

# UNDERSTANDING YOUR S-METER

by Tony Martin, G4HBV

One of the intriguing aspects of radio is the tremendous variation that can exist in the paths between two stations in communication with each other. This is evident in the ability of low-power HF transmitters, using powers of 1W or so and in the hands of skilled operators, to cover what seem like amazing distances. In an article such as this, it would be impossible to describe in detail all the mechanisms involved, some human, some part of our natural environment and some determined by the equipment we use.

However, a simple review may reveal aspects that we have never thought deeply about. So what lies behind the S-meter reports that feature so prominently in amateur contacts?

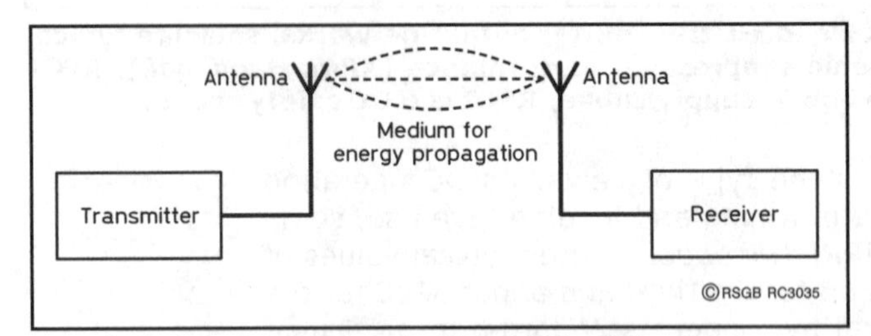


**Fig 1: The basic communications system.**

## SOME THEORY

The block diagram of **Fig 1** shows a radio system, with a transmitter, receiver, aerials and the transmission medium. For simplicity, let us consider a modulation system using SSB voice or Morse, a single path between transmitter and receiver, and an omnidirectional aerial erected in free-space (the transmission medium). We can think of free-space simply as a place where there are no obstacles or complications to the uniform expansion of the radiation fields (thus ignoring the effects of the ground and the ionosphere, for example).

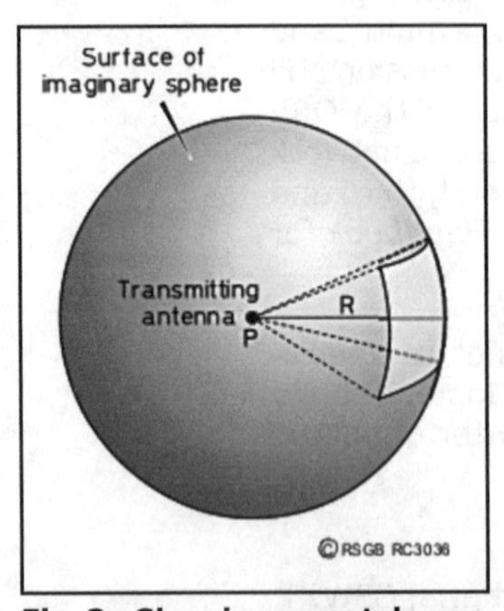


**Fig 2: Showing a patch one metre square on the surface of a sphere of radius $R$ and surface area $4\pi R^2$, receiving power from the transmitter at the sphere's centre.**

We can imagine the radiated power, P, from the transmitting aerial as spreading out in all directions. Thus, at a point where the receiving aerial is located, which is a distance, R, from the transmitter, the total power available is spread over the surface area of an imaginary sphere of radius R, as shown in **Fig 2**. In simple mathematics, the power in a unit area of such a surface is the total radiated power divided by the surface area of the sphere:

$$\text{Power per unit area} = \frac{P}{4\pi R^2},$$

where $P$ is the total radiated power.

Outside the immediate vicinity of the aerial, the radiated power in a unit area can also be calculated as the product of the $E$ (electric component) and $H$ (magnetic component) field strengths at that point. In free-space, the field strengths $E$ and $H$ of this radiated energy have a fixed relationship, known as the plane-wave impedance. This impedance, defined as $E/H$, we can call $Z$, and so we can also represent power in a unit area at a receiving point as being equal to $E^2/Z$, where $Z$ is a constant. Thus, the power per unit area is given by

$$\frac{P}{4\pi R^2} = \frac{E^2}{Z}.$$

By making $E$ the subject of the equation, we find that

$$E = \sqrt{\frac{PZ}{4\pi R^2}} = \frac{1}{R}\sqrt{\frac{PZ}{4\pi}},$$

showing that E:

- varies with the square root of the power;
- varies with the reciprocal of the distance from the transmitting aerial.

These two factors are represented conveniently by the graph of **Fig 3**, which relates power, field strength and distance, under the free-space conditions we are considering here [1]. For the moment, ignore the right-hand decibel scale.

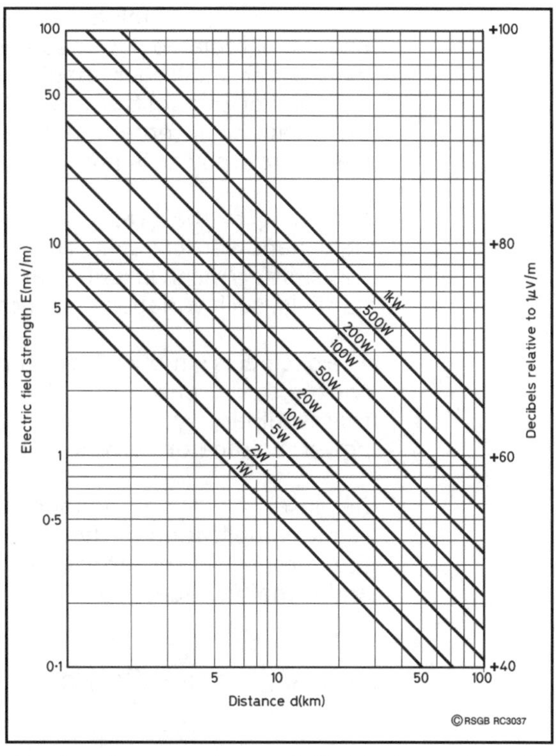


**Fig 3: The variation of field strength with distance and power for an omnidirectional aerial in free space [1].**

## VARIATIONS

The field strength and distance axes are drawn on logarithmic scales, which allow the graph to show the variations as simple proportional ones, which otherwise, with linear scales, would be curves. The essential points to grasp are:

- For a fixed value of power, the field strength, $E$, follows a linear characteristic according, as we have noted, to the reciprocal of the distance between aerials. Thus, for a 1W transmitter at 1km distance, $E \sim$ 5mV/m. At 10km, $E$ becomes $^1/_{10}$ of this at ~0.5mV/m.
- The graph also shows how the field strength, $E$, changes according to the square root of the power. On the graph, this is evident by examining the vertical line at the appropriate distance, and noting that this line intersects the two different 'power' straight-line characteristics at different points on the field strength axis.

Thus, for 10W at 10km, $E \sim 1.8$mV/m, and for 100W, $E \sim 5.7$mV/m. Note that the power changes by a factor of 10, and that the new value of $E$ is found simply by multiplying its original value by the square root of 10 (which is 3.16):

3.16 x 1.8mV/m = 5.7mV/m.

To fix this in our minds, we have to return to the concept of a fixed amount of radiated power spreading outwards from the transmitting aerial, over larger and larger imaginary spheres as the distance increases. Also, in terms of transmitter power, only a small part of the spherical surface area can be intercepted by the receiving aerial, so that increasing transmitter power is not as effective as might at first be thought.

Returning to the graph, we can now consider the decibel scale on the right-hand vertical axis. The tremendous variations that can exist in the transmitter/receiver path have already been shown. Without the use of decibels, such variations are inconveniently large, and calculations difficult.

The decibel is simply a logarithmic measure of the ratio of two quantities. For the same reason that the logarithmic scales of Fig 3 made it easy to represent large values, the decibel is used to express changes in electrical power, voltage or current as a logarithm.

Because of the way voltages and currents relate to power, there is a slight complication in the use of decibels for these three quantities. For the purposes of this article, we are interested in changes of field strength and signal voltages, so:

$$\text{Change} = 20\log\left(\frac{A_2}{A_1}\right) dB,$$

where $A_2$ and $A_1$ can represent either field strength or signal voltage. Since the decibel is a ratio, $A_1$ can be given a specific value (eg 1mV/m) which, in turn, enables the decibel to represent specific values. Hence, the right-hand vertical scale of Fig 3 shows decibels relative to a field strength of 1mV/m.

Now let us consider what I call the 'QRP equation', which indicates that quadrupling transmitter power results in only about one S-point increase at the distant receiver. The decibel scale on Fig 3 is not precise enough to check this directly.

Instead, using the graph, check that a 5W transmitter produces almost 2mV/m at 6km distance. Now check that a 20W transmitter produces almost 4mV/m at the same distance and convince yourself that we have doubled the field strength (remember it varies as the square root of the power). The increase in field strength works out at 6dB, often quoted as the 'standard' S-point. We have just established the viability of QRP operation.

In theory, if your 100W signal is being received at S9, reducing the power to 25W should produce S8. Reducing it to about 6W should produce S7, and 1.5W should produce S6 - in theory, not necessarily in practice!

## COMPLICATIONS

At the receiver end of the path shown in Fig 1, we have now to translate field strength at the receiving aerial into signal voltage at the receiver input terminals. This is rather complex because, although the value of the induced voltage per metre of the receiving aerial in our system does correspond to the field strength, other complications arise. Firstly, the induced voltages are distributed along the aerial and cannot just be added together. The induced voltages cause currents to flow in the aerial, which then re-radiates because of them. This means that some of the intercepted energy cannot be delivered to the receiver, and can be accounted for by assuming that the aerial's equivalent radiation resistance appears in series with the receiver input. We may simplify matters, however, by just accepting that signal voltage, delivered to the receiver, is proportional to the incident field strength, *E*.

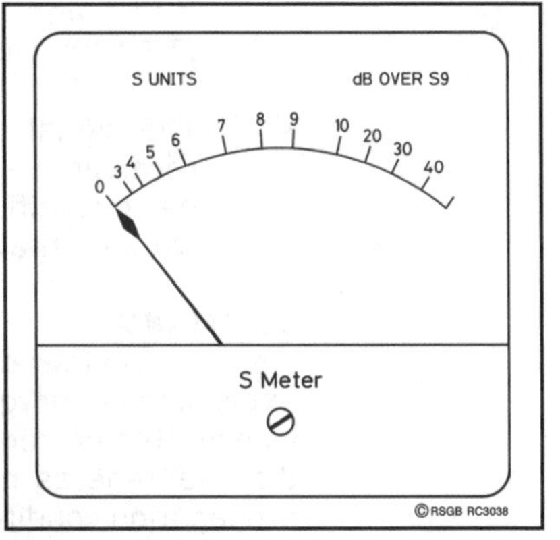


**Fig 4: The logarithmic S-meter display of signal voltage, based on decibels, allows what would otherwise be a gigantic linear variation to be conveniently displayed. Calibration of the meter requires a signal voltage to be assigned to one of the S-points, typically 50mV at S9 [1].**

It should now be apparent why traditional S-meter scales always look like **Fig 4**, because we are measuring in logarithmic terms.

The free-space conditions and simple aerials that we assumed, however, do not describe what actually happens in real life: obstacles complicate the fields; receiving aerials intercept the wavefront obliquely; the polarisation of the field may change.

The other assumption we made, that of considering the transmission path only one way, also needs further comment. We would normally expect similar performance of the system if the transmitter and receiver positions were reversed. There are some reasons why this basic property of radio-wave behaviour seems not always to apply, and this is sometimes evident to operators at the ends of a path.

- We can use the term 'reciprocity' to describe the comparative performance of two aerials at the ends of a communications system. Put simply in this context, reciprocity means that if the two aerials only in Fig 1 were swapped, and that the electrical conditions at the newly-assigned transmitting aerial were as before, the performance of the system would remain the same. The proviso of electrical conditions at the transmitting aerial remaining unchanged rules out as absurd any idea of interchanging a ferrite-rod receiving aerial with a medium wave broadcast aerial. Reciprocity of two aerials cannot be assumed under conditions of ionospheric propagation, where path conditions may differ in each direction. Reciprocity will become most evident in the VHF and UHF bands by similar S-meter reports being given both ways.

- With high-performance transceivers commonly in use at most amateur stations, the question of compatibility between receiver and transmitter capabilities at each end of a path is less likely to be evident. Such compatibility can be upset, either by addition of receiver pre-amps or transmitting linears; this situation often applies if VHF and UHF stations are upgraded, when both a linear and a pre-amp will be needed.

- The signal-to-noise ratio, determined by local conditions at the receiver, may differ significantly at each end of the path. This may be so significant that, for instance, stations attempting difficult DX working may need to erect separate receiving aerials.

## SUMMARY

We have reviewed a basic feature of radio-wave behaviour, the spatial expansion of energy from a transmitting aerial. We also saw that reducing transmitter power down to QRP levels of a watt or so was not such a disadvantage as might at first have been thought. This means that, if propagation conditions are good and QRM is low, successful QRP operation is possible and will depend on the effectiveness of the aerials and the skill of the operators.

We saw how logarithmic S-meter scales are necessary in order to display the tremendous variation in signal levels possible at the receiver. Finally, it was briefly explained why disparities sometimes arise in the S-meter reports exchanged by stations in a QSO.

## REFERENCE

[1] *Radio Communication Handbook*, RSGB.

# Chapter 5 GENERAL INFO CHOOSING COAXIAL CABLE

**by Paul Balaam, G4LNA**

There are various ways that an antenna can be fed. The simplest method is via direct connection from the rig, ie plugging an antenna straight into a transceiver.

This may be acceptable on a VHF handheld or for listening around the HF bands, but not for any *serious* receiving or transmitting. In fact, it is particularly risky when transmitting, because a large amount of RF can be induced into nearby objects, including the rig and the operator. This can cause problems such as burnt ears if you are using headphones, a burnt lip from the microphone, and RF getting back into the audio. This does not mean to say that such a system cannot be used at all, but it requires a lot more knowledge of how antennas function to make sure the RF goes where it's supposed to.

## CONSTRUCTION

Coaxial cable is made by surrounding one conductor with another, as can be seen from **Fig 1**. An insulating material is used to space the conductors a constant distance apart. Coaxial cables have characteristic impedances, and these are defined by the relative diameters of the inner and outer conductors, and the distance and material between them.

Unfortunately for the newcomer, several words are used interchangeably to refer to the various parts of coaxial cable.

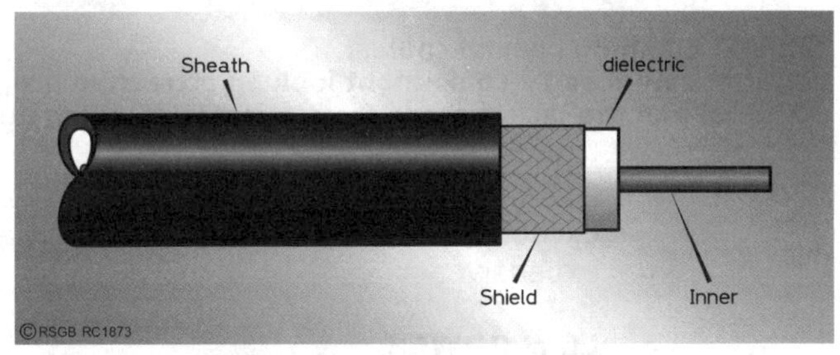


**Fig 1: The construction of coaxial cable.**

The 'shield' is often referred to as the 'outer', the 'braid' (even though it is not always braided), or the 'screen' (or 'screening').

The inner is often referred to as the 'core'.

The dielectric is often referred to as the 'insulation'.

The sheath is often referred to as the 'jacket'.

A good-quality coax will have a thick, closely woven shield; a poor-quality coax will have a thin, wispy shield, which offers less protection from the escape or ingress of signal along its length.

**GETTING STARTED**

So, you want to keep the RF well away from you, the rig and the house. This has the added advantage of keeping the antenna away from a lot of man-made noise. In turn, keeping the antenna away from the house will help to reduce EMC problems.

If you are using a dipole for one particular HF band, or one of the many multi-band antennas (eg a G5RV, trapped dipole, or trapped vertical), the easiest way to feed it is with coaxial cable. The problem a newcomer is likely to encounter is trying to decide which coax to purchase from the large range available. **Table 1** details some common types, but is by no means an exhaustive list.

| Cable type | Diameter (approx mm) | Nominal Impedance (Ω) | Attenuation per 10m(approx dB) | | Cost per metre (approx p) |
|---|---|---|---|---|---|
| | | | 100MHz | 1GHz | |
| URM-43 | 5 | 50 | 1.3 | 4.5 | 25 |
| RG-58CU | 5 | 50 | 2.0 | 7.6 | 25 |
| URM-6 | 10 | 50 | 0.68 | 2.5 | 94 |
| RG-213AU | 10 | 50 | 0.62 | 2.63 | 83 |
| RG-59CU | 5 | 75 | 1.3 | 4.6 | 32 |
| Low-loss TV | 6.7 | 75 | 0.82 | 2.8 | 29 |
| CT-100 (satellite TV) | 6.7 | 75 | 0.6 | 2.0 | 40 |
| UR-203 | 7.3 | 75 | 0.75 | 2.7 | 30 |

**Table 1: Common coaxial cables.**
**Caution: Low-loss TV coax might look an attractive proposition, but the screening is often marginal and the jacket made from a material which deteriorates after a few years.**

Like anything in this world, you get what you pay for. Coax cable is no exception; the lowest loss *and* heaviest duty is going to be the most expensive.

**LOW POWER**

Let's put things into perspective. If your interest is only in receiving, transmitting low power at HF, or if your pocket is not very deep, why spend vast sums of money when there is a cost-effective alternative?

If you are using a dipole, CT-100, UR-203 or RG-59CU will be a close match to the 72Ω impedance of the antenna. Even at VHF I have used TV coax on simple dipoles with great success.

**HIGH POWER**

If you intend running high power, say 100W or more, you will have to invest in an appropriate coax. Coax which can handle the power, without going up in smoke. This is particularly true if the antenna-to-coax impedance match is not to good, which is likely to cause higher voltages to appear on the coax. Certainly the cellular polythene used for the dielectric of TV coax will not withstand very high voltages or being heated to any extent.

Nearly all commercial rigs these days need to be matched to 50Ω. This should not pose a problem if the system is for receive only, as a slight mismatch will almost certainly not be noticed. Some QRP rigs have built-in antenna matching units, but if they don't, these units can be easily built using receiver-type variable capacitors.

However, if you are intending to run a reasonable amount of power, consideration should be given to obtaining a matching unit if the rig doesn't have one.

RG-59CU would be a good choice at HF, for feeding a home made dipole with 100W. However, bear in mind it is a balanced antenna, so a 1:1 balun is required at the feed-point. Baluns can be purchased from various suppliers, or built quite easily and inexpensively.

A good source of reference for their construction *is HF Antennas For All Locations*, by the late Les Moxon, G6XN [1]. For longer runs and/or higher powers, consider using UR-203 or something even more substantial.

**50Ω ANTENNAS**

So, what if you have a commercial HF antenna? Most of these are 50Ω impedance. Again, Table 1 shows that URM-43 or RG-58CU would be good choices of coax, as these are competitively priced and have an acceptably low loss. They are perfectly good for carrying up to 100W. The only exception to their use would be if you needed to carry the RF a particularly long distance, in which case consider URM-67 or RG-213AU.

On VHF, if you use RG-58CU you will have a loss of nearly half an S-point at 144MHz for every 10m of coax; worse on 432MHz. Consequently, URM-67 or RG-213AU should be used for long runs.

Whatever the frequency, if you run high power, use heavy duty coax (URM-67 or RG-213AU).

**REFERENCE**

[1] Available from RSGB.

# EXCERPTS FROM 'IN PRACTICE'

**by Ian White, GM3SEK**

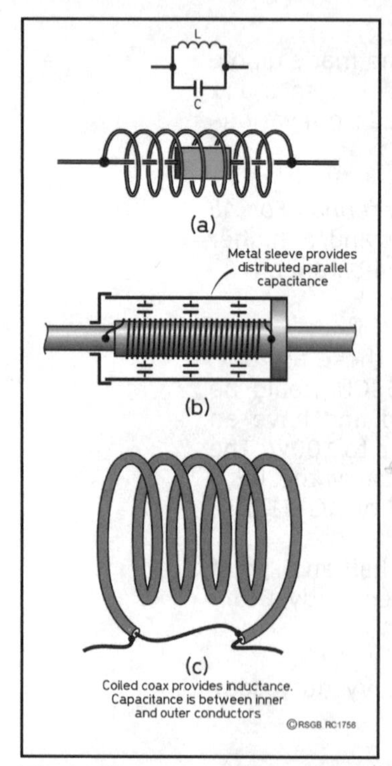


**Fig 1: Three examples of trap construction: (a) normal inductor and capacitor, (b) tubular construction, using distributed capacitance from metal sleeve, (c) trap made from coaxial cable.**

## WHAT DO TRAPS DO?

**Q:** ***I have a trap dipole which works at two frequencies. Is it legitimate to view the trap as a low-pass filter for the lower frequency, and a load on other bands?***

**A:** The trap is not a filter in the classical sense, because there is no common ground connection between input and output which a filter would require. The trap is better regarded as a load impedance, which will differ between the two bands. Traps are parallel connected L/C circuits in many different kinds of construction, for example as shown in **Figs 1(a) – (c)**.

On the higher-frequency band, the trap is usually designed to be parallel resonant, so it acts as a near-resistive load of hopefully several thousand ohms. Like any parallel resonant circuit, it will look inductive at a lower frequency so, on the lower band, the trap will provide loading inductance, which will make a shorter resonant aerial. On bands higher than the trap's parallel resonance, the trap will look like a capacitive load, which will make the resonant length longer. In all cases, there will be some resistive losses because the *Q* of the components is not infinite.

**Q:** ***If the trap is really a 'trap', is it fair to assume that it simply stops (truncates) the current distribution at the higher frequency, as opposed to tapering it?***

**A:** Typically, the traps are inserted at what would be the ends of a half-wave dipole for the higher-frequency band, so the current at those locations is naturally heading towards zero anyway. Although the trap is parallel-resonant on the higher-frequency band, there is a small level of current in the ends beyond the trap. Aerial modelling programs, experiments with a current probe, and plain common sense all agree on that. The current tapers away to exactly zero at the physical ends of the aerial - it obviously has to - but with good traps (that present a very high impedance at or near resonance) the currents in the end sections are so small that the radiation is negligible (**Fig 2(a)**).

On the lower-frequency band, where the traps are acting as loading inductors located some way in from the ends, the level of current at the traps/inductors is quite significant. The models predict a downward kink in the current profile at this point, again tapering away to zero at the physical ends of the aerial (**Fig 2(b)**). The effect of the kink is to shorten the normal smooth quarter-wave current profile to make it fit into the physically-available length of wire.

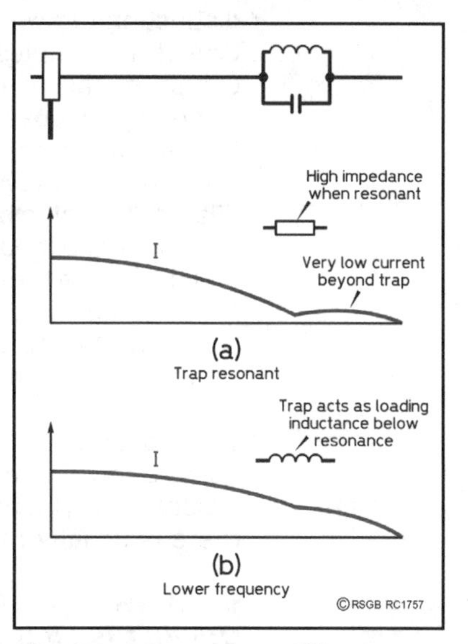


**Fig 2: Current profiles on a trapped element, (a) at trap resonance, (b) on a lower-frequency band where the trap acts as a loading inductance.**

## DESOLDERING ICS – AND SAFETY

***Here's yet another episode in the saga of desoldering multi-leg ICs. Hopefully, one of the techniques described will be the one for you. Also, keep control over the risks involved.***

This tip is from the TV and video repair business (Roy, G0FDU, and his partner). Take a $^3/_4$in (19mm) paint brush and cut the bristles down until they are only about $^3/_4$in long. If possible, stand the circuit board vertically and spray the connections of the IC to be removed and the area around it with something oily such as switch cleaner or WD40. This is to prevent the hot solder from sticking to other parts of the print - but don't overdo the spray. With a hot soldering iron, heat each pin of the IC to be removed until the solder is well molten, then immediately brush it away. Give each pin in turn the same treatment, then with a little leverage under the IC it should come away easily. After a little practice they come out cleanly, with no damage to either the printed circuit board or the IC. After removing and replacing the IC, brush away all the bits of solder, then clean the spray residue off the board. Thanks to Ron, G3CUR, for passing on this trick.

Incidentally, this method would not apply to circuit boards that have plated-through holes, because solder needs to be removed from inside the holes to avoid damaging the fragile plating when you extract the IC. One of the best methods for deep solder removal is to use a vacuum solder-sucker, as described in 'In Practice', August 1997. G8BMI passes on a useful hint to improve the vacuum seal around the solder pad (August 1997, Fig 4): a short piece of silicone rubber sleeving over the tip of the nozzle. You can probably salvage many of these sleeves from good-quality hand-wired junk - silicone rubber sleeves being the soft, flexible ones that can withstand soldering-iron temperatures. You can also buy them, of course, but only in large quantities and at high prices. They are meant to be slipped over the stripped ends of wires, using a lubricant and a special three-pronged tool to stretch them open. Without these you may have an interesting time stretching the sleeve over the solder-sucker's nozzle, but it can be done. Finally, trim the sleeve back with a scalpel to leave a short, flexible end which seals well onto the printed circuit board.

G0XAU rightly points out the hazards of some of these component removal techniques, especially the ones that require drastic measures such as hot-air guns or blowtorches. These methods can generate some nasty fumes. Equally, it is very easy to inhale lead and resin flux fumes in ordinary soldering operations, especially when working close-in with miniature components. DIYers of all types tend to do risky things that would be illegal in the workplace, but we also need to make a sensible distinction between occasional activities and repeated day-long exposure. One of the simplest, cheapest methods for clearing fumes is to use an old computer fan to direct a current of air across the area where you are soldering. It doesn't have to be a strong blast, just enough to send the fumes away from your face. For larger-scale or heavier-duty requirements, use a room fan. However, this has only a short-term benefit if it merely fills up a small closed room with fumes. G0XAU also points out that you can use a household vacuum cleaner fitted with hoses to suck away fumes and discharge them outdoors, but be careful where you direct the exhaust. If you're generating a lot of fumes, it's preferable to work outdoors or in a well-ventilated area (eg a garage or workshop with the door open and a through draught), or decide to do the job a different way.

The most effective safety precaution of all is to be aware of what you are doing and think ahead, to minimise the risks involved. Don't let 'tunnel vision' suck you into doing something stupid: whatever you want to do, there is usually more than one way to do it. Choose the safer way.

## S-UNITS

***Q: what is an S-unit? Is there any standard?***

**A:** The original definition of signal strength was in words. The original RST (Readability, Strength, Tone) system of reporting defined nine levels of signal strength, as shown in **Table 1**. Signal strength reports were largely guesswork... and who knows how those nine carefully-graduated English descriptions came out in other languages? When S-meters began to appear, they gave a slightly more objective indication based on signal strength alone - but the calibration was still guesswork.

| | |
|---|---|
| **S1** | Faint signals, barely perceptible |
| **S2** | Very weak signals |
| **S3** | Weak signals |
| **S4** | Fair signals |
| **S5** | Fairly good signals |
| **S6** | Good signals |
| **S7** | Moderately strong signals |
| **S8** | Strong signals |
| **S9** | Extremely strong signals |

**Table 1: The original signal strength scale.**

In an SSB/CW/AM receiver, the S-meter is connected to the AGC circuit which holds the audio output reasonably constant by controlling the gain of the IF and RF amplifier stages (**Fig 3**). The stronger the signal, the more AGC voltage it generates, and the S-meter indicates this. (In an FM receiver, the S-meter often works on a different principle, but still indicates signal strength.)

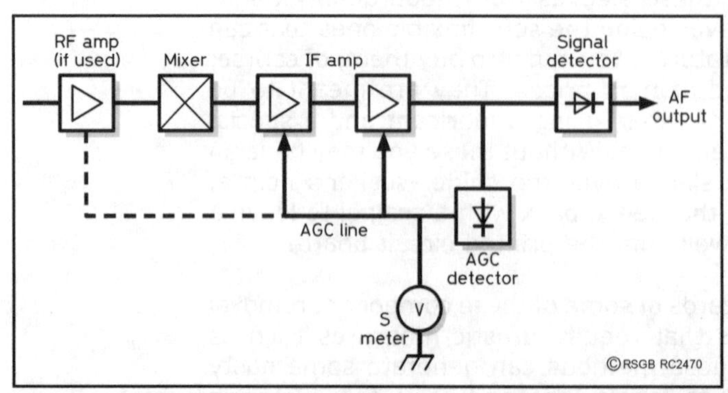


**Fig 3: The AGC loop in a receiver keeps strong signals at a constant output level. The S-meter measures signal strength in terms of the AGC control voltage that needs to be applied.**

There are two big problems with this. One is that AGC is usually not applied on very weak signals, because turning down the IF/RF gain will degrade the signal-to-noise ratio. This can lead to interesting situations where a signal is perfectly copiable in the absence of QRM, but the S-meter hasn't started to move off the stop - it doesn't seem right to give someone a report of 'Readability five, strength zero', does it? Manufacturers usually avoid this problem by being very vague about what happens below S1, rather like your car's speedometer below 10mph. The other problem is what to do with signals that are stronger than S9: what does 'Stronger than extremely strong' mean... if anything?

The interesting property of most S-meters is that they give a fairly linear decibel scale. This is because the AGC-controlled amplifiers have a fairly linear relationship between gain in decibels and the applied AGC voltage that is driving the S-meter. The accuracy of this relationship is by no means guaranteed, because it depends on the design of the controlled amplifiers and the way that AGC is applied, but it does lead to the notion that each S-unit should represent the same number of decibels increase in signal strength. And as we know, after reaching S9 the scale then continues in plain decibels: S9 +10dB, +20dB and so on, up to maybe +60dB. However, decibels are always a *relative* measurement, ie a measurement of power *ratio*, not simply power. This means that the entire S-unit scale also needs to be referenced to some absolute signal level.

So how many decibels *is* an S-unit, after all that? The answer is: it varies! It varies between manufacturers, it almost always varies along the scale of the S-meter, and quite possibly it varies between different examples of the same receiver. There is an IARU standard that each S-unit represents a received signal level change of 6 dB, but a glance at Peter Hart's receiver reviews shows that real-life S-units vary enormously [1]. Typically, the manufacturer aims to get the 20dB step between S9 and S9+20 about right, but below S9 the 'value' of one S-unit gets smaller and smaller, and can be less than 2dB per S-point at the bottom of the range.

The IARU recommendation is that S-meters should be referenced to S9 = 50µV for HF receivers and S9 = 5µV for VHF receivers. The difference reflects the greater sensitivity of VHF receivers, for a 5µV signal is indeed 'extremely strong' at VHF. However, Peter Hart's review measurements once again show that the 'S9' levels of commercial HF receivers can be anywhere from 250µV down to less than 20µV. The other important factor is whether the internal preamp is on or off - if it is on, all signals jump up the S-meter scale. Should you then modify the signal strength report you give to the other station? The answer of course is no, because the other person's signal strength cannot rationally depend on which buttons you choose to push at the receiving end.

Why are commercial amateur S-meters so bad? Well, first of all because the IARU has no authority to enforce a standard on manufacturers. But the manufacturers themselves will justly blame the market. It costs money to produce an S-meter with a truly linear decibel scale (although the more 'digital' the receiver becomes, the easier it would be). Also, an S-meter with a genuine 6dB per S-point calibration would seem very sluggish compared with existing meters, promoting the entirely mistaken rumour that the receiver is 'deaf'. Finally, there's the human factor that nobody likes to seem mean about the signal reports they give.

The only logical conclusion is that none of this makes sense! If you look for deep inner truths from your S-meter, it will drive you crazy. In the end, most experienced amateurs almost ignore the S-meter and give subjective reports based on some personal version of Table 1... so much for progress.

## AERIALS IN SUBURBIA

***Q: How can I fit a multi-band HF aerial into my small suburban plot of land?*** **A:** RSGB receives many pleas for help of this kind. While the Society can't give too much individual detailed advice, and neither can 'In Practice', here are a few ideas that will help. **Fig 4** is a typical small suburban plot, with both a front and a back garden but with the house set closer to the front. Fig 4 also shows a typical attempt to fit a multi-band HF aerial into the available space. What's good is that somebody has already learned the first lesson: Don't insist on any particular length of wire; just put up as much as you reasonably can, and centre-feed it with open-wire through a balanced ATU.

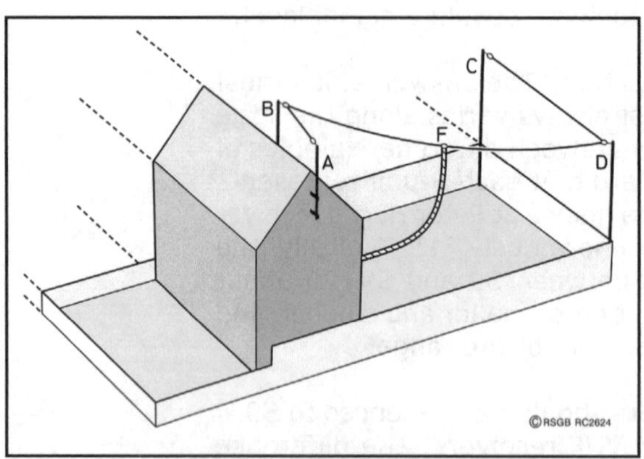


**Fig 4: Typical small suburban plot with problems in erecting a multi-band HF aerial.**

Too many amateurs – especially beginners – are fixated on certain 'magic' lengths of wire, such as 102ft for a G5RV, or 51ft for a 'half-size G5RV' with loading coils for 80m. This may indeed be a viable approach if 51ft is all you have space for, but it can also lead to silly situations where people who already have limited space put up *even shorter* aerials than they could actually fit in. Just put up what you can, feed it with open-wire, and away you go. With a balanced ATU, almost any length can be made to work on all bands, or at least down to the frequency where the total length is about 0.2λ. 'In Practice' for August and September 1999 gave several ideas for handling difficult feed impedances. Don't worry about the radiation patterns on different bands; the main point is to be on the air!

Unfortunately the example in Fig 4 doesn't succeed quite so well with lesson 2: Get the wire as high as you reasonably can – and especially at the centre. Worry less about the ends, because they don't contribute much to radiation. Fig 4 does not make good use of the available supports A, B, C and D. The poles A and B are the best assets, because they take advantage of the height of the house, but in Fig 4 they are rather wasted in supporting the end of the wire. This also leaves the feed-point F unsupported and sagging, which is exactly the opposite of what you should be aiming for. For cramped sites like this, where you can't afford the space to put back-guys on the support poles, the wires are often quite slack, so an unsupported feed-point can easily be dragged down several feet by the weight of the feeder.

But the most obvious problem with Fig 4 is that it only uses half the property - the front garden is completely wasted! Of course there are limitations on the use you can make of a front garden, especially if the fronts of neighbouring properties are relatively open, but don't assume that it's a complete no-go area. A 'stealth' wire running down from pole A to the corner of the front garden transforms the whole situation - see **Fig 5**. Now the feed-point F can be directly supported on pole B, and the central length A-F-C is high and in the clear.

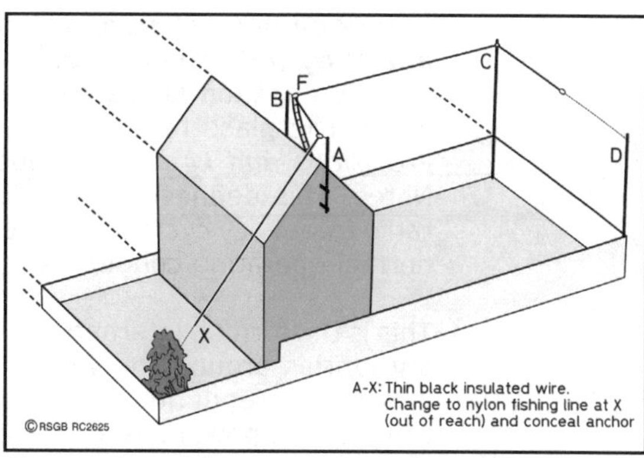


**Fig 5: A 'stealth' wire into the front garden can transform the situation.**

Moving the next support along from D to C means that the whole aerial in Fig 5 is much less severely Z-folded than in Fig 4. The potential benefits of adding that extra wire AX are so great that it's worth trying really hard to do it. The most important requirement is safety. The open end X is a high-voltage point, so it needs to be well above head height, but at the end of the wire you can change over discreetly to nylon fishing line with no need for an extra insulator. The combination of thin black-insulated wire and almost transparent fishing line is very hard to spot. Even though the rest of the aerial is fairly conspicuous, that doesn't ruin your chances of adding extra wire using 'stealth' techniques – in fact, it rather improves them! The possibilities for anchoring the line at the front corner of the plot depend very much on your situation. For example, you could plant a young tree (which will of course need a stake to help it grow straight), or erect a trellis. It's amazing what can be done with a little imagination... and also a little nerve.

The exact dimensions for the aerial will depend on the individual plot and the supports available. In Fig 5 you'd probably need to start from wherever you could position point X. Then work upwards and determine first how much wire you could fit into the length X-A-F. Then you'd need an equal length of wire in the opposite leg, so you may or may not need the whole of the length F-C-D. As most people already know, it's also possible to drop the ends of wires directly down to save a little more space. With such highly asymmetrical layouts as Fig 4 or Fig 5, you should expect some problems with common-mode radiation from the feed-line, which could bring EMC problems unless you choose the feeder length and/or the ATU configuration appropriately - see the September 1999 column. Finally, it should go without saying that you must never run aerial wires anywhere near overhead power lines, for reasons of both safety and EMC. Likewise you shouldn't run aerial wires close to overhead telephone lines, because that too is inviting EMC problems. Even so, I hope this example has given you some ideas how to get out better on HF from a typical small British plot.

**PEP TALK**

***Q: I have been away from amateur radio for 10 years, and the power limits have changed. I am no longer clear what 'Peak***

***Envelope Power' means, but I also asked several active amateurs and they didn't seem too clear either!***

**A:** This question keeps coming around - or rather, it never really goes away. The place to go for a definition of PEP is *BR68,* Ofcom's *Terms, Provisions and Limitations* booklet, currently paragraph (e) (iii) of the Notes. This defines PEP as: "The average power supplied... during one radio frequency cycle at the crest of the modulation envelope taken under normal operating conditions".

This is the formal, internationally-agreed definition used by many licensing authorities around the world. To explain that definition, note first that 'average power during one radio frequency cycle' is what we more familiarly call just 'RF power'. That means the average over the *whole* RF cycle.

There is a common misconception that PEP has something to do with the instantaneous peak RF voltage - it hasn't! The word 'peak' applies to the modulation envelope, not to the RF. We need to identify the moment during the transmission where the RF power is highest - that's the peak of the modulation envelope - but then we average the RF power over the *whole* of that biggest RF cycle.

One reason for the misconception is that you can measure RF power into a known load resistance by measuring the peak RF voltage as viewed on an oscilloscope, applying a factor of 1/"2, and then doing a $P = V^2/R$ calculation. But that's just a rather indirect technique for measuring RF power – it doesn't change what PEP *means*.

There was a long, long Internet discussion about the meaning of PEP, which still left some people complaining that PEP has something to do with peak RF voltage or the 'instantaneous' power at the peak of the individual RF cycle.

Where all efforts at explanation and persuasion fail, the only remaining answer is the brutal one: the definition that the licensing authorities use is correct and *it's the law*, so you'd better change your mind, because they certainly aren't going to change the definition!

| Mode | Duty cycle | Peak Envelope Power |
|---|---|---|
| FM (F3E, F1B, F2B, J2E etc) | 100% | Same as carrier power |
| Keyed CW (A1A) | 50% typical | Key-down carrier power |
| SSB (J3E) | Depends on voice waveform and degree of compression | RF power* at highest peak of audio modulation |
| AM (A3E) | Depends on voice waveform, degree of compression, also percentage modulation | RF power* at highest peak and of audio modulation |

* RF power is defined as the average power supplied by the transmitter during one complete RF cycle.

**Table 2: Duty cycle and PEP.**

**Table 2** shows how the definition of PEP applies to common transmission modes. On CW and FM, every RF cycle is the same, so PEP is just the RF power. In a mode such as SSB or AM, where the RF power varies with the modulation, the meaning of the PEP is the RF power at the highest peak of audio modulation. My interpretation of a 26dBW (400W) PEP power limit is thus very simple: never let the RF power exceed 400W at any time during the transmission.

If you have a monitor oscilloscope, it can be useful for confirming compliance with the power limit. Under the present regulations, where you can transmit the same PEP on any mode, the best way to adjust to the legal limit is to tune-up on CW and adjust the output power to 400W as measured on a calibrated wattmeter. I don't recommend measuring RF power using the osciloscope – it's nowhere near as accurate as you'd like to think. Having set up a 400W carrier as measured by a wattmeter, simply mark the screen along the top and bottom of the ribbon of light.

Then switch to SSB, and adjust the modulation so that the spot never goes outside
of your marks.

## USING OPEN-WIRE FEEDER

***Q: 'AERIALS IN SUBURBIA' (see previously) was very interesting – but how do you bring the open-wire feeder into the house?***

**A:** It's much easier than the old aerial textbooks might suggest. Open-wire feed-line is sensitive to proximity effects, and needs to be kept several line-widths away from ground, walls or metal objects. But that doesn't mean you have to drill a pair of holes through the window or use those enormous old-fashioned feed-through insulators – it can all be done with coax instead. The vast majority of ATUs used with open-wire or 300/450Ω ribbon are basically unbalanced circuits (eg the familiar C-L-C T-network) with a ferrite toroidal balun built into the back. However, these baluns are subject to losses when terminated in reactive loads that are far from the design impedance - which is sure to be the case on some bands. In such situations, you'd do better to use an external balun. Therefore, you might as well use coax to bring the feed-line out of the house without the installation difficulties of open-wire, and then have the balun just outside (**Fig 6**).

The reason why an external balun is usually better than the one built into your ATU is that size is much less of a constraint. You can make excellent baluns for all HF bands using coax, with no ferrite core to saturate or overheat. Let's just recall what the balun is there for [2].

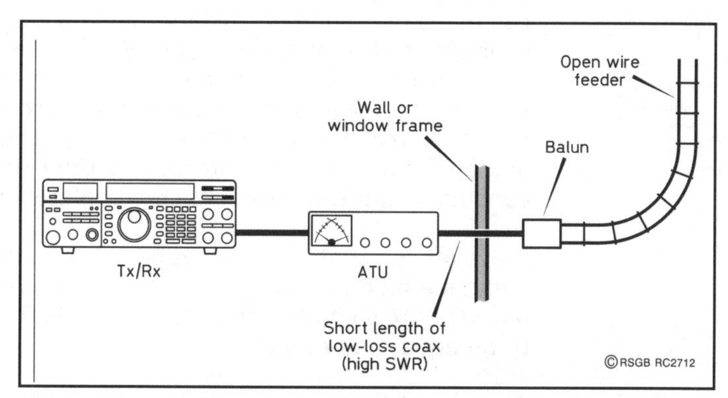


**Fig 6: Instead of bringing open-wire feed-line into the house, you can use coax and mount the balun just outside.**

'Balun' is short for 'balanced-to-unbalanced transformer' and that's basically all it

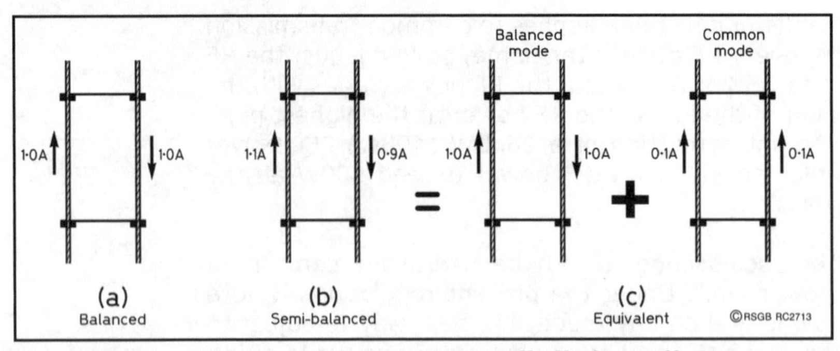


**Fig 7: Open-wire feed-line is not necessarily 'balanced'. It can carry both balanced currents and unwanted common-mode currents at the same time.**

does. The open-wire feed is a balanced system, meaning that both sides have – or *should have* – equal impedance to ground. Coax is an unbalanced system where one side generally is – or rather, *should be* – at ground potential. The balun simply converts between these two systems. (Note that a balun does not have to be an impedance transformer, although baluns can do this also.)

A truly balanced feed-line has equal and opposite currents in its two conductors (**Fig 7(a)**) but it's a fallacy to assume that this will happen automatically. Unless you actually take steps to *make* it balanced, a more realistic situation might be two conductors carrying unequal currents of say 0.9A and 1.1A (**Fig 7(b)**). This can be separated into balanced (equal and opposite currents) of 1.0A and a current of 0.1A that flows in the same direction in both conductors (**Fig 7(c)**). This current in the same direction is called the common-mode current, and your so-called 'balanced' feed-line will happily support both the balanced mode and the common mode together. The common mode current *shouldn't be there* because it brings a high risk of EMC problems and generally erratic behaviour. The purpose of the balun is to force this common-mode current as near to zero as possible, and there are two basic methods to do this [3, 4]. One is to force equal voltages at the terminals of the open-wire line, which is how the transformer-type 'voltage' baluns work. The other method, more direct and generally more effective, is to create a high impedance against common-mode currents while letting the balanced currents though. This latter type of balun is called a 'current balun' or 'choke balun' [4].

I'd recommend the choke balun for transitions from open-wire line. You can make one quite easily by coiling up the end of the coax, just before it connects to the open-wire (**Fig 8(a)**). Because of the skin effect which confines RF currents to the surfaces of conductors, separate currents flow on the inside and outside of the coax shield, and coiling up the coax doesn't affect the currents inside [4]. These inside currents are equal and opposite, strongly enforced by the close coupling between the inner conductor and the outer which completely surrounds it.

However, currents on the outside see the coil as an RF choke, which creates a high impedance. When open-wire feed-line is connected to the output of this choke balun, the currents in its two conductors are forced to be equal and opposite.

Any would-be common mode current sees only the high impedance of the RF choke, so common mode currents are effectively suppressed.

Enforcing zero common-mode current at the bottom of the feed-line still doesn't guarantee the same all the way up to the aerial, especially if the aerial is markedly asymmetrical and the feed-line is an odd number of quarter-wavelengths long [4, 6], but it's the first and biggest step towards a truly balanced feed system.

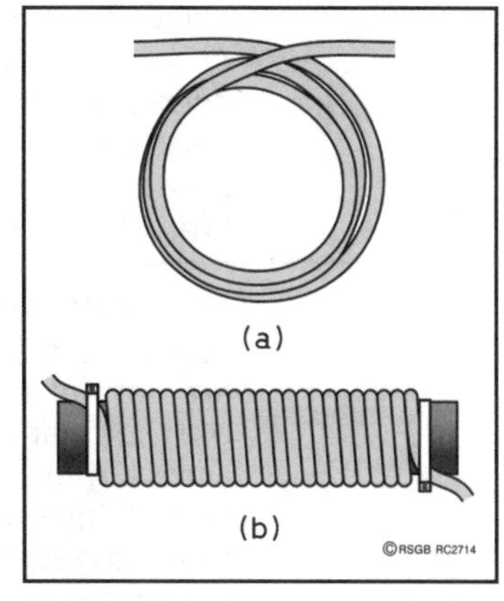


**Fig 8: Two ways to wind a common-mode choke (or choke balun) out of coax.**

So how do we make a choke balun from coax? Physical size is important here, because you probably should be using large coax. The impedance at the bottom of open-wire feed-line can be a lot different from the 50Ω characteristic impedance of coax, so the SWR on the coax will generally be very high – in other words, there will be unusually high currents or voltages. This means the resistive or dielectric losses of the coax will be much higher than the normal matched loss. Therefore you should be using low-loss coax such as RG-213/URM-67 (or better) for the balun and then all the way to the ATU, and also the total run of coax to the ATU should be as short as possible. (QRPers, this means you too! *Nobody* can afford the extra losses in thin coax carrying a high SWR.)

There are two basic ways to wind a common-mode choke out of coax: in a flat coil (Fig 3a 8(a)) or as a solenoid on a former (**Fig 8(b)**). In both cases the inductance needs to be sufficient to create a high impedance against common-mode current at the frequency of operation, but the self-capacitance will be different for the two forms of construction. The self-capacitance appears in parallel with the inductance and, at a certain frequency, the choke forms a parallel-resonant circuit with extremely high impedance. Above this frequency the choke appears capacitive, but

| MHz | RG213/ URM67 | | RG58/ URM76[1] | |
|---|---|---|---|---|
| 3.5 | 22ft, | 8 turns | 20ft, | 6-8 turns |
| 7 | 22ft, | 10 turns | 15ft, | 6 turns |
| 10 | 12ft, | 10 turns | 10ft, | 7 turns |
| 14 | 10ft, | 4 turns | 8ft, | 8 turns |
| 21 | 8ft, | 6-8 turns | 6ft, | 8 turns |
| 28 | 6ft, | 6-8 turns | 4ft, | 6-8 turns |
| 3.5-30[2] | - | | 10ft, | 7 turns |
| 3.5-10[2] | - | | 18ft, | 9-10 turns |
| 14-30[2] | - | | 8ft, | 6-7 turns |

Notes:

[1] This thinner coax is not recommended for chokes connected to open-wire line, where the coax often has to operate at a very high SWR.

[2] The multi-band chokes are less critical in construction than the monoband chokes, but being a compromise they are also less effective.

**Table 3: Self-resonant choke baluns.**

still with quite a high impedance up to 2 – 3 times the resonant frequency. At lower frequencies, the choke appears inductive and its reactance decreases. With the right size and shape of winding you can thus make a very effective monoband choke balun. The resonance is actually quite broad, so the choke will cover several amateur bands with reasonable effectiveness.

The *ARRL Antenna Book* gives details of various choke baluns for the bands from 3.5 to 30MHz, reproduced in **Table 3**. The monoband chokes are very effective on the design frequency and on adjacent bands, while the multiband chokes are compromises aiming to spread the performance across more bands. In all cases the required length of coax is simply wound into a flat coil of the specified number of turns, which are then taped together and the coil allowed to hang freely. Some users favour the alternative solenoid type of construction, for example winding the coax on to a length of large-diameter plastic water pipe (the material is not at all critical for this application). I haven't seen any tables of data for various amateur bands using this type of construction, but if you short the ends of a trial coil you can easily find the resonant frequency using a GDO or SWR analyser. Yet another trick for a single lower-frequency band, where the size of a self-resonant choke might be too large, is to resonate it deliberately using a tuning capacitor (**Fig 4a 9(a)**) [6]. This is a narrow-band solution, but there's nothing to prevent you from adding another multi-band choke directly in series to cover the higher bands (**Fig 4b 9(b)**). Just take care to avoid magnetic coupling between the two chokes, by separating them by a few diameters or orientating them at right angles.

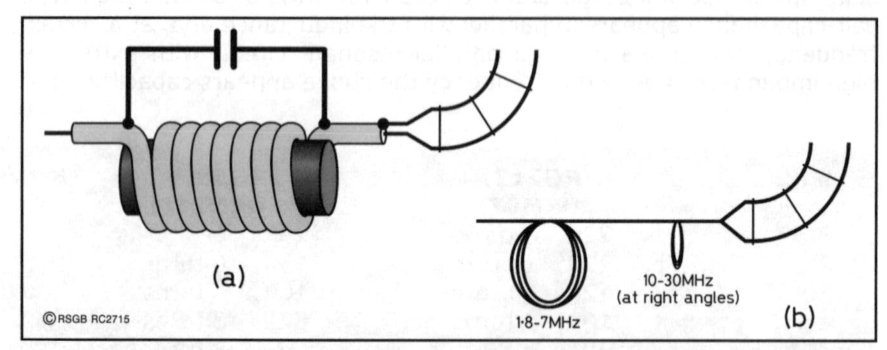


**Fig 9: (a) Monoband common-mode choke using a resonating capacitor. (b) Common-mode chokes for different bands can be cascaded (but avoid mutual coupling).**

Finally, remember that you can use these common-mode chokes as baluns for aerials with direct coaxial feed, or anywhere else that outer-surface currents on coax need to be suppressed.

### UP THE DOWNLEAD, DOWN THE UPLEAD

***Q: I recently mounted a multi-band HF vertical on the roof of the house, and am now finding considerable noise pickup from the shack PC. The PC is directly connected to the transceiver, but there is no interference when I use the horizontal G5RV, down the garden.***

**A:** It's always hard to diagnose RF interference (RFI) problems from a distance, because so much can depend on small details of the setup, but here goes. With luck, we can probably cure your received interference problem, and at the same time drastically reduce the risks of your transmitter causing RFI to other systems.

There are potentially several routes for interference to get from the PC into your receiver, but you can probably check - and dismiss - most of them fairly quickly.

1. Direct path from the PC to the radio, either by radiation or by conduction along leads. This is the least likely, because you say there is no interference when using the G5RV.
2. Direct radiation from the PC to the vertical. This is possible, but relatively unlikely - if the PC was radiating significantly, you'd probably be hearing something on the G5RV too.
3. Conduction from the PC along various leads, *up* the coax to couple into the HF vertical, and then back *down* the coax to the receiver. This is my best guess, so let's look at it more closely.

If you're not already familiar with these problems, the first question you'll ask is: 'What do you mean, the interference is going both up *and* down the coax?'. The answer is in the skin effect, which makes RF currents (at HF and above) flow exclusively on the *surfaces* of conductors. This means that a shielded cable such as coax has two personalities. On the inside, it's a coaxial two-conductor cable, just as you thought. But on the outside surface, quite separately, it behaves like a single fat wire.

The skin effect also applies to the metal case of your transceiver; although your transmitter generates quite intense RF currents, the skin effect confines them to the *inside* surface of the case. Likewise, all kinds of RF currents can flow on the *outside* of the case, but they won't get inside unless they're offered a way in. The coaxial aerial connector is *not* a way for surface currents to get either inside or out, because it is always securely grounded to the chassis on the rear panel, and this maintains continuity for the two separate RF paths. The currents on the inside of the coax are connected only to the inside

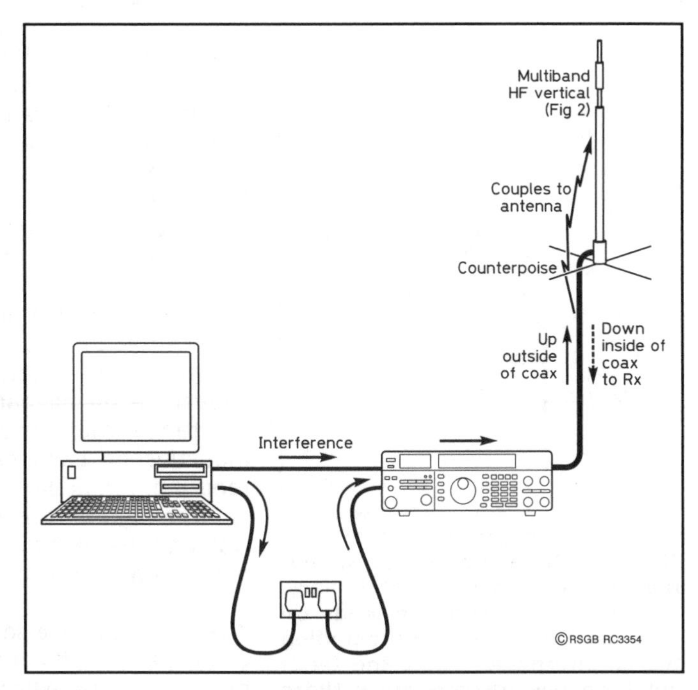


**Fig10: How interference can flow from the PC to the receiver, up the outside of the coax and then down the inside. The place to stop it is on the outside of the coax, by winding the cable to make RF chokes.**

of the transceiver, while the currents on the outside stay outside. The December 2001 column [8] showed what happens if those precautions are ignored: if a shielded microphone lead is not grounded where it enters the transceiver case, it can be a major cause of RF feedback problems. The cure, of course, is to ground the shield to the case, right at the connector.

Returning to the main question, the absence of problems with the second aerial shows that this particular transceiver is pretty RF-tight. What I think is happening is illustrated in **Fig 10**. The computer is somehow leaking its internal signals, quite probably through the monitor or its external leads. These signals are then flowing on various other leads in the shack, and on the outside of the transceiver case. But they aren't getting inside the transceiver directly; instead they flow *up the outside* of the coax and then jump across to the aerial. From there on they behave like normal received signals, and come back *down the inside* of the coax and into the receiver.

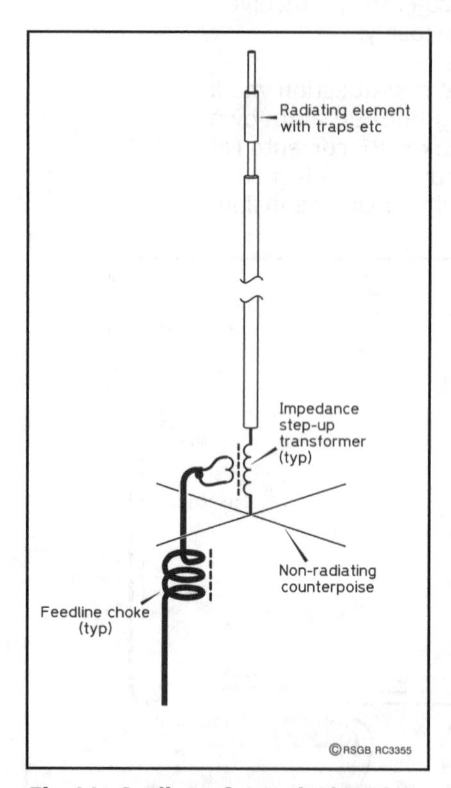


**Fig 11: Outline of a typical HF 'no-radials' vertical aerial – details vary, but the basic resonance mode on each band is a half-wave including the counterpoise. Since the feed-point is a long way off-centre, there is a significant risk of induced currents on the coax feed-line.**

How the signals 'jump across' from the coax to the aerial needs more detailed examination. The mechanism will be the same for both reception and transmission, and although we're interested mainly in the reception case (where the interference signal jumps across from the feed-line to the aerial), it's much easier to understand it in terms of transmission. So let's look how unwanted RF currents can get from the aerial to the outside of the coax feed-line. A typical 'no radials' HF vertical is in fact a multi-band half-wave dipole, fed a long way below its electrical centre [9]. The vertical element is the radiating part; the fan of short horizontal radials forms part of the half-wave resonance on each band, but is arranged to act as a non-radiating counterpoise (**Fig 11**). The counterpoise works because the RF current divides itself equally among all the radials, which are in pairs pointing in opposite directions. In the far field, radiation from each rod is cancelled by the radiation from its opposite number. The counterpoise is electrically short, so it looks capacitive on all bands. To cancel this capacitive reactance, the vertical radiator is arranged to look inductive on each band, giving a purely resistive feed-point impedance at mid-band. This impedance is quite high because the 'dipole' is being fed some way from its centre, so the designer usually provides a broadband step-down transformer to 50Ω.

The feed-point is also the place where the RF current can cross from the inside to the outside of the coax, because that's where the inside and the outside meet. So the designer also provides a feed-line choke – probably some thin coax wrapped round a ferrite toroid – which creates a high impedance to RF currents flowing down the outside of the main feed-line.

Ideally, this creates a clear separation; the coax acts only as a feed-line, and if its outer surface is dead to RF it will not try to behave as part of the aerial. But, in practice, this separation is not so clear-cut. The difficulty is that the aerial itself will induce considerable currents directly onto the outer of the coax. Unlike a centre-fed dipole, a 'no-radials' HF vertical is highly asymmetrical, so there is really *no* 'best' direction to run the coax away from the aerial to avoid induced currents. If you want to clean-up the outside of your coax, the only solution is to add your own feed-line chokes at some distance down from the aerial. Feed-line chokes are very easy to make – simply wind some coax into a small, flat coil (dimensions below) and tape it up.

But where should the chokes be? If the built-in choke at the feed-point is working well, this creates a high-impedance point which is a good place to start reckoning from. Remember that high-impedance and low-impedance points alternate every quarter- wavelength down the line, and that current maxima like to sit at low-impedance points. Therefore, the first maximum of the induced current would want to be at the first low-impedance point, a quarter-wavelength down from the feed-point. Since your objective is to make life as awkward as possible for these unwanted currents, you place your first choke exactly where the current maximum would want to be – an electrical quarter-wavelength down from the feed-point. How far from the feed-point is that?

Resonances on the outside of the feed-line will be greatly modified by the presence of nearby objects, but you have no way to quantify this, so you might as well relax and use the uncorrected free-space values.

Starting at the shortest wavelength of the multi-band HF aerial, 10m, a quarter-wavelength at 28MHz is about 2.7m, so that's where to place your first choke. Dropping down to the 20m band, a quarter-wavelength here is about 5.4m, so the best place for the second choke is another 2.7m down the line from the first one. This treatment will also be quite effective for the bands between 20m and 10m, because the presence of the two chokes makes it very awkward for induced currents to build up at those wavelengths either. Unless your vertical also covers bands below 20m, just those two chokes near the aerial should do the job. Another choke in the shack is always a good idea too [10].

Dimensions for the choke coils are given in the *ARRL Antenna Handbook*, and are intended so that the coil's self-capacitance forms a parallel resonant circuit with its inductance. If you're planning to cover the upper HF bands, a good choice would be 2.5m of cable, wound into 6 – 8 turns, like coiling a rope. Just tape it together and let it hang. If you're winding two identical chokes only 2.5m apart, as suggested above, wind the second one in the opposite direction from the first, so that you cancel-out the twists in the rest of the cable.

Now that your coax is hostile to outer-surface currents, either going up from the shack or coming down from the aerial, your whole RF environment should be much cleaner. On receive, you should be at much less risk of RFI from local sources. On transmit, you should also be at much less risk of RF feedback into your own equipment, or of injecting your RF into the overhead mains or phone lines if they pass near your coax.

This discussion has wandered in several different directions from the main question... but 'not all those who wander are lost'. To solve practical RF problems, you often need to understand several different topics before you can see how they all fit together.

In this case we've needed to know something about the skin effect, how RF shielding works, how 'no radials' HF verticals work, and how unwanted currents get on your coax... and then we can finally understand how to stop them.

**BEST VHF POLARISATION?**

***Q: Why is horizontal the standard polarisation for SSB/CW on VHF? I have heard that horizontal is better than vertical - so why is vertical used for FM and packet?***

**A:** The answer lies in the different uses that we make of these modes. VHF SSB and CW are generally 'DX' modes - they can of course be used for local contacts, but their useful range generally extends far beyond the horizon. DXing is best suited to fixed stations, where you can settle down and concentrate fully on the task of operating.

And when you are operating from a home station, or have gone to a portable site to operate for several hours, you also have the opportunity to use beam aerials – almost invariably Yagis – to extend your range further. For this style of operation, horizontal polarisation has significant advantages. VHF signals propagate beyond the horizon by a variety of means, but often the path involves some element of diffraction over the horizon [11]. Normally, radio waves propagate in straight lines but, when a wavefront strikes the horizon, diffraction is what allows it to bend 'around the corner' and continue to follow the earth's curvature instead of going straight on into space. There is some loss involved, but a horizontally-polarised wave suffers slightly less loss than a vertically-polarised wave. This gives horizontal polarisation a slight advantage.

As the wave continues, it probably uses many other propagation mechanisms, but almost all of these preserve the polarisation intact. Therefore, a second horizontally-polarised aerial at the far end of the path will receive the maximum possible signal.

There are a few exceptions, but not enough to make a difference to the standard use of horizontal polarisation for VHF/UHF DX. In particular, backscatter modes such as auroral reflection may scramble the polarisation, and so too may ionospheric propagation modes on 50MHz. The same applies to space communication, because the plane of polarisation is rotated as the VHF/UHF waves pass through the ionosphere. Circular polarisation is often used for space communication because it is unaffected by rotation, but it is not favoured for terrestrial communication because it involves a 3dB loss in communicating with any linear polarisation.

FM is used in a very different way, generally for local communication and often between mobile and even hand-portable stations. These stations are often in very poor locations with lots of nearby obstructions, so the plane of polarisation is likely to be altered by local reflections. In addition, beam aerials are often undesirable because you want to work in all directions. These features make vertical polarisation the best choice because the vast majority of FM stations can use simple omnidirectional whip aerials.

The most important thing of all is to use the same polarisation as all the other people you want to work. You'll lose a lot by going against the standard. Theoretically, the loss between perfect plane polarisations exactly at right-angles is infinite! In practice, you can expect losses of at least 20dB - and that makes all the difference between a station that works and one that doesn't.

## PASSIVE GRID

***Q: I'm trying to build a tetrode HF amplifier with a 50Ω 'passive grid' resistor, but I can't get a good input match at 30MHz. How can I cure this without any band switching? The tube is a Russian 4CX800A.***

***A:*** The 'passive grid' circuit is a very good choice for tetrode amplifiers, at least as high as 50MHz, because it provides a good input match to the exciter, and in principle it requires no band-switched input network. **Fig 12** shows the basic circuit, which has appeared in many handbooks. If the tetrode is operated in Class AB1 with no grid current, it essentially requires only a voltage swing and absorbs very little drive power.

The necessary voltage swing is developed across the 50Ω resistor, R1, which absorbs almost all the input drive power and provides a stable load to the transceiver. Even if the tube is driven into a small amount of grid current (Class AB2), causing its input impedance to fall to maybe a few kilohms, R1 still dominates and stabilises the load seen by the transceiver. This low value 'swamping' resistor, connected directly across the input, also gives stability against oscillation and usually avoids the need for neutralisation.

A 50Ω grid resistor is the right load for the transceiver, but the voltage swing developed by the full output of a 100W transceiver may be too high for many modern ceramic-metal tetrodes. For example, the 4CX800A / GU74B requires a negative grid bias of about 56V, and this peak voltage can be obtained across 50Ω at a power level of only 31W [1 12]. The data sheet suggests using resistive negative feedback in the cathode (about 30&!, as shown dotted in Fig 1 12) which increases the RF voltage swing required [2 13], and thus increases the drive requirement to 45 – 50W. This is actually quite convenient, because Peter Hart's reviews show

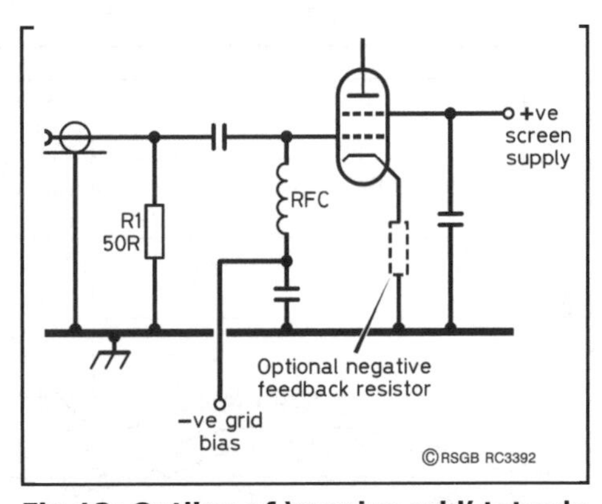


**Fig 12: Outline of 'passive grid' tetrode input circuit. Grid voltage swing is developed across R1, which also provides a stable load for the exciter. Also shown is the optional negative feedback resistor in the cathode.**

that many transceivers, particularly the ones with 12V transistor finals, have rather poor intermodulation distortion figures at the 100W level. However, the distortion often improves dramatically when you ease back the RF POWER control to 50W output or less. With a power amplifier that doesn't need a lot of drive and also doesn't add too much distortion of its own, the result will be a signal that is both stronger *and cleaner* than it was with the 'barefoot' 100W. The penalty is that you *must* use Automatic Level Control (ALC) fed back from the amplifier to the transceiver, to make certain that the tube is never overdriven with the full 100W. This applies particularly to modern tetrodes whose delicate grid structures must be protected from excessive power dissipation.

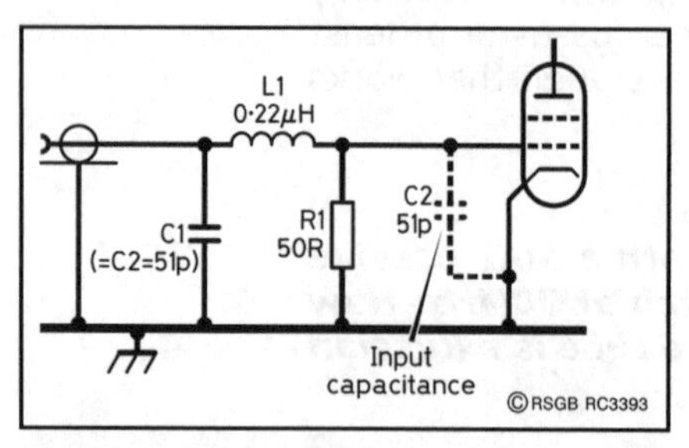


**Fig 13: Additional components C1 and L1 form a low-pass p-network which 'absorbs' the parasitic input capacitance of the tube (other components from Fig 12 omitted).**

Now what about the input matching problem? The clue is that it is only serious at 30MHz, and not at lower frequencies. Whenever there is a problem that increases with frequency, you can bet that some parasitic capacitance or inductance is to blame. ('Parasitic' is one of those fancy engineers' terms for something they'd forgotten to account for!) In this case it's the grid-cathode input capacitance of the 4CX800A which, according to the data sheet, is 51±5pF. This appears in parallel with the 50Ω load resistor. At low frequencies its reactance is so high that it makes no practical difference – the SWR at 1.8MHz is only 1.03. But at 30MHz, the reactance is much lower, so the SWR has risen to 1.6, a value that will make many transceivers complain.

So what's the solution? Obviously you could parallel-resonate the capacitance with a shunt inductor, but that would only fix the problem on a single band. The broadband solution is to use the tube's input capacitance as the 'C2' of a π-network at the input, as shown in **Fig 13**. Electrically, this is an ordinary three-component π-network, but you only need the two extra components, C1 and L1. Another way to think of a π-network is as a low-pass filter. If we design this little filter to be matched somewhere near 30MHz, it will be almost transparent to lower frequencies, so that will achieve our goal of a 'flat' input match across the whole of HF.

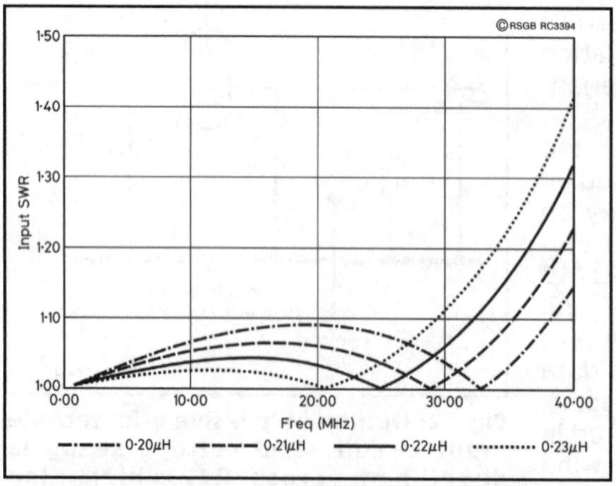

**Fig 14: Effect of different value of L1 on the input SWR. The optimum value is around 0.22μH, but is not critical.**

We're looking for a 1:1 impedance transformation, so straight away we know that C1 = C2. It only remains to calculate L1. I won't trouble you with the calculation here, because the value can easily be scaled from the 0.22μH value shown in Fig 13. **Fig 14** shows a set of computed SWR curves for C = 51pF and various values of L1 from

0.20μH to 0.23μH. Increasing the inductance will move the frequency of the perfect match downwards, and at the same time reduce the small SWR bump in the middle of the HF band. Clearly, the value of L1 is not critical, because all the values tried give an SWR of 1.1 or below, right across the HF band. The optimum value is about 0.22μH, which sacrifices matching at the top end of 10m band for the sake of better matching on all lower bands.

The component values in Fig 2 13 are specific to our example of 51pF input capacitance for the 4CX800A. To scale these values to a different input C, simply make C1 equal to that new value, and multiply L1 by (51/C). That doesn't give the mathematically correct new value for L1, but it's close enough because, in practice, you'll need to make some small adjustments. The recently-introduced TO-220 thick-film power resistors are ideal for a passive grid input load ('In Practice', June 2002).

Two 100Ω 50W units in parallel should handle the average drive power very well, if you bolt them down to an area of the chassis that is well-cooled by the blower. However, these resistors will add some input capacitance of their own, and that's why you need to make some adjustments. C1 can be a combination of fixed silvered-mica capacitors and a mica 'postage-stamp' trimmer, and L1 can be a small self-supporting coil which you can squeeze, stretch or change. You can safely measure the input SWR with the tube in its socket but completely unpowered. Adjust C1 and L1 in turn until the SWR is a minimum at a spot frequency, either 28.0MHz or, even better, around 25MHz if your transmitter isn't locked out. You should be able to get a perfect match at your chosen frequency. Then switch to the other bands and confirm that the SWR is still good.

Obviously there are some interesting variants on this little circuit. For example, you could extend the SWR compensation up to 50MHz, though at the expense of a larger SWR bump at lower frequencies (probably the best frequency for the perfect match would be around 40MHz). For tetrodes that require less grid bias, and therefore also require a smaller grid voltage swing, you might find yourself having to turn the drive power down very low. If this is too finicky to adjust, a solution would be to build a power attenuator as detailed in the June 2002 column, and/or to tap the grid connection down the load resistor.

### HARMONICS FROM VSWR METER?

***Q: In my HF station, should I put the low-pass filter between my in-line VSWR meter and the aerial, or between the transmitter and the VSWR meter? Local club members are split on the issue!***

***A:*** It matters less than some people think. There is an old, old story that the rectifier diodes in the VSWR meter cause harmonics, and that therefore the low-pass filter should come after the VSWR meter to prevent these harmonics from reaching the aerial and being radiated. That may have been true, but it required some special conditions that hardly apply any more.

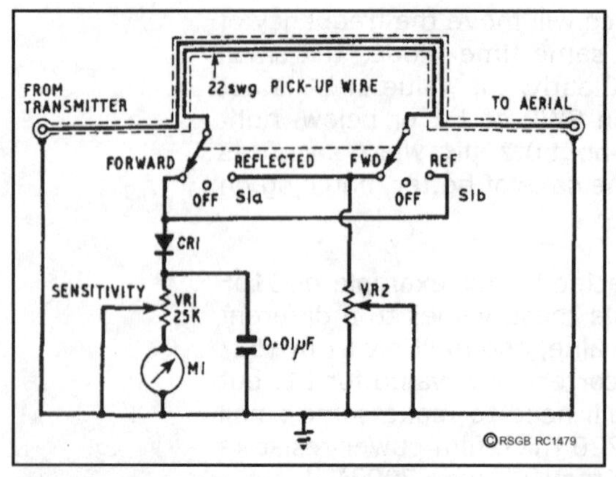


**Fig 15: This old-fashioned type of HF VSWR meter sometimes caused harmonic TVI on Band I.**

The older style of HF VSWR meter used coupled sampling lines, parallel to the main conductor of the coaxial cable (**Fig 15**). The coupling between the main line and the two sampling lines increases with frequency, so a sensor designed to cover all HF bands would be very strongly coupled on 21 and 28MHz - all the more so if the moving-coil meter was not very sensitive and the line coupling had to be increased to compensate. On the higher HF bands, the sensitivity control, VR1, would be turned down so as not to 'pin' the meter, but there is no doubt that the diodes were being driven quite hard. Because the coupling between two short parallel lines increases with frequency, any harmonics generated in the sampling lines would be coupled even more effectively back into the main line. Under these circumstances, interference to the old Band 1 TV on the second or third harmonic would be quite possible.

Modern HF VSWR meters are quite different. They use a transformer coupling system that is essentially independent of frequency, and in most instruments the moving-coil meters are also more sensitive than in days of old. Therefore the diodes are driven nothing like as hard, and there is much less likelihood of significant harmonic generation. Even with the older-type of VSWR meter, harmonic generation was by no means universal. G3RZP has grossly overdriven his old 'Labgear' meter until the terminating resistor, VR2, bubbled hot wax, but still found no sign of harmonics above 70dB with respect to the main carrier. It therefore seems that, especially in modern times, the EMC hazards of connecting a VSWR meter straight to the aerial are over-rated.

## ATU POWER RATINGS

***Q:*** *I just burned up a so-called '3kW' aerial tuner with only a few hundred watts! What does this power rating mean? [From the Internet newsgroup* ***rec.radio.amateur.antenna****]*

***A:*** One of the best features of Internet newsgroups is that you can sometimes get an answer straight from the horse's mouth, as this message from Tom Rauch, W8JI, will show.

"I'm an independent RF designer, and a year or two ago I was asked by MFJ to look at their existing range of tuners and make technical suggestions. Over the years I've done the same for other well-known tuner and amplifier manufacturers who always want to keep improving their products. I feel that insight into engineering facts is important, and that people should understand a few things about tuners in general, and the '3kW' rated models in particular. I've yet to find any currently available tuner that always meets its advertised power specifications, at every frequency and under all possible load conditions.

"The only exceptions were the ATR-15 as advertised by Ameritron in the 1980s, which was rated at the RF output power over a specific load impedance and frequency range; and the old Johnson KW Matchbox, which was rated for a 1kW AM transmitter (about 750W continuous carrier and 3kW PEP on modulation peaks) so it easily handled 1500W CW or PEP output.

"Power ratings for tuners are an offshoot of when PAs were measured by plate *input* power. For example, the Heathkit SB-220 was advertised as a '2kW PEP Input Power' PA, but in modern terms the SB-220 was really rated at about 1kW SSB PEP output. The matching tuners were called '2kW tuners'. Enter Dentron: they called their PAs '3kW', even though they only put out the same reliable power as the SB-220. Likewise Dentron called their matching tuners '3kW tuners'. That rating of '3kW' has since become attached to any tuner with the same basic component quality as Dentron's, and it seems to have stuck. So, a typical so-called '3kW tuner' is actually a 1200W PEP tuner by today's *output* power standards.

"The '3kW'-rated MFJ-989 used a roller inductor produced by Oren Elliot Products. It is an industry-standard roller, and is similar to or the same as rollers used in other high-dollar tuners. Roller inductor *Q* is absolutely *not* a problem of cabinet size or anything external to the roller, it is a problem of materials and the basic design. Although the Oren Elliot roller has a Delrin form whose losses are an additional weakness, it is the overall roller design that seriously lowers the power rating. Even the expensive ceramic form roller from Cardwell barely offers an improvement; it has almost the same loss, but the ceramic form stands the heat much better. The new design of air-core roller used in the MFJ-989 helps to solve *Q* and heating problems, and the manual was also re-written to correct earlier misinformation on how to adjust the tuner for the lowest matched loss.

"It is *not* correct to specify a tuner power rating without specifying both a load impedance and a frequency. Any manufacturer who says 'this is a 3kW tuner' without specifying a frequency and a load is just giving a number that may not – and probably will not – mean anything to you. The only way to know for sure if a tuner will work is to try it with your own aerials. T-network tuners (**Fig 16**) can handle the least power on 1.8MHz and / or loads of low resistance and high capacitive reactance. T-network tuners can handle more power into higher-resistance loads or loads with some amount of inductive reactance. In a T-network tuner, maximum efficiency and power handling generally occurs with the least amount of inductance and roughly equal amounts of capacitance. This is true even though many other settings will produce a low SWR."

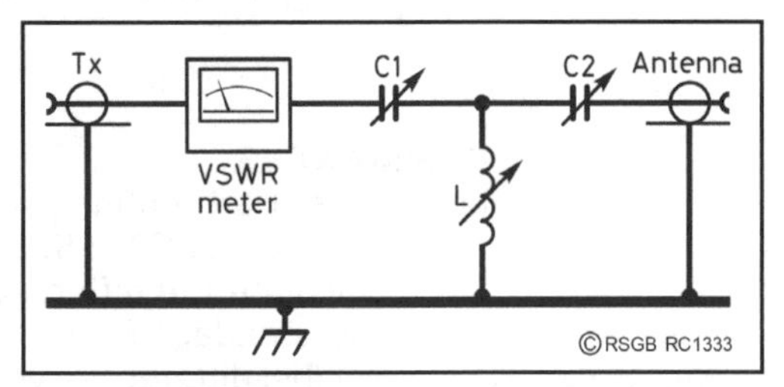


**Fig 16: The T-match ATU circuit. Inductor L1 can either be a roller-adjustable type or use switched taps. For lowest losses in the ATU, use the LOWEST inductor value and the HIGHEST capacitor values that will produce a 1:1 VSWR.**

W8JI concludes, "I hope this dispels some of the hyperbole about tuners." In other words, he has explained that the '3kW' rating developed from an upward spiral of marketing hype, and is technically meaningless. Remember that the engineers who designed a product have very little influence on the way that it is marketed and advertised, or even the printing on the front panel!

Dividing everything by 10, you could probably say the same about the popular '300W' ATUs. These ATUs can handle the rated power in favourable circumstances, but not into a severe load mismatch. This is also the reason why the small automatic ATUs in many 100W solid-state transceivers are only rated to handle very modest levels of VSWR, up to maybe 3:1. However, a '300W' manual ATU will match a much wider range of impedances at the 100W level if it is used intelligently. With a three-knob ATU of any type, many combinations of settings will give a 1:1 VSWR into the transmitter, but there is only *one* combination that gives optimum efficiency. With severe load mismatches and inappropriate adjustment, the wrong settings can melt the inductors and arc the capacitors in almost any ATU! Here's an effective adjustment procedure for the conventional T-network ATU (Fig 16), starting with a completely unknown load:

1. Set both C1 and C2 to MAXIMUM.
2. Adjust L1 for maximum receiver noise.
3. Transmitting low power on a clear frequency, adjust C1 and C2 for a VSWR of 1:1.
4. Reduce the inductance of L1 and repeat step 3.
5. Repeat step 4 until a match cannot be achieved; then reverse the last change you made to L1.
6. Note the settings of C1, L1 and C2 for quick and easy band-changing. You can mark them directly on the front panel with felt-tip pens, using different-coloured dots for each band.

This procedure gives you the optimum ATU efficiency, which comes from using the lowest possible value of L1, and the maximum values of C1 and C2, that will achieve a match. If the ATU shows signs of distress when transmitting at full power on some bands, think about changing the aerial before upgrading the ATU. For example, you could relieve some of the strain on the ATU by using a pre-matching network at the aerial itself, or by changing to an aerial design that already has a "tendency to match", leaving only a moderate VSWR for the ATU to deal with.

**REFERENCES**

[1] Peter Hart routinely measures the signal levels in microvolts required to give S1, S3, S5, S7 and S9, and then S9+20, 40 and 60dB. To calculate the difference in decibels between any two voltages, use the standard formula: $dB = 20 \log_{10}(V2/V1)$. Remember that the decibel differences you calculate from Peter's tables will be for steps of *two* S-units.

[2] This topic connects with several previous *In Practice* columns, especially September 1994, May 1997 and September 1999; and also with Reference 3. Even so, I make no apology for repeating

this information, simply because the same questions about baluns keep coming back!

[3] Tony Plant, G3NXC, 'An Introduction to Baluns', 'Down To Earth', *RadCom,* July 2000 – an excellent introduction indeed.

[4] Roy Lewallen, W7EL, 'Baluns: What They Do and How They Do It'. This article is a classic combination of theory and experimental work, and also the source of the terms 'voltage balun' and 'current balun'. Although the conclusions are summarised in the *ARRL Antenna Handbook*, the complete original article is hidden away in *The ARRL Antenna Compendium, Volume 1* (1985) which, frankly, only a dedicated enthusiast would buy at this late date.

[5] 'Coax - Inside and Out', 'In Practice', May 1997.

[6] 'Best Feeder Length?', 'In Practice', September 1999.

[7] Rick Littlefield, K1BQT, 'Light-Weight Resonant-Trap Baluns', *Communications Quarterly*.

[8] 'In Practice', December 2001 contains a lot more references to previous columns – this is a popular topic!

[9] Product Review: Cushcraft R7000, by Peter Hart, G3SJX, *RadCom* January 1997.

[10] As explained in December 2001 and previous columns, attempting to connect an 'RF earth' in the shack is often a *bad* idea. Instead of curing RFI problems, it can just as often make them *worse*.

[11] For a detailed explanation of propagation modes, see *The VHF/UHF DX Book*, Chapter 2.

# HOMEBREW HINTS

**When cutting horizontally, angle the saw blade slightly down. This will prevent it wandering away from its guide.**

## STRAIGHT CUTS

It's very easy to end up with a wavy line when cutting PCB and other soft material. It's equally easy to prevent it!

Simply place the material in the vice, so that you will be cutting horizontally rather than vertically. The top of the vice then acts as a guide for the saw. If you want to protect the vice, sandwich a piece of angle iron in the jaws along with the material you wish to cut.

## A THIRD HAND

To make an inexpensive 'third hand', you need nothing more than an offcut of wood (the heavier, the better), two clothes pegs, and some wood screws or epoxy cement.

When the pegs are fixed to the wood, it makes an ideal method of holding a printed circuit board, a cable or connector, etc.

## MINI-WORKSTATION

To prevent burning a table top and to provide a convenient method of putting things away at mealtimes, build yourself a mini-workstation.

This need consist of nothing more then a piece of hardboard (mine is 35cm x 35cm), with wooden battens fitted along three sides to prevent components slipping off the edge. Margarine tubs make ideal component holders, and a treacle tin provides an ideal place to stow your hot soldering iron. To prevent the tubs and the tin moving, they should be stuck down with impact glue.

## RF INDICATOR

You can build a very simple RF indicator for a transmitter, simply by connecting a coil of wire to an LED. See **Fig 1** for details.

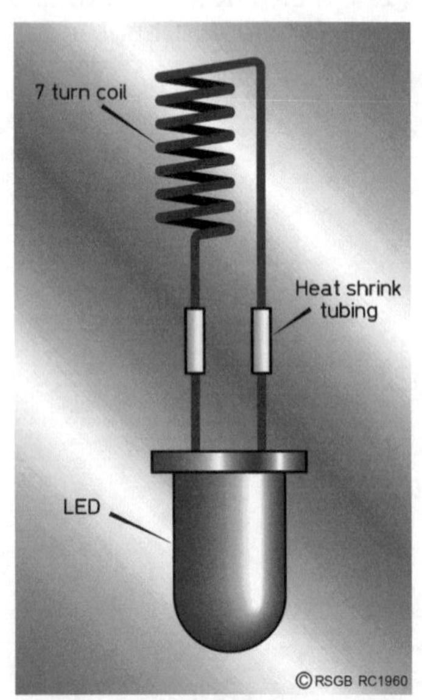


**Fig 1: A seven-turn coil of enamelled copper wire connected to an LED makes an RF indicator. Is this the simplest design ever?**

Depending on the power level in use, the coil will need to be located closer to, or further from, the RF source.

**SOLDER DISPENSER**

If you buy solder in small quantities, it can often end up getting screwed up into a dirty, tangled mess in the bottom of your toolbox – assuming you can find it at all when you want it! Alternatively, if you buy solder in large quantities, it can be inconvenient to carry around a large, heavy reel. A 35mm film canister can be used to stow a metre or two of solder, help to keep it clean, and prevent it from getting knotted. What's more, you are more likely to be able to find it when you want it.

Wind some solder into a helix (eg on a short, wooden dowel), pop the whole lot into the film canister. Feed one end of the solder through a small hole in the cap, snap the cap back on, and you'll have a handy solder dispenser.

**Ready for action, the mini-workstation can be put away and brought back out with everything exactly the way you left it.**

# LONG-DELAY ECHOES ON TAPE

**by Pat Hawker, G3VA, from 'Technical Topics'**

Almost 30 years ago a 'TT' item 'Long-Delay "Cosmic" Echoes' noted that a team at Stanford University was planning a new investigation into "the baffling, 40-year-old puzzle of long-delayed echoes (delay times measured in seconds) that have been reported only extremely rarely".

It was noted that these strange Long-Delay Echoes (LDEs) were originally reported in 1927 in a series of experiments by Professor Stormer at Oslo and Dr Balthazar Van der Pol at Eindhoven. The 1960 translation of *Radio Wave Propagation and the Ionosphere*, by the Russian scientist Ya L Al'pert includes a summary of this work: "The first cosmic echo was observed at the end of the summer of 1927. Searches were organised on 31.4 metres from March 1928, but it was not until near noon on 11 October that a large number of echoes were received in Oslo with delay times between 3 and 15 seconds (most about 8 seconds)... during the night 120 echoes were observed at Eindhoven... in 1930, echoes were reported in Indochina... in 1934 more than 70 echoes were observed between 30 May and 8 July."

Post-war investigations showed LDEs being heard only very rarely, and some researchers concluded that all reports must be mistakes, hoaxes, subjective effects originating within the listeners own brain, etc, etc. Al'pert distinguished LDEs from the far more common HF Round-The-World echoes, on which a good deal of research was carried out by the Germans during WWII in the course of devising a means of locating a distant transmitter by measuring the difference in time of arrival of the short and long path signals. It was shown that the average delay time of more than 750 RTW echoes was 137.78ms, with a spread in the individual measurement of 218 echoes less than $10^{-4}$s. In the following 11 years, I included at least eight further items on LDEs, including several reports of such echoes being heard, or not heard despite long periods of listening, and also several hypothesises of how they might occur.

Two of the more convincing reports of LDEs in the 1930s - not only heard but also recorded - were included in 'TT' (November 1975, p846). R L A Borrow, G3ZTK, who was one of those who carried out listening tests together with Professor Appleton at King's College, London, during the original transmissions by Van der Pol from The Hague in 1928 – 29 wrote: "In addition to hearing a number of echoes, I did succeed in obtaining what was then the only known photographic record of an echo. This record was examined by Appleton and he was quite satisfied that it was undoubtedly genuine (see his letter to *Nature*, Vol 122, December 1928)... On a very few occasions I distinctly heard the signal (an X)... on one occasion an echo was heard by me in the laboratory at King's and also by Appleton at his home in Potters Bar."

L A Reeves, G4CEM, also verified that even if LDEs were not being heard in the 1970s, they were heard and recorded by himself and many members of the commercial telegraph fraternity. He wrote: "The last time I heard this phenomenon was around 1938 on the Cable and Wireless London-to-Bangkok route... The Bangkok station when 'idling' had the habit of putting on a call-band of spaced Vs and callsigns. The space between each V was about 5 seconds, and following each V there could sometimes be heard a perfect echo-type signal approximately 2s behind the original. It was nearly always strong enough to be recorded on the pen-undulators used at that time... These echoes were recorded on the daytime frequency of 19MHz during winter, when the whole route was in daylight... I can assure you that what we heard and recorded on these tapes was tangible and certainly not subjective." The last time I included an item on LDEs was in 1980, but I have now received a most interesting report from Peter Martinez, G3PLX, who has recently observed and recorded echoes while using his electronic Hellschreiber equipment. The delay time of (209 ± 5)ms is clearly too long for an RTW echo, although less than what has generally been regarded as the delay time of an LDE. He writes:

"The subject of 'long-delayed echoes' is a bit like the Loch Ness Monster or the Abominable Snowman. However implausible their existences may seem, reports still appear from time to time describing them. Here is another, but this time I captured it 'on tape', as it were, that I can show to those who may be interested. I was in contact with PA0OCD on 3.5MHz at 1930UTC on October 25, 1997, using the 'Hellschreiber' mode (described in detail by the late Stan Cook, G5XB, 'Hellschreiber - What it is and How it Works' (*Radio Communication*, April 1981, pp320 – 323). In this mode, letters are transmitted by sending patterns of dots in on/off keying mode, the patterns representing directly the letter shapes as they appear on the printed tape. The original German direct-printing telegraph equipment used mechanical methods, but several amateurs are using DSP software and computers. My own set-up displays the incoming signal on the computer screen as black letters on a white background, stronger signals giving a blacker image. The sending software works so that whenever I hit a key, the transmitter immediately transmits the letter shape, then drops back to receive, in quick-break fashion.

"Imagine my surprise when, over a period of about one minute, I started to hear short 'pips' in the receiver speaker at the ends of my words. I then realised that I could also see the right-hand edges of the final letters of my words appearing on the receive screen, but shifted out-of-phase so that they did not line up with the transmitted letters.

"This appeared to be due an echo, so I did a quick test by sending a series of full stops in rapid sequence. Sure enough the dots showed up on the receive screen, appearing just after the transmitted dots but at the top of the text line rather than the bottom. I quickly 'saved' the entire page. The effect faded and did not occur again.

"**Fig 1** shows a fragment of the screen-shot. At the top, the signal from PA0OCD; then my own test, with band-noise appearing between my words as a result of the quick-break operation. On the bottom few lines can be seen my remarks about the presence of this echo, followed by my tests

GIVING SMALL SIGNALS BUT I THINK THAT ON SUNDAY
NALS N GREECE. OK BTU PETER, G3PLX DE PA0OCD PSE
GUESS SIGNALS WILL BE TOO WEAK FOR CREECE IN TH
ORNING. THAT TIME WE HAVE WORKED HIM ON PSK31. I
K SPACE EITHER. HI. NOW I CAN HEAR MY
AND I CAN SEE A BIT OF THE END OF MY LAST CHA
MAYBE I CAN WORK OUT THE DELAY FROM THE WAY IT

**Fig 1: Screen-shot of a part of G3PLXs Hellschreiber-mode contact with PA0OCD.**

with the string of dots. The echoes can be seen on some of the letters at the ends of words on the lines above, for example the 'K' at the beginning of line 5 and 'N' and 'CAN' in line 6. Where you see my text, this is 'forced' onto the screen by the software equivalent of 'sidetone', and are not signals received through the receiver which is muted while transmitting.

"I have spent some time analysing this screenshot, picking out the actual sequence of received signals. The Hellschreiber signal scans in columns, bottom-to-top, the columns being transmitted at a precise rate of 17.5 columns per second. Each column is divided into 14 pixels or dots, each of 4.08ms duration. A full stop consists of 4 pixels in a square, and the waveform of a repeated string of dots is thus a pair of pulses 8.16ms long, separated by 57ms. They repeat at intervals of 457ms.

"I processed the screen-shot to unravel the rectangular characters into strips 457ms long, repeating down the page to show the repeated full stops. This is shown in **Fig 2**. I have added a line which is 49 pixels long, which represents exactly 200ms. As a separate exercise, by listening to my own transmission on a separate receiver and displaying that on the screen while I was transmitting, I measured the delay from the 'sidetone' image on the screen-shot to the actual transmission. This was 55ms, formed mostly from a fixed software delay and a few milliseconds of delay through transmitter and receiver SSB filters. This enabled me to move the sidetone image from Fig 3 back by 55ms so that it represents the transmitted signal. The distance between the transmitted signal and the echo is seen to be about 209ms, with an uncertainty estimated at ±5ms.

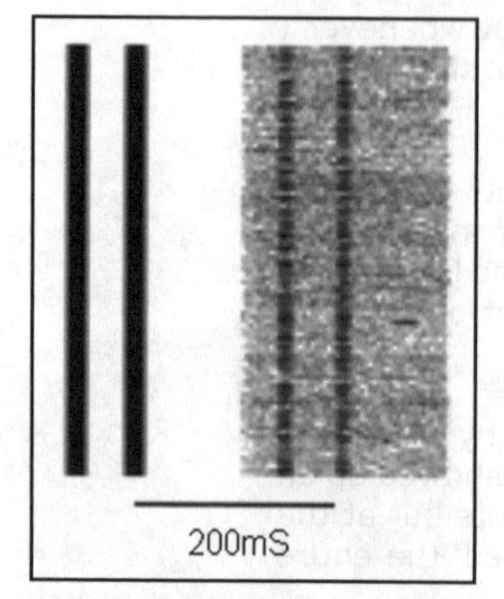


**Fig 2: Processed version showing more clearly the echoes received 209ms after the transmission of the series of dots shown at the bottom of Fig 1.**

"I have heard a similar echo once before on 3.5MHz, while transmitting AmTOR but, on this occasion, I was fortunate in being able to record it for posterity. To put 209ms into perspective, it represents some 1.5 times the delay of a Round-The-World echo. Remember, this is 3.5MHz not 21MHz!"

[A check of DK0WCY beacon propagation forecasts (see 'TT', October 1997, pp80 / 81) shows that, during 25 October, there was a warning of an expected minor geomagnetic storm, and this must have been pronounced since, most unusually, I did not hear the beacon on 26 / 27 October. There must be some suspicion that the echoes and the magnetic storm may have been connected, with the echo possibly arising from a storm-enhanced Van Allen belt - *G3VA.*]

# MAKING A CERAMIC COIL FORMER

**by Ted Garrott, G0LMJ**

A ceramic article is made by forming it in clay and then firing (heating) to a high temperature. It then takes on the appearance of unglazed porcelain. Ceramic material has the following characteristics that make it suitable for use in electrical and electronic components:

- it can be produced in any shape;
- electrical insulation is excellent at all frequencies;
- it has low dielectric loss;
- it can be used at high temperatures without losing strength.

Ceramics are not used by amateurs as the material is not available on the commercial market. As the use of a kiln was available to me, I decided to make a ceramic coil former for my balanced-line aerial tuning unit. This tuning unit was described in the July and August issues of *RadCom* in 1998.

**The plaster mould in two halves.**

## SLIP-CASTING

Many methods of forming clay into various shapes can be used and, after initial experiments, I decided to use a procedure called 'slip-casting'. Here, a plaster mould is filled with slip (liquid clay). The porous plaster removes water from the slip and a layer of clay forms on the inside of the mould. When the surplus slip is poured out, this layer of clay remains, thus forming a hollow object in the shape of the mould.

The mould, which is in two halves, may then be split and the clay model removed. This principle is now described in detail.

## MAKING THE MOULD

The method of making the mould is shown in **Fig 1**. A box is made out of the plastic coatedboard, the sides being screwed together. A model of the ceramic article is supported in the box with pieces of clay and the clay smoothed out along the model centre line. This is shown in the photograph.

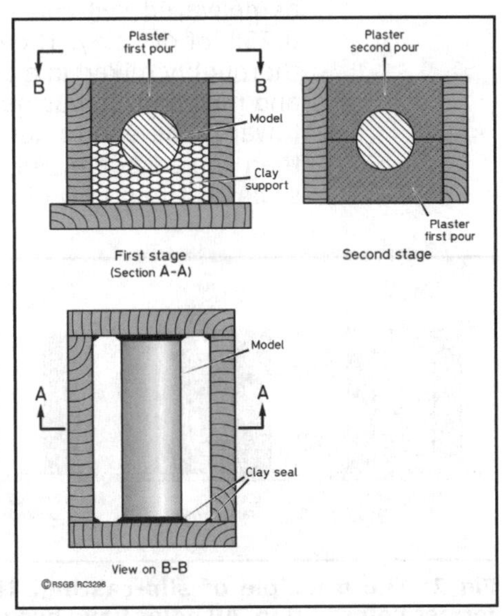


**Fig 1: Method of making the plaster mould.**

**The model is set in the mould on a clay support and the first pour of plaster can be made.**

The model should be non-absorbent. I used a plastic tube, but wood may be used if it is given two coats of primer. The model may be solid (not hollow).

Fine casting plaster is poured in the top of the mould; when the plaster has set the mould bottom and the clay are removed. The exposed plaster around the model is given three coats of soft soap and water (1:1 ratio). The whole mould, with the model still in place, is turned upside down and the other half filled with plaster.

When the plaster is set, the mould box is removed, the mould halves separated and the model removed. The photograph shows this.

## MAKING THE SLIP

I used an industrial porcelain clay. The slip is made using dry powdered clay mixed with water and a deflocculant. The deflocculant separates the clay particles and keeps them in suspension so that the clay is fluid.

Experiment showed that a slip density of 33oz/pint was suitable. This slip was made from 21.67oz of dry powdered clay and 11.33oz of water, 2.6g of sodium dispex (the deflocculant) was added. I calculated that 1.25 pints of slip would be plenty and increased the quantities accordingly.

The clay as supplied was rolled out with a domestic wooden (not glass) rolling pin into slabs about 4mm thick and dried in a domestic oven at a temperature of 100°C. The dry clay was then crushed to a powder using a rolling pin. A face mask was used to avoid inhaling clay dust. The clay as delivered will contain 22% water, so 1lb of wet clay will yield only 0.78lb of dry clay. The deflocculant was added to the warmed water and thoroughly mixed in a plastic bowl. The clay powder was gradually added and mixed in. It was necessary to use a hand-held electric kitchen mixer towards the end to get a smooth mix. The slip was left for 24 hours and then mixed again. All materials were weighed using domestic kitchen scales with an LCD display.

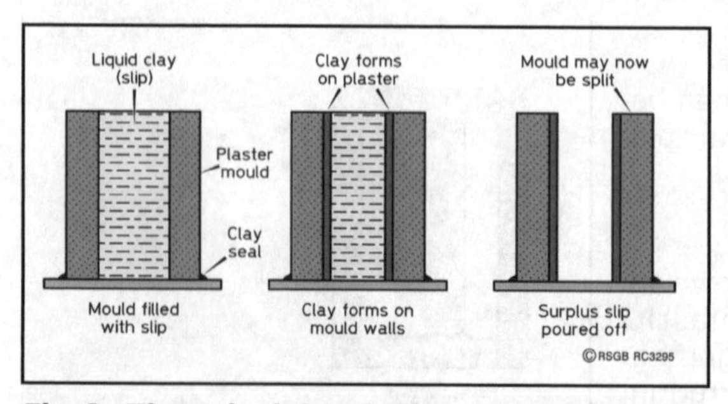


**Fig 2: The principle of slip-casting. The coil former before firing. All holes have been drilled and wire guide-grooves cut.**

## POURING THE SLIP

The two halves of the mould were held together with elastic bands and the mould mounted upright on a piece of plastic-coated board as shown in **Fig 2**. A fillet of clay around the bottom of the mould prevented leakage. The mould was filled with slip and left for one hour.

The slip needed topping up during this period. The liquid slip was then poured out. This left a clay layer about 5mm thick on the inside of the mould.

When the clay was seen to shrink away from the mould, the bottom board was removed and the clay trimmed neatly at teach end with a hobby knife. The time to separate the mould needs a bit of judgment, but it should be done when the clay is clearly stiff enough not to distort too much.

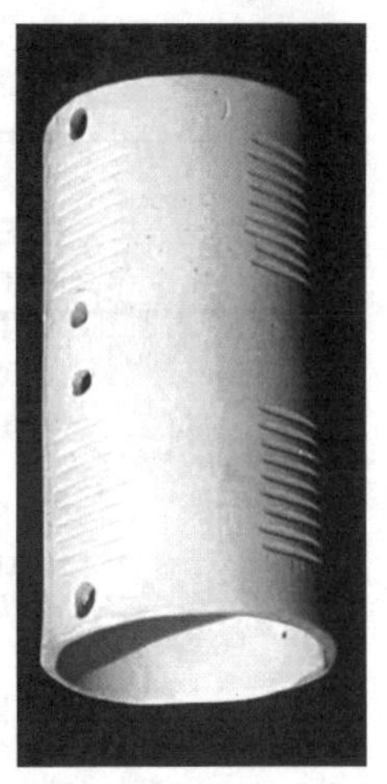

**The coil former before firing. All holes have been drilled and wire guide-grooves cut.**

### DRYING THE CLAY TUBE

This must be done slowly to avoid cracking or distortion. I took a week to dry mine and controlled the drying rate by keeping a plastic bag over the clay most of the time. Early in the drying it is possible to square off the ends of the tube with a sharp hobby knife.

The clay is dry when it has the same colour as a sample piece of clay taken from the drying oven. The dry clay is rather fragile and must be treated gently. At this stage, I drilled all necessary holes and cut grooves to position the coil windings. The grooves were cut using an 'Abrafile' tension file, which was just the right diameter for the 1.25mm wire.

Very careful setting-out was required for these grooves. The photograph shows the tube at this stage.

### FIRING

The clay tube was fired in an electric kiln fitted with an electronic controller. The temperature was increased at 40ºC per hour to about 600ºC with the ventilation bungs removed. This slow heat build-up allows chemically-combined water to escape as it changes into steam. The organic matter in the clay also decomposes.

The ventilation bungs were then replaced and the temperature allowed to rise at a fast rate to 1210°C. The kiln was then switched off and, when the temperature reached 200°C, the kiln lid was raised, and the tube removed when cool enough to handle.

### SHRINKAGE

The clay will shrink during drying and firing. I found that the shrinkage during firing was 8% and the total shrinkage from the model stage to the end of firing was 12%.

The last photograph shows the plug-in ceramic former with all windings in place. The plugs were glued in with Araldite. Slip-casting is one of many ways that can be used to shape clay and I recommend that anyone wishing to take up the challenge should read books on the subject first.

**The ceramic coil former with all wire and plugs fitted.**

### KILNS

An electric kiln with a programmable electronic controller is best. My wife uses one for china painting. The kiln was made available to me after negotiation.

Even a small kiln will cost over £500, so it is best to make enquiries amongst friends or the local colleges, where one may be available for firings at little more than the cost of electricity.

**BIBLIOGRAPHY**

Many books are available at public libraries under the subject heading 'Pottery'. I consulted a number of them and eventually bought the following: *The Potters Manual*, by Kenneth Clark and *Slip-casting*, by Sasha Wardell.

**SUPPLIERS**

Both these books give lists of suppliers. I located my suppliers from the *Yellow Pages*: Clayman, Morells Barn, Park Lane, Lagness, Chichester, West Sussex PO20 6LR, tel: 01243 265 845. This company was most helpful in all respects and obtained the industrial porcelain clay from Valentine Clays of Stoke-on-Trent ST1 6BA, Tel 01782 271 200.

# OBTAINING NON-STANDARD VOLTAGES FROM TRANSFORMERS

**by Ian White, GM3SEK, from 'In Practice'**

***Q1*: I need to supply a valve heater with 12.6V AC at 2.9A, but the transformers in the catalogues only have 'standard' voltages like 12V and 15V. How can I obtain the correct voltage?**
***Q2*: Me too! I need 8V AC at 0-3A, and the catalogues only show 6V and 9V.**
***A*:** There are several ways to do this. Let's see which are the best options for the two applications above.

**Option 1.** Adjust primary voltage taps This is a time-honoured way of obtaining a different voltage, by connecting the 230V AC mains to the 'wrong' primary tap... that's assuming the transformer has at least one primary tap in addition to the start and finish of the winding. However, the scope for variation is not large, and there can be safety implications. For example, if a transformer is marked as 6.0VAC and it has primary taps at 210-230-250V (**Fig 1**), you could connect the 230V mains to the 210V tap and obtain 6.0 x (230 / 210) = 6.6VAC. Connecting to the 250V tap would give you 5.5VAC. Note that the scope for varying the output voltage is usually only about ±10%. Note also that if you connect 230V to a 210V tap, you would be overrunning the transformer. Most transformers should stand this, but there is a risk of core saturation and overheating. Don't even think of connecting 230V to a 115V tap!

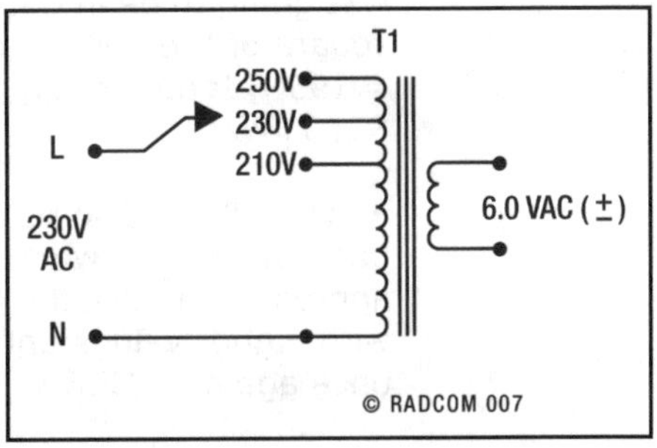

**Fig 1: Using the 'wrong' primary tap will change the output voltage by a few percent.**

I don't recommend using a 'spare' secondary winding to add to the primary voltage or subtract from it. Modern practice is to keep the primary and secondary completely separate - often with the primary on its own fully insulated bobbin (see photograph). The insulation between windings on the secondary side of a low-voltage transformer may not be up to mains standards, and it's risky to rely on it.

**Option 2.** Series resistor in secondary circuit This is one way of reducing the voltage for a valve heater - but only if the load is constant. **Fig 2(a)** shows how this can be done for the valve in Question 1, starting with a 15VAC transformer. That voltage at 2.9A represents 43.5VA, so you should be looking in the catalogues for a transformer with a 50VA rating. You need to drop (15 - 12.6) = 2.4V at 2.9A, so the resistance, R1, needs to

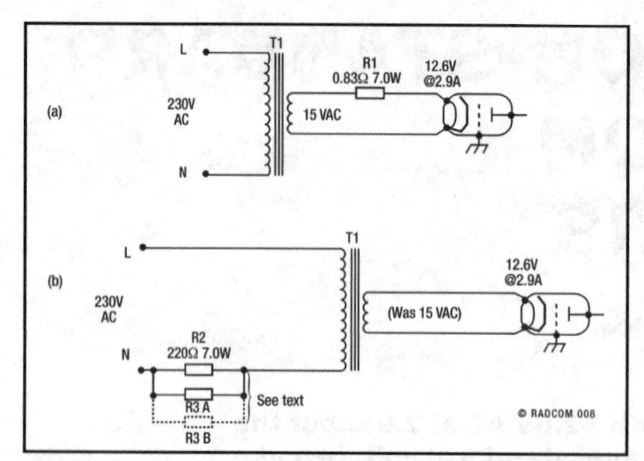


**Fig 2: Reducing output voltage with a series resistor: (a) in the secondary, (b) in the primary.**

be 0.83Ω and the power dissipation will be 7.0W. And there's your problem: 0.83Ω represents the *total* resistance of R1 plus the wiring, which may itself be a few tenths of an ohm. In order to obtain exactly 12.6V AC, right at the valve's heater terminals – which is the only valid place to measure – you'd probably have to experiment with power resistors of uncommonly-low values, or else make something out of resistance wire. There is a better way.

**Option 3.** Series resistor in primary circuit At the primary side of the mains transformer, voltages are higher and currents are correspondingly lower, so the value of resistor R2 in **Fig 2(b)** will be much more convenient. You'd use the same mains transformer (230V primary, 15V secondary, rated 50VA). The easiest way to work out the resistor value is to do the calculation exactly as above, as if the resistor was going to be in the secondary, but finally multiply the R1 value by the *square* of the voltage ratio (230 / 15). So now R2 = 0.83Ω x $(230 / 15)^2$ = 195Ω. Interestingly, the power dissipation of R2 is exactly the same as R1: 7.0W.

R2 will still need some adjustment to allow for secondary wiring resistance, but now you're working with much more convenient values. The best approach is to use the next-higher preferred value, and shunt that resistor with higher values until the voltage at the valve heater terminals arrives once again at 12.6V.

The next preferred value above 195Ω is of course 220Ω, and a good practical choice would be one of the metal-clad power resistors that bolts to the chassis, because this also provides the tags for mounting wire-ended shunt resistors.

The shunt resistor, R3, needs to be... well, whatever value in parallel with 220Ω gives about 195Ω, so that would be

$$R3 = \frac{1}{\frac{1}{195} - \frac{1}{120}} = 1716\,\Omega.$$

This is obviously not a preferred value so, once agai, we apply the same trick. Use the next-higher preferred value – 1.8kΩ or 2.2kΩ – and call it R3A. Then be prepared to add a second shunt resistor, R3B, if necessary. Since R3 is about 10 times higher than R2, the power dissipation will be less than 1W, so a 1W metal-oxide resistor will do fine for R3A. You probably won't have the correct resistors to hand, but at least you can buy these two components R2 and R3A with confidence. You may have to add an R3B as well, to get the voltage exactly right. If you do, you'll probably find the correct component in your stock of 0.25W preferred-value resistors.

This approach is so much more convenient than trying to find suitable low-value power resistors for R1. Note that in Fig 2(b), R2 and R3 (A, B...) are shown connected at the *neutral* end of the transformer winding.

This is important, because the smaller metal-clad resistors do not have high-voltage insulation between the wirewound element and the grounded metal case. The 10W range only has 165VAC rating and, even though you'd probably choose to buy the 15W range for only a few pence more, these are only rated at 265VAC. Having heard rumours about long-term deterioration of the insulation in this type of resistor, I'd always opt to give it an easy life.

A few small points to complete the answer to Question 1...

Judging by the voltage and current, this looks like a transmitting valve with an indirectly-heated oxide cathode. The presence of the series resistor has two useful side effects. First, it reduces the switch-on current surge into the heater, when it is cold and its resistance is low (this is not critical for indirectly-heated valves, but it always helps to treat them kindly).

Second, if you measure the voltage at the valve heater terminals with a high-resolution digital AC voltmeter, you can watch the voltage creep up very slowly, finally stabilising at the correct value. In effect, the heater is acting as its own resistance thermometer, as it brings the oxide cathode up to its correct operating temperature. This allows you to check the valve manufacturer's recommended warm-up time.

But the series-resistor methods of Fig 2 also have drawbacks. The added resistance makes the delivered voltage more sensitive to variations in mains voltage; and more seriously, these methods are suitable only for constant loads. This makes series resistance the wrong choice for the person who wants 8.0V at 0 – 3A from a 9V transformer (see Question 2 above).

The voltage would only be 8.0V at maximum current, and at lower currents the voltage would rise *above* 9V because the secondary voltages are specified at full load. It would be so much better if the transformer really did have an 8V secondary. The next option allows you to make it so.

**Option 4.** Change the number of secondary turns This is easier than you may think, especially if you pick the right kind of transformer to modify. A typical modern transformer will have its secondary winding on a separate bobbin, with some spare space around it. That space is all yours to play with – but leave the primary alone, and watch out for the mains voltage while experimenting.

The first thing to do is to estimate the number of volts per turn, usually by adding about five turns of insulated wire and measuring the voltage on this new temporary secondary winding. That will give you a fair estimate of how many turns to add or remove from the real secondary. If you're removing turns, you need to find which end of the secondary is the finish of the winding, where it's easy to remove the turns (Murphy's Law dictates that the end you want will always be the *second* one you try). If you're adding turns, use at least the same wire diameter as the existing secondary, take care not to kink or scratch the wire. Actually, another option to reduce the output voltage is to add more turns, but to reverse the connection of the new winding; but this also adds to the secondary resistance so it isn't always a good choice.

Threading or removing wire through the small 'window' around the secondary winding can be an awkward job, so this brings us to the final option.

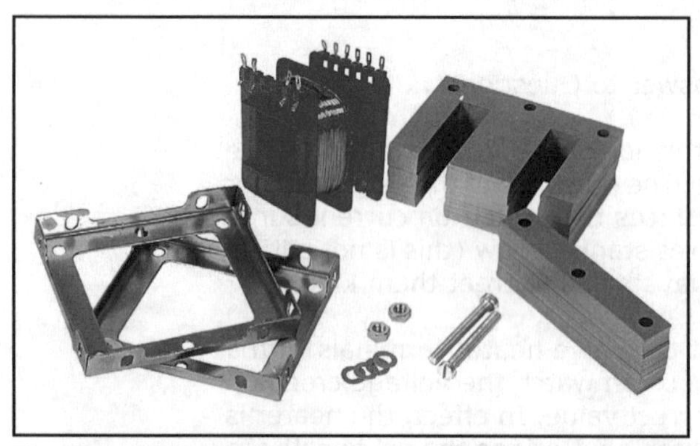

The Maplin transformer kit (insulation removed to show ready-wound primary).

**Option 5.** Wind a completely new secondary This involves starting without the transformer laminations, so that you have free access all around the bobbin. I honestly cannot recommend stripping the laminations off an existing transformer, because it's almost impossible to do that without buckling the laminations... and then it's *totally* impossible to get them all back in again without damaging the bobbin. The result is a loose stack of laminations that will almost certainly buzz.

Another transformer, showing the primary winding on a separate bobbin, below which is the secondary.

Fortunately, Maplin sell transformer kits that have a pre-wound primary, and are supplied with a stack of new, flat laminations (see photograph). Go to **www.maplin. co.uk** and search for YJ63 – this will illustrate the three sizes (20VA, 50VA and 100VA) and you can also view the full winding instructions. It still isn't easy to make a secondary winding with neat, flat layers that completely fill the bobbin, but at least the Maplin kit gives you a sporting chance.

# RHO ($\rho$) – THE ALTERNATIVE TO SWR

**by Bob Pearson, G4FHU**

It is easy to make a reflectometer (or SWR meter) for the HF bands, but the resulting instrument may disappoint if it does not register a minimum reflected signal when connected to a good quality dummy load of the chosen design resistance.

A minimum other than zero may be tolerable, but what is not is a minimum that occurs with anything other than the correct load. Such problems are most likely to arise when working at the lowest or highest frequency bands.

After studying and building a number of traditional designs and testing out some error calculations via practical measurements, it was concluded that to avoid disappointment one really has to take quite a bit of trouble, both in the essential design, in the layout and in the final adjustment. The instrument described here includes compensation for the most common frequency-dependent errors. It also includes a remarkably simple but useful way of testing whether a load impedance magnitude is higher or lower than it should be, which is something a conventional SWR meter cannot do.

A *RadCom* article [1] showed some of the undesirable effects of an incorrect load impedance on a transmitter and gave reasons for wanting to know more about that impedance than is revealed by the Standing Wave Ratio alone. Also, when adjusting and developing aerial systems (ie aerials, feeders and tuning units) it is useful to be able to monitor how impedance magnitude varies with frequency.

An ordinary SWR meter shows when the impedance of a load is a resistance of the required design value, (usually 50Ω). If there is a mismatch it does not reveal whether the load impedance is too high, too low, or a mixture of the wrong magnitude and phase. For example, an SWR of 2.0 might be the result of a 100Ω or a 25Ω resistive load. Equally, it could be the result of any one of an infinite number of other impedances that are not purely resistive.

The design shown can be switched to act as voltmeter and ammeter. From this the load impedance magnitude can be estimated as the ratio of |V| / |I|.

**The complete unit in its case.**

Traditional SWR meters, like the design shown, do not measure SWR *directly*. What they actually measure is the magnitude of the 'Reflection Coefficient'. This is the ratio of the amplitude of the reflected wave ($V_r$) divided by the amplitude of the incident or forward wave ($V_f$) at an interface or other discontinuity (such as that at the connection between a transmitter and the aerial system it feeds). The shorthand for reflection coefficient is often the Greek letter 'ρ' (pronounced 'rho'). Its magnitude (ie after discarding any information about its phase) is represented by '$|\rho|$'. This contains exactly the same information as SWR because of the following simple relationships:

(i) $|\rho| = \left|\frac{V_r}{V_f}\right|$,

(ii) $\text{SWR} = \frac{1+|\rho|}{1-|\rho|}$, or conversely

(iii) $|\rho| = \frac{\text{SWR}-1}{\text{SWR}+1}$.

For most amateur radio checks, a measured SWR figure is noted with interest, but not used in further calculations. For practical purposes, it is just as useful to know $|\rho|$ as it is to know SWR. If this seems unfamiliar at first, the user may refer to the conversion chart shown in **Table 1** or make a calculation using equation (ii).

| 100\|ρ\|% | SWR | SWR | 100\|ρ\|% |
|---|---|---|---|
| 0 | 1.000 | 1.00 | 0.00 |
| 2 | 1.041 | 1.10 | 4.76 |
| 5 | 1.015 | 1.20 | 9.09 |
| 10 | 1.222 | 1.30 | 13.04 |
| 15 | 1.353 | 1.40 | 16.67 |
| 20 | 1.500 | 1.50 | 20.00 |
| 25 | 1.667 | 1.60 | 23.08 |
| 30 | 1.857 | 1.70 | 25.93 |
| 35 | 2.077 | 1.80 | 28.57 |
| 40 | 2.333 | 1.90 | 31.04 |
| 45 | 2.636 | 2.00 | 33.33 |
| 50 | 3.000 | 2.50 | 42.86 |
| 55 | 3.444 | 3.00 | 50.00 |
| 60 | 4.000 | 3.50 | 55.56 |
| 65 | 4.714 | 4.00 | 60.00 |
| 70 | 5.667 | 4.50 | 63.64 |
| 80 | 9.000 | 5.00 | 66.67 |
| 90 | 19.000 | 10.00 | 81.82 |
| 100 | infinity | infinity | 100.00 |

**Table 1: Conversion of |ρ| to SWR and SWR to |ρ|.**

As far as the home constructor is concerned, an important practical advantage of indicating $|\rho|$ rather than SWR is elimination of the tedious and risky procedure of opening up a meter and trying to add the non-linear SWR scale. After all, it takes only one slip of the fingers or one dropped scrap of iron or steel to ruin a moving coil meter.

## CIRCUIT DESCRIPTION

The RF part of the circuit is shown in **Fig 1**. R1 and R3 form a potential divider, and T1 is a current transformer with R2 as the low resistance 'burden' on its secondary, L2. R2 is also arranged to provide a centre tap, to avoid having to make L2 a bifilar winding. To obtain satisfactory HF performance with adequate power and voltage ratings from readily available and cheap, close-tolerance resistors, R1 consists of three resistors and R2, four resistors.

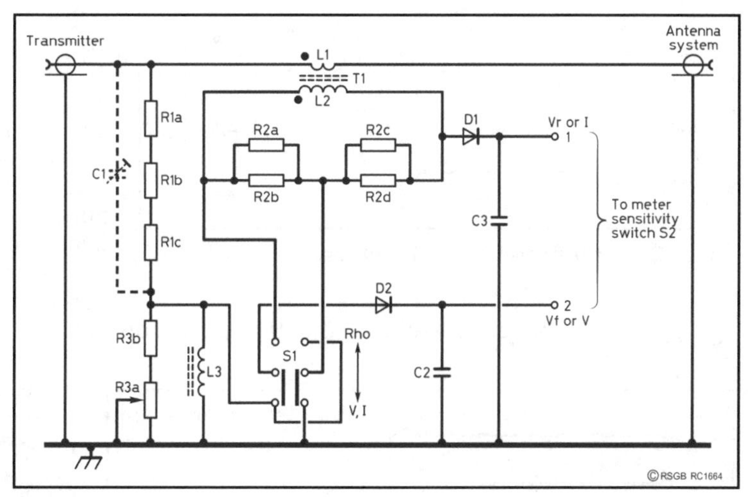


**Fig 1: RF Section circuit diagram.**

In most traditional designs, it is difficult to make transformer T1 work well over the whole frequency range. If L2 is large enough to minimise low-frequency phase shifts, the excessive number of turns and consequent inter-turn capacitance spoils the high frequency performance.

In the design shown, fewer turns than usual are used to make L2, but the consequent low-frequency phase shift is compensated by adding inductor, L3, to the potential divider. With careful choice of values, the resulting low-frequency phase shift in the potential divider can be made to compensate for the phase shift in the current transformer T1. This results in good accuracy, even on the 1.8 – 2.0 MHz band. L3 also serves usefully as a DC path for rectifier currents.

Another weakness of some designs is the lack of provision for high-frequency compensation of the potential divider. In the design shown, this is achieved by adding a tiny capacitor across R1. Since the requirement is for only about 1pF, it consists simply of a pair of well-insulated stiff wires hooked over each other for a distance of about 5mm.

When switch S1 is in the 'Rho' position, voltages proportional to load current and to load voltage are combined in series so that diode D1 detects their difference, representing the reflected amplitude $|V_r|$ and diode D2 detects their sum which represents the forward (incident) wave amplitude $|V_f|$. The ratio $|V_r|/|V_f|$ is the reflection coefficient magnitude, $|\rho|$, as

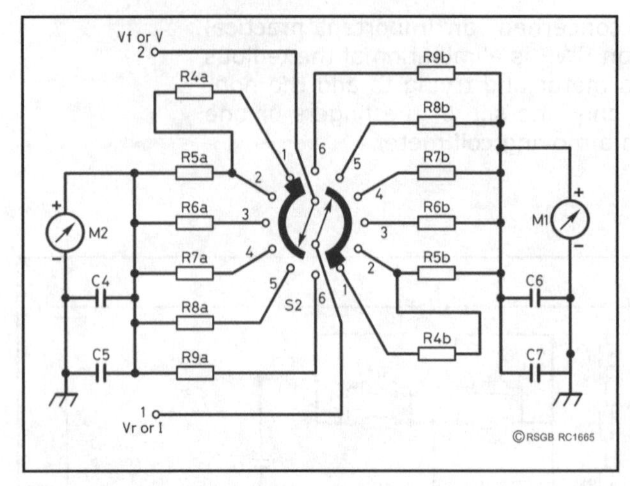


**Fig 2: Sensitivity switch and meters.**

mentioned earlier. This can conveniently be read directly as a percentage, if the forward amplitude is set to indicate 100%.

The 100μA FSD moving coil meters receive the rectified currents from the diodes via close-tolerance fixed resistors, selected in pairs by S2, as shown in **Fig 2**. This unusual arrangement avoids the errors and irritations that arise with the use of continuously-variable potentiometers, as shown in **Fig 3**. The only ganged pairs of potentiometers readily and cheaply available are often ill-matched and get worse with wear. However, for use with a transmitter that has no panel control for adjusting power output, the traditional arrangement of Fig 3 would have to be employed instead of that of Fig 2. In such an event, the V and I ranges will have to remain uncalibrated.

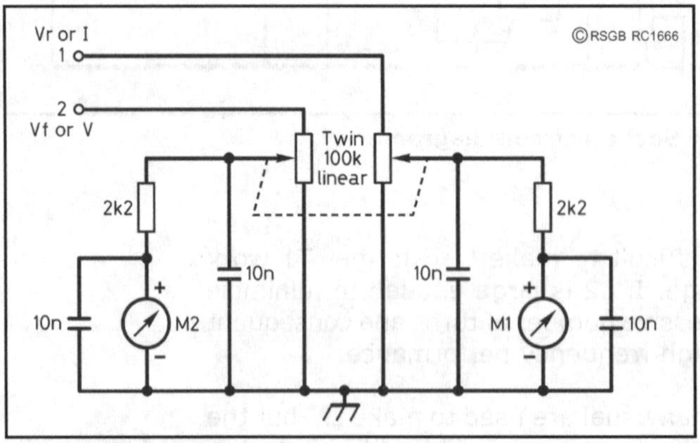

**Fig 3: Alternative sensitivity control.**

When S1 is in the other position (labelled V, I), D1 derives its input only from the current transformer secondary, so meter M1 now indicates load current instead of reflected wave amplitude. D2 receives an input only from the potential divider, so meter M2 shows the load voltage instead of forward wave amplitude.

The magnitude of the load impedance (ie the aerial system impedance) can be calculated by dividing the voltage indication by the current indication. If the actual 0 – 100 meter scale readings are used, rather than converting them to volts and amps, simply multiply the result by $Z_0$ (which is 50Ω resistive in the present design). If the aerial system impedance magnitude happens to be precisely $Z_0$, then the V and I meters show the same angular deflection (see **Table 2**).

As in all passive SWR meters, needing no power supply and containing no complex electronic circuitry, the accuracy deteriorates at low powers due to the significant forward voltage drop of the diodes and, to a lesser extent, by mismatch between the two diodes. Nominal voltage and current ranges (ie calculated neglecting these errors) are shown in **Table 3** but, in practice, the meters will indicate less. Nevertheless, the much sought-after and critical indication of minimum $V_r$ and therefore $|\rho|$, showing

**MATHEMATICAL EXAMPLE**

Set to sensitivity range 2, suppose the following figures are obtained:

| | |
|---|---|
| S1 set to 'Rho' | $V_f$ set to 100% on meter M2 by adjusting transmitter output, |
| | $V_r$ indicates 30% on meter M1. |
| S2 set to 'V, I' | V indicates 26% on meter M2 |
| | I indicates 68% on meter M1 |

***Calculate as follows . . .***

| | |
|---|---|
| Reflection coefficient magnitude | $\|\rho\| = 30 / 100 = 30\%$ or 0.3 |
| Standing Wave Ratio | SWR = (1 + 0.3) / (1 - 0.3) = 1.3 / 0.7 = 1.86 |
| Load impedance magnitude | $\|Z\| = 50 \times 26/68 = 50 \times 0.382 = 19.1\Omega$ |
| Approximate load voltage and current | $V_{rms}$ = 26% of 200 = 52 |
| | Irms = 68% of 4 = 2.72 |

**Table 2: Calculating Reflection Coefficient |r| from meter readings.**

| **Switch range** | | **1** | **2** | **3** | **4** | **5** | **6** |
|---|---|---|---|---|---|---|---|
| 100% | $V_{rms}$ | 400 | 200 | 100 | 50 | 20 | 10 |
| 100% | $I_{rms}$ | 8 | 4 | 2 | 1 | 0.4 | 0.2 |

**Table 3: Nominal voltage and current ranges for each position of the range switch S2.**

that the load is close to 50Ω and zero phase, is more reliable than for some other designs, owing to the compensation techniques used here. Also the ratio $Z=|V| / |I|$ is much more reliable than the indications for $|V|$ and $|I|$ separately.

## CONSTRUCTION

RF connections need to be short and direct. This is more important than elegance. Switch S1 must not introduce long paths internally. A common type of slider switch is shown, but push-button and small wafer switches were used equally successfully in earlier versions. The prototype layout is shown in **Fig 4**. For clarity, the wiring is shown laid out flat and tidy, in the style of a circuit diagram but, in practice, most routes can be shorter. The resistors connected to the sensitivity switch can be wired directly or on strip-board. The solder blobs in Fig 4 refer to connections made on small stand-off

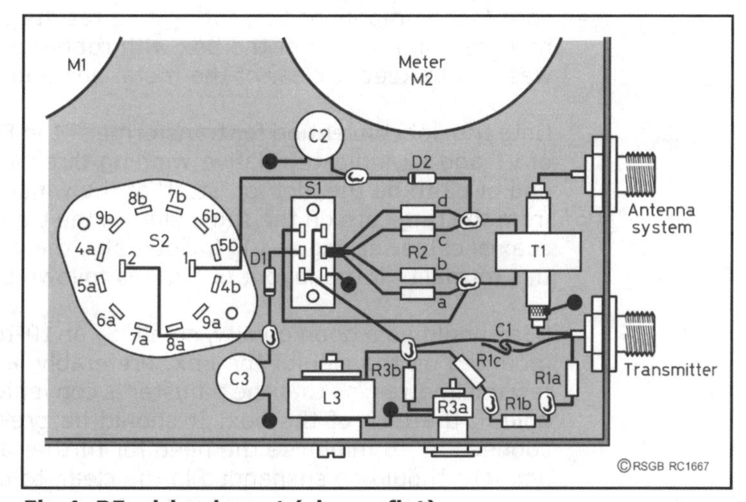


**Fig 4: RF wiring layout (shown flat).**

insulators or insulated solder tags. The large black circles refer to earth connections to the box. If these are made via solder-tags fixed by bolts, it is convenient to join them by some plain copper wire to form a network of places where one can make an earth connection. Alternatively, to avoid drilling the box for earth tags and stand-offs, glue some copper-clad board inside the box, to allow soldering directly to an earth plane at any point required. In that case, mount the stand-offs on the board first. Connect the copper foil to the box, preferably at or very near to the transmitter connection socket.

BNC sockets would be more elegant, but crude and robust SO-239 sockets were used to withstand the many prototype rebuilds during development. They should be located close together, allowing just enough room to permit convenient connection of external leads and to accommodate a short, straight piece of coaxial cable, threaded through the current transformer ring core to serve as the one turn primary, L1. The use of coaxial cable is not, as sometimes claimed, to prevent a discontinuity in the characteristic impedance. It should be earthed at only one end (the transmitter end). It is just a convenient and practical method of supporting the ring core and providing an electrostatic screen between L1 and L2. It should not be longer than necessary, because it adds shunt capacitance to the effective transmitter load.

The secondary of the current transformer, L2, comprises 13 turns of insulated 'hook-up' wire. The compensating inductor L3 has 30 turns of enamelled wire of any convenient gauge, such as 24SWG or 0.5mm diameter (or thinner). L2 and L3 are wound so that the turns uniformly occupy nearly the whole circumference of the ring core. The specified number of turns and core type should be sufficient to define essential parameters, but if you have access to an inductance bridge and wish to make a further check, then the nominal specification of L2 is 9.7μH and L3 is 52μH. In the unlikely event that L2 requires adjustment, do this by altering the spacing of the turns, rather than by changing the number of them (and thus the turns ratio of T1). L3 can have any number of turns, so long as the inductance is correct. It is *vital* to obtain the right kind of core (see component list), otherwise results will be disappointing. L3 is mounted on the side of the box with rubber grommets and/or insulating washers to keep it clear of the metal box and the securing nut.

Note the dot convention for transformer T1 in Fig 1. The dot-marked ends of L1 and L2 indicate relative winding directions. Arbitrarily choose one end of L2 to be the 'dot' or 'start' of the winding. That end *must* emerge from the ring core on the same side as the transmitter end of the piece of coaxial cable serving as L1. Incidentally, there is no need to bother about this formality if the layout of Fig 4 is followed.

R3a should be a good quality, small, 5- or 10-turn pre-set potentiometer, mounted on the side of the box. Preferably, it should be held clear by a few millimetres, so that the adjuster is conveniently accessible via a small hole in the face of the box. It should be pre-set to about 120Ω before connection, to minimise the need for further adjustment later. R1a, R1b and R1c should be suspended in the clear, to minimise stray capacitance to earth and to other components. One or two small stand-off insulators should suffice. Do *not* mount these resistors on a neat printed circuit board, as this would add stray capacitance. Similarly, the four resistors comprising R2 can be supported on three stand-off insulators, though they are less likely to be upset by stray capacitance to earth. Further stand-off insulators may be used for the components in Fig 2 but, in the prototype, the alternative method of using the bypass capacitors as connection points was preferred.

It may surprise those who have built such instruments before to see the potential divider wired to the transmitter side of the reflectometer. This means that its loading effect on the transmitter is not indicated by the instrument. The effect of this is rather small, about 1% when using a well-matched load, and even this is partially offset by a smaller opposite error introduced by the effective primary impedance of the current transformer. The important practical reason for this configuration is that when testing or setting-up the instrument on a high-quality dummy load, it is much more convenient to set up the instrument for minimum reflected reading than to try and set the correct offset. For the same reason, the coaxial screen should be connected to the transmitter side earth, so that the slight error caused at HF by its shunt capacitance is also excluded from the set-up procedure.

## SETTING-UP

1. Connect the transmitter to the 'Tx' side and a 50Ω dummy load in place of the aerial system on the 'Ant' side. Use a good quality load and good quality 50Ω cable.
2. Select a frequency, preferably in either the 7MHz or 10MHz band, but lower if necessary.
3. Adjust the transmitter power level and meter sensitivity for convenient operation well within the capabilities of the transmitter and the dummy load.
4. Switch the meter to 'Rho' mode and adjust the transmitter power so that meter M2 reads a forward wave amplitude of 100% (ie full scale).
5. Minimise the $|\rho|$ reading on M1 by adjusting R3a. It should be possible to get this to below 10% of full scale, even if the inductors are not quite correct. (Note: Even quite expensive dummy loads are not perfect. Neither are connecting cables, plugs and sockets.)
6. Check at the very lowest frequency to be used, and readjust R3a if necessary. If this upsets the 7 or 10MHz zeroing, choose a compromise setting. If all is well and the diodes are well matched, the 'V, I' mode should show almost identical meter deflections for voltage and current.
7. If the results are disappointing, ie it is not possible to get below say 5% for the reflected amplitude, add home-made capacitance C1 and adjust the amount of overlap between the wires. Use wires with good insulation to withstand the full aerial system voltage. *Do not* attempt to adjust the wires or poke about inside the box while the Tx is on. Only make adjustments between trials, with the Tx off.
8. Check what happens on the 28MHz band. If the minimum reading is higher than 3 or 4%, go back to step 7 and make further adjustment to C1. It may require some patience to achieve really

good error compensation, and it is best to make only very small changes between trials.

### OPERATION

The prototype model was intended for use with power outputs from about 20 to 150W. As when using any similar meter, it should be kept in mind that a poorly-matched load can cause excessive voltage to arise in some circumstances and excessive current in others. Consequently you should *always* begin with the sensitivity set to minimum (ie Range 1, using the *highest* resistance in series with each meter). *Never* apply full power until load matching adjustments have proved satisfactory at low power levels.

### REFERENCES

[1] 'How Big is a Bad SWR', *RadCom* March 1993 and April 1993.
[2] Farnell Electronic Components, Canal Road, Leeds LS12 2TU. Tel: 0113 263 6311.

### COMPONENTS LIST

**Resistors**

All fixed resistors are metal film 0.5W 1% tolerance.

| | |
|---|---|
| R1a, b, c | 2k2 |
| R2a, b, c, d | 47R |
| R3a | 2k0 or 2k2 or 2k5 pre-set miniature, 5- or 10-turn potentiometer |
| R3b | 120R |
| R4a, b | 100k |
| R5a, b | 100k |
| R6a, b | 47k |
| R7a, b | 22k |
| R8a, b | 6k8 |
| R9a, b | 2k2 |

**Capacitors**

| | |
|---|---|
| C1 | See text |
| C2 – C7 | 10nF ceramic, 20% tolerance |

**Miscellaneous**

| | |
|---|---|
| T1, L3 | Each on a Philips 4322097180 ring core (type RCC14/5, violet colour ferrite, Grade 4C65, AL=55 ±25%). Order Code 180-008 from Farnell [2]. See text for winding details. |
| D1, D2 | OA91 |
| M1, M2 | Moving-coil meter, 100μA full scale, scaled 0 to 100 |
| S1 | Two-pole two-way, break-before-make, slide, push button or rotary. |
| S2 | Two-pole six-way, break-before-make, rotary. |
| Sockets | To suit |
| Knob(s) | To suit |
| Case | Die-cast (or other metal) box. Prototype size was ample at 170 x 120 x 50mm. |

# THE TWILIGHT ZONE – just what is 'grey-line' propagation?

**by Steve Nichols, G0KYA**

World-wide communications using the MF and HF bands are dependent on radiation coming from the Sun. But twice a day, at sunrise and sunset, the ionosphere undergoes dramatic changes, giving enhanced propagation in some directions.

In terms of radio propagation, the D- and E-layers are responsible for most of the absorption of radio waves that pass through them, but the absorption is frequency-dependent. The D-layer can completely absorb signals on 160, 80 and 40m during the day, and can attenuate signals on 20m too. Hence the reason you don't hear much, if any, DX on the low bands during the day as sky-wave signals are absorbed before they can reach the reflective (more correctly, refractive) E- and F-layers.

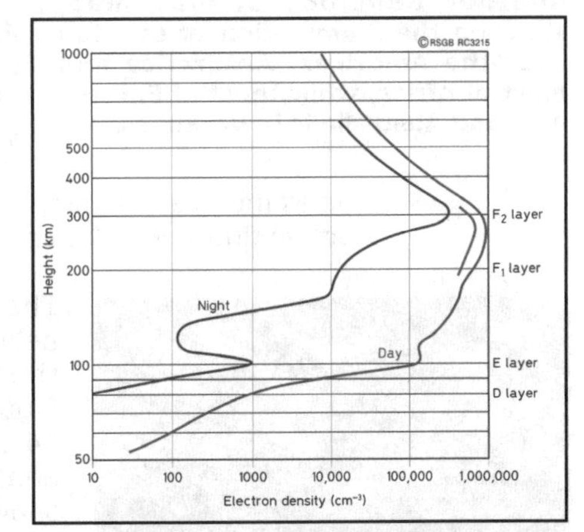


**Fig 1: Electron density during day and night as a function of height above the earth's surface, showing how the ionosphere undergoes dramatic changes in ionisation at the transition from day to night.**

The ionosphere undergoes a dramatic change in ionisation at the transition from day to night. The electron (and ion) density in the E-layer decreases by a factor of 200 to 1 and the F1 by nearly 100 to 1 (see **Fig 1**). At sunset, the D-layer disappears rapidly.

## ENTER THE TERMINATOR

Around the other side of the world other regions that are entering into daylight have yet to form any significant D-layer and the E-layer has not built up from its night-time low. Therefore, for a short period, propagation between two regions simultaneously experiencing sunrise and sunset can be highly efficient. Signals on the lower bands can theoretically travel over great distances with little attenuation.

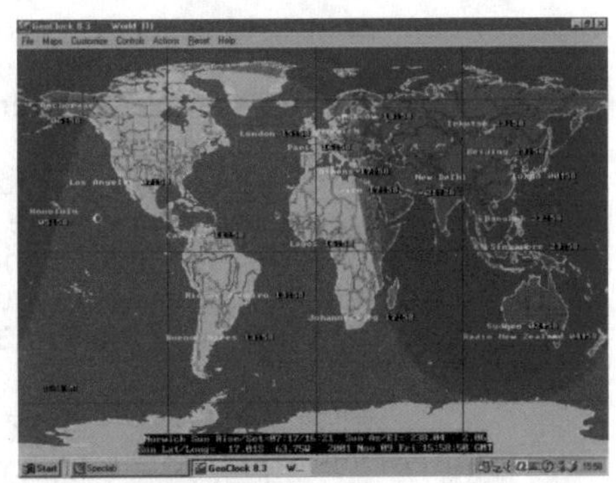

**Darkness, daylight and the terminator for any time of day and any day of the year can be displayed using the *GeoClock* programme.**

This is well documented, with many examples of such propagation being logged on 160 and 80m over the years. Many amateurs will be familiar with this so-called *grey-line* propagation (the term

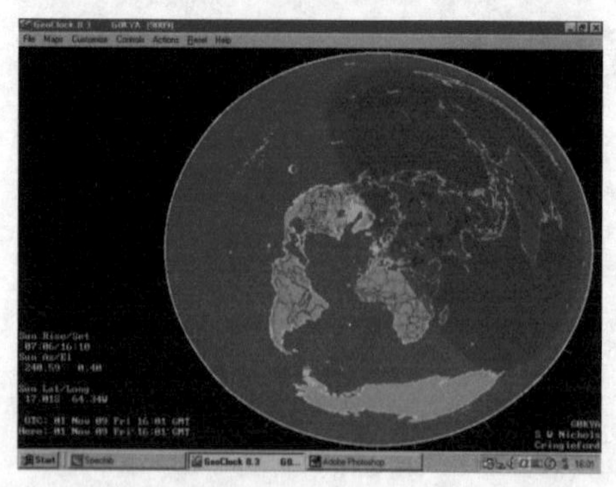

**Another *GeoClock* display map, showing the illumination of the Sun over the Americas, Antarctica and most of Africa, while most of Europe, Asia and Australia is in darkness.**

was coined in 1975 - see [1]) - propagation that occurs along a line separating night from day. The line is called the *terminator*, but it is diffuse, due largely to the earth's atmosphere that scatters the light over a large area. In radio terms, the radio terminator is not the same as the visual one. The latter refers to the point when we see the sunrise or sunset at ground level on the earth and the period of visual twilight that either precedes or follows. The former is related to the way the sun illuminates the ionospheric D-, E- and F-layers.

For example, the PC program *GeoClock* defines the point at which the sun starts / stops illuminating the D-layer as being offset from the visual sunrise / sunset by 6.596° longitude. Because the earth rotates 15° per hour, this could be as much as 24 minutes before or after sunrise or sunset, although the actual figure will depend upon the time of year and latitude.

**Fig 2: The twilight zone - the line shows the areas experiencing physical sunset and the two shaded areas show the loss of first the D-layer and then the F-layer illumination.**

The radio 'twilight zone' - the region on earth between the creation / loss of the D-layer and where the sun starts / stops illuminating the F-layer (roughly defined as being offset from sunset by 14.165° longitude) can be almost one hour before and after sunrise and sunset. See **Fig 2** - the line shows the areas experiencing physical sunset and the two subsequent shaded areas show the loss of D-and then F-layer illumination at that time. E-layer illumination starts / finishes somewhere in between these two, but the average height is much closer to that of the D-layer.

To confuse matters, these values are based on average D- and F- layer heights, but the apparent heights of these can change too. The conclusion is that it is no good looking for grey-line DX just at your visual sunrise / sunset - you could be out by up to an hour depending on the band, your respective locations, and the time of year.

And - even worse - for signals at an angle to the terminator, we are interested in where the first ionospheric refraction (or hop) actually occurs once you radiate a signal, which is likely to be many hundreds of miles to the east or west of you - where the sun may still be illuminating the E- and F-layers (see **Fig 3**).

Most books relating to HF propagation give a brief description of grey-line propagation, and how and why it works. What they don't tell you is the actual frequencies affected, other than a vague idea that 80 / 160m are definite bands for grey-line, and 'some' HF bands also exhibit grey-line enhancements. Either way, all these books tell you that grey-line enhancements occur along the terminator, ie when both stations are at the sunrise / sunset condition.

The book *Low-Band DXing*, by John Devoldere, ON4UN [2], shows that paths perpendicular to the terminator may enjoy the greatest signal enhancement. That is, on the low bands, as sunset occurs at the receiving station, you may get enhancements at right angles to the terminator in the direction towards the dark side of the earth – and *not* along the terminator itself. He also points out that the width of the terminator will vary according to the season and your position on the earth, and cannot be thought of as a fixed entity – the grey-line will be narrower at the equator and wider at the poles. So the time-span available for grey-line conditions will also vary depending upon the time of year, and the locations of the two stations.

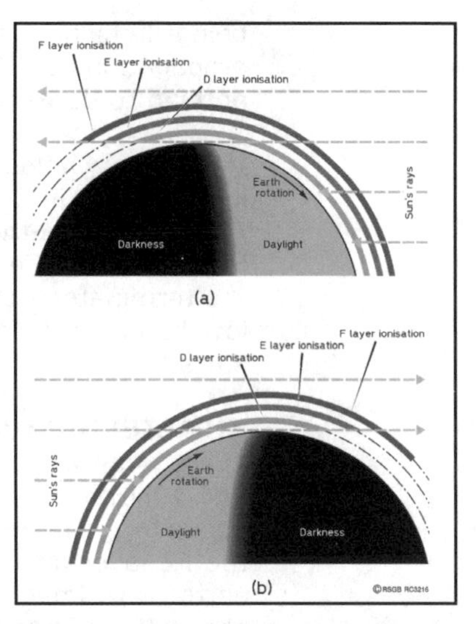


**Fig 3: The effect of ionisation of the D-, E- and F-layers, (a) at dawn, and (b) at dusk (scale exaggerated for clarity).**

Likewise, the width of the grey-line will depend upon frequency as D-layer absorption is frequency-dependent – you may be able to work DX on 40m 24 hours a day in mid-winter, while DX on 160m will fade out quite quickly after sunrise due to the greater D-layer absorption.

## WHAT ABOUT HF?

There appears to be little hard research of grey-line propagation at higher frequencies. The vague suggestion in most books appears to be that grey-line enhancements can and do occur on 20m. 10m is theoretically too high for the effect to appear, as D-layer absorption is virtually non-existent normally at these high frequencies, although I have read more than one article about how to work grey-line on 10m! See the graph of frequency-dependent D-layer absorption predictions on the Internet.

My own studies show that twilight enhancements on 10m *do* occur, but not necessarily along grey-line paths. On many occasions, I have heard signals from Indian, Indonesian and other stations on 10m just after their local sunset – these stations were not audible before. I have also worked a Brazilian station (PT2GTI) on 10m just after his local sunrise, receiving a 59+ report using just 10W into an indoor dipole. He was still audible later that morning, but at reduced signal strengths - down 10 – 15dB. The same has occurred with KP4NU in Puerto Rico. These were definitely not grey-line paths, but still showed definite enhancements.

Reports of sunset / sunrise enhancements at 50MHz over long distances have also been logged, notably between the UK and USA. One suggestion [3] is that this is due to E or Es enhancements as the E-layer increases in altitude at sunset.

The increase in altitude of the E-layer needs further explanation. As the sun sets, the lower regions of the E-layer are not illuminated, so the effective height of the reflecting layer appears to increase. Likewise at this time, we can imagine the radio ionosphere as being tilted as it is being illuminated at an angle. This is probably the vehicle for the enhanced propagation at 28MHz and 50MHz – the loss of the D-layer probably has nothing to do with it.

If my theory holds, look for enhanced signals on 28MHz and 50MHz during local daytime in Great Britain from stations experiencing their local sunrise / sunset - from the west at their local sunset and from the east at their local sunrise. The signals should be strongest at roughly right angles to the terminator - the same as ON4UN's prediction of propagation on the low bands, but from the illuminated parts of the globe, not the dark parts.

## AN ALTERNATIVE VIEW

There is an alternative way of looking at grey-line conditions on 7MHz and 10MHz connected with the critical frequency ($fo_{F2}$). At frequencies above $fo_{F2}$, a radio wave travelling vertically upwards would pass through the F2 layer into outer space. Below $fo_{F2}$ it would be reflected back to earth. Now imagine a radio wave hitting the ionosphere at an angle of about 75 – 85° to the earth – a near-vertical incidence sky-wave (NVIS). Below the critical frequency, it would be returned. If it is some way above $fo_{F2}$, it will pass into space. At some frequency close to $fo_{F2}$, it could be refracted through a large angle and end up travelling almost parallel to the earth, giving a very long first skip distance.

This is the condition for the Pedersen [4] or critical ray, discovered in 1927 and characterised as being high-angle, long-distance, close to, and probably above, the $fo_{F2}$ frequency. As there would be no intermediate ground hop,s the signal strength could be very high indeed.

It is likely that these conditions exist around local sunset / sunrise as $fo_{F2}$, passes through the two bands and could account for long distance communications under grey-line conditions on 7MHz and 10MHz. A near-real-time $fo_{F2}$ map can be found on the Internet.

Either way, there is more to grey-line and twilight propagation than meets the eye. The effects and the mechanisms behind the propagation are probably different on each band too. What we *can* say is that twilight propagation is not always best along the terminator, hence I try to avoid the term 'grey-line' where it is not applicable. There may not be any enhancement at all on some bands. Some books would have you believe that you can just tune up on 20m at sunset and work ZL at 59+20dB every day - if you can I would like to hear about it!

## A RESEARCH OPPORTUNITY

I am currently doing some research into twilight propagation on many of the amateur bands, starting with 10m. The early results confirm that we can and do see twilight enhancements from signals originating from areas experiencing sunrise / sunset.

The graph of the beacon SV3AQR on 28,182kHz shown in **Fig 4** is typical. This was produced using *SpectrumLab* software connected to the audio output from a Yaesu FT-920. With the AGC turned off, the vertical scale indicates signal strength while the horizontal scale shows time. You can quite clearly see a 10dB increase in signal strength near the beacon's radio 'dusk'.

The effect has been seen on other beacons, but like all ionospheric effects, it doesn't occur every day and is virtually impossible to forecast.

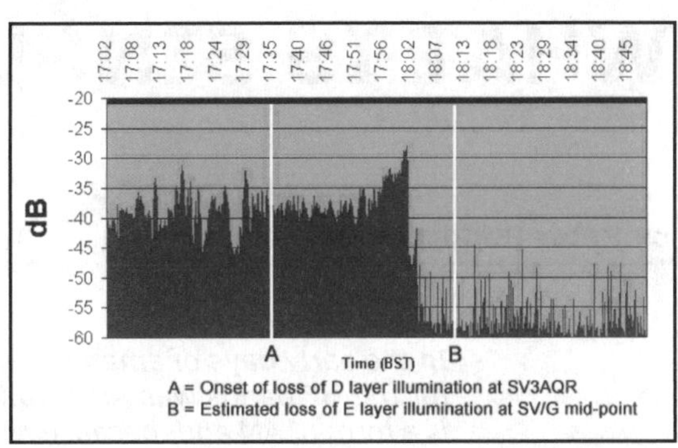


**Fig 4: The SV3AQR beacon on 28,182kHz, as received at G0KYA, on 8 October 2001.**

More monitoring work needs to be done before we can write the definitive guide to grey-line and twilight propagation and this is where I need readers' help. If you have a PC with a soundcard, can run the *SpectrumLab* software, have a very stable receiver (the software needs stability in the order of a few Hz), and can leave your system monitoring for an hour or more at a time, then I would like to hear from you. As part of the Propagation Studies Committee's work I plan to look at twilight and grey-line propagation on all the HF and LF bands systematically, using known, quantifiable signal sources such as beacons and broadcast stations. This is a long-term project though, but is essential if we are finally to clear up what has been a grey area of propagation research for a long time - every pun intended!

**REFERENCES**


[1] 'The Grayline Method of DXing', Dale Hoppe, K6UA, et al, *CQ* Sept 1975.

[2] *Low-Band DXing*, John Devoldere, ON4UN, ARRL, 1999.

[3] 'E layer and Sporadic E – Two Modes of Propagation at 50MHz', Ken Osborne, G4IGO, UK Six Metre Group website.

[4] *An Introduction to Radio Wave Propagation*, J G Lee, Babani BP293, 1991.

[5] *Your Guide to Propagation*, Ian Poole, G3YWX, RSGB, 1998.

[6] *The New Shortwave Propagation Handbook*, Jacobs, Cohen and Rose, *CQ* Communications, 1995.


**WHERE TO LOOK ON THE INTERNET**

| | |
|---|---|
| *Geoclock* software | www.geoclock.com |
| D-layer absorption predictions | www.sec.noaa.gov/rt_plots/dregion.html |
| Critical frequency ($fo_{F2}$) predictions | www.spacew.com/www/fof2.html |
| *SpectrumLab* software | www.qsl.net/dl4yhf/ |
| UK Six Metre Group (UKSMG) | www.uksmg.org |
| *Beacon Time Wizard* by Taborsoft | www.taborsoft.com |

# WHAT IS SINGLE SIDEBAND?

**by Peter Dodd, G3LDO**

*In the early days of amateur radio, the main telephony modulation method for the HF bands was Amplitude Modulation (AM). Single Sideband (SSB) is a form of AM and, because AM is easier to understand, it will be described first.*

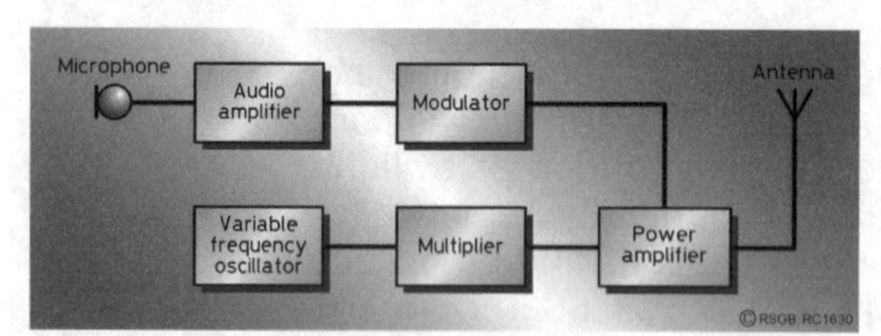


**Fig 1: Block diagram of a simple amplitude modulated transmitter.**

## AMPLITUDE MODULATION

A block diagram for a transmitter suitable for AM is shown in **Fig 1**. With early amateur AM transmitters, the frequency of the RF signal was obtained by multiplying from a lower-frequency oscillator. Because the 3.5, 7, 14, 21 and 28MHz bands are all multiples of 3.5MHz, an 'all-band' transmitter (before the 10, 18 and 24MHz 'WARC' bands were introduced) could be constructed using a switched multiplier to obtain the desired band. The voice audio frequencies were introduced to the radio frequency (RF) power amplifier using a high power modulator.

To show the signals produced by the arrangement in Fig 1 it is, for simplicity, assumed that the modulator is producing a 1kHz audio tone. If the transmitter RF output (known as a 'carrier') is at 14.2MHz (14,200kHz), two additional signals (sidebands) are produced when the audio tone is superimposed on the carrier. One signal is 1kHz above 14,200kHz, and the other 1kHz below (ie, at 14,201 and 14,199kHz). This is shown in **Fig 2**, which shows in (a) a three-dimensional diagram of a signal (f0) modulated by an audio tone. The same waveform is viewed in the time domain in (b). In practice you don't see each individual waveform, but the effect of the three waveforms mixed together. This is how the signal would look if viewed using an oscilloscope. If the signal in Fig 2(a) were viewed in the frequency domain, as seen when using a spectrum analyser, the display would appear as shown in Fig 2(c). These signals arrive at the receiver and the three signals are mixed (demodulated) together in the detector stage to produce a 1kHz tone in the loudspeaker.

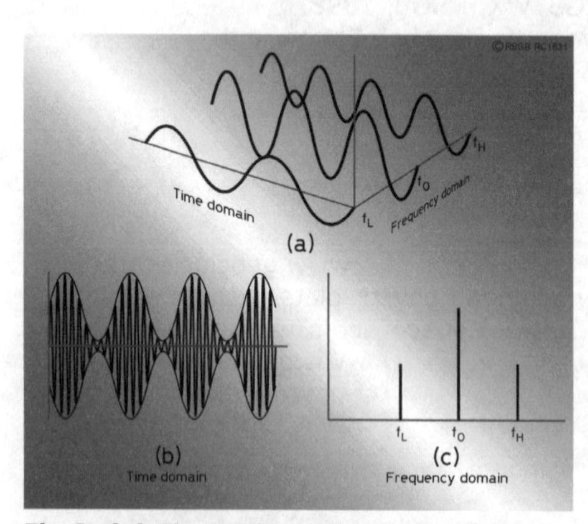


**Fig 2: (a) Three-dimensional diagram of a signal (f0), eg one at 14200kHz, modulated by a single audio tone. (b) Viewed in the time domain. (c) The same signal viewed in the frequency domain.**

If the transmitter were modulated with two tones, one at 1kHz and the other at 2kHz then there would be four sideband signals, two either side of the carrier.

When the transmitter is modulated with a voice signal, all the multiple tones that make up the voice signal are produced in multiple tone sidebands either side of the carrier. These sidebands and carrier arrive at the receiver and are demodulated together in the detector stage to produce a voice signal in the loudspeaker.

## DOUBLE SIDEBAND SUPPRESSED CARRIER

The carrier does not contain any of the transmitted tone or voice information, and its only purpose is to demodulate the signal at the receiver. The signal could be transmitted without the carrier - in the case of the 1kHz tone this would be just the two sideband signals. The carrier for demodulation could then be inserted in the receiver using an oscillator. This method of modulation is called Double Sideband Suppressed Carrier.

The problem with this arrangement is that inserted 'carrier' must be *exactly* on the right frequency, otherwise there would be two sets of conflicting tones in the sidebands, which would make a voice signal unintelligible.

## SINGLE SIDEBAND

The difficulty of accurate placing of the carrier can be overcome by transmitting only one sideband. If the inserted carrier is not exactly on frequency all that happens is that all the frequencies in the sideband are shifted slightly in frequency. This gives the voice a strange 'Donald Duck' sound, but the voice is still, in the main, intelligible. The operator can then adjust the receiver to improve the quality of the voice signal. This method of modulation is called Single Sideband. Because the carrier and one of the sidebands are removed, all the transmitter power is confined to this one transmitted sideband. The transmission also takes up less spectrum space.

## GENERATING AN SSB SIGNAL

The method of generating an SSB signal described here is called the 'filter method'. There is also a 'phasing method' and the 'third method'. The filter method is described here because it is the one in most common usage - and it's also the easiest to understand.

An SSB signal is generated in a similar manner to AM, ie by using a modulator. A carrier is generated at, say, 9MHz (the reason for using this frequency is described later). If an audio signal of 1kHz were used, the output signal from the modulator would contain three signals, one at 9000kHz, and sidebands at 8999kHz and 9001kHz. Some modulators (a balanced modulator) also balance out (suppress) the carrier, thus part of the job of creating an SSB signal is accomplished.

The remaining step to create an SSB signal is to eliminate one of the sidebands. This is achieved using a filter. The passband of the filter should be just wide enough to pass voice intelligence, while at the same time removing the un-wanted sideband and any residual carrier that may be output from the balanced modulator. For voice communication, such filters are about 2.5 or 3kHz wide.

The filter can be used to remove either the upper or lower sideband, depending on the relationship of the frequency passband of the filter to the carrier frequency. In the amateur radio bands, by convention, upper sideband (USB) is used above 10MHz and lower sideband (LSB) on the lower frequency bands, so the ability to switch from USB to LSB is important.

In practice, this is done by changing the frequency of the carrier oscillator by switching crystals; placing the carrier on one side or the other of the filter passband. It could be done by having the carrier oscillator on one frequency and by using two switched filters, but this is more costly.

The unwanted carrier and sideband is absorbed by the filter and modulator and converted to heat, so it is necessary that the above process take place at a relatively low signal level, of the order of a volt or two, so that power dissipation is low.

The SSB signal must now be changed to the required frequency band. It cannot be multiplied, as in Fig 1, so one or more mixer stages are used to translate the SSB signal to the required band.

## SIMPLE TWO-BAND TRANSMITTER

To reduce circuit complexity, early amateur filter-method SSB transmitters generated the signal at 9MHz to make a two-band transmitter. If the 9MHz SSB signal is mixed in a balanced modulator with a VFO having a frequency coverage of 5.0 to 5.5MHz, two SSB signals are generated as shown in **Fig 3**. As the VFO is varied from 5.0 to 5.5MHz, the one output from the balanced modulator will vary from 14.0 to 14.5MHz, while the other output will vary from 4.0 to 3.5MHz. All that is required are switched filters to select the desired band.

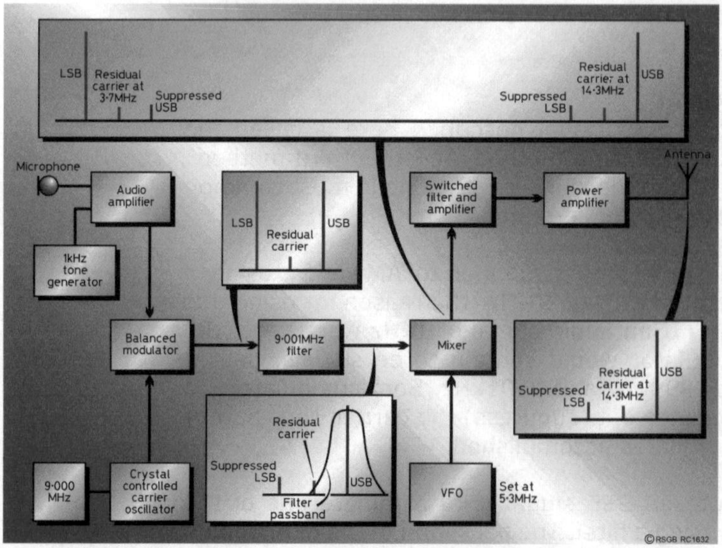

You will probably have noticed that the 9MHz SSB signal, when translated to the 3.5MHz band, tunes from the top of the band to the bottom as the oscillator frequency is increased. This downward translation also has the effect of reversing the upper and lower sidebands, so that an USB generated at 9MHz

**Fig 3: Block diagram of a simple two-band (80 and 20m) SSB transmitter with frequency domain diagrams of the signalsat various points.**

becomes LSB when translated down to the 3.5MHz band. This is the probably the reason for the convention of using USB on the higher frequency bands and LSB on the lower frequency bands (as already described), despite the flexibility of most modern amateur SSB equipment.

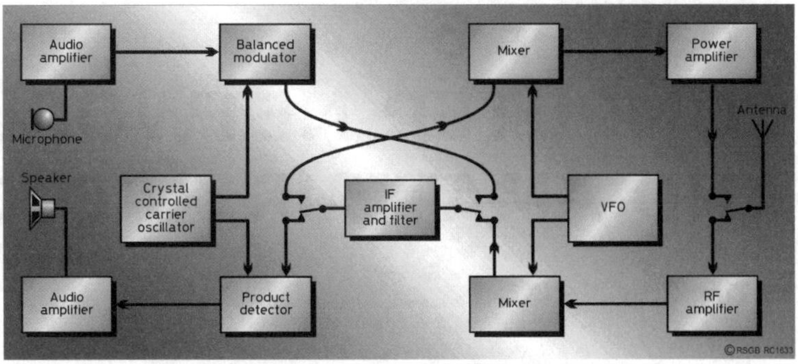

**Fig 4: Block diagram of a single band transceiver, showing that many blocks of the diagram can be shared with the transmitter and receiver.**

## THE SSB TRANSCEIVER

With the SSB transceiver, the functions of transmission and reception are combined, allowing a substantial reduction in cost and complexity, along with greatly increased ease of operation. A single-band filter-type transceiver is shown in **Fig 4**. It shares a common carrier oscillator, IF amplifier / filter and VFO. The transceiver is designed to communicate on a single frequency selected by the VFO.

Transfer from receive to transmit is carried out by one or more relays and by application of blocking voltages to unused stages, where savings in size, weight, and power consumption are important. Dual usage of components and stages in the SSB transceiver permits a large reduction in the number of circuit elements and facilitates tuning to the common frequency desired for two-way communication.

Common mixer frequencies are used in each mode and the high-frequency VFO is used to tune both transmit and receive channels to the same operating frequency. In addition, a common IF system and sideband filter is used.

The transceiver is commonly switched from receive to transmit by a multiple-contact relay which transfers the commonly used modules and the antenna.

# Chapter 6
# COMPUTER NETWORKING AND AMATEUR RADIO

by Tim Kirby, G4VXE

***A computer network at home? Perhaps even five years ago, this would have been unusual for the vast majority of people. But now, many homes have several computers in them and it makes sense to share resources (a printer, scanner, Internet connection - or amateur radio transceiver) between them. This article explains how to go about creating such a network and what it can do for you.***

The first thing you need is at least two computers! We're going to assume that you have *Windows*-based computers. Your task will be easier if you have *Windows 95, 98, ME, 2000 or XP* (it can be done with *Windows 3.1*, but it takes a bit more effort).

Next, you need some network cards. Ethernet cards which operate at 10 or 100Mb/s are very cheap, for both desktop machines or laptops, which use PCMCIA cards. You can pick them up very cheaply at rallies or through computer suppliers: don't expect to pay much more than £25 for such a card.

The simplest method to connect two computers together is to use a crossover cable. This means you connect the transmit line of one network card to the receive of the other. You can buy crossover cables, or an adapter to convert a standard CAT-5 network cable into a crossover cable. Or, armed with a cable, plugs and a crimping tool, you can make your own (see 'Web Search').

**BASIC NETWORKING**

Installing network cards in the machines will prompt you with a few questions about how you want to set up your network. You'll need to supply a name for each machine. Something like 'SHACK' or 'TIM' will do fine. You will need to install a Network Protocol that your network card will use to talk to other machines. You will need to install TCP/IP if you ever want to use the Internet or pretty much anything else on your machine.

You may be prompted about a TCP/IP address. What's this? Well, each computer that uses an Internet connection, or talks to another computer will almost certainly have one. Think of it like a phone number for computers. Assuming that you are setting up a network which won't, for the moment, talk to the outside world, you should use a range of private IP addresses. I'd recommend being logical about it. For example, the first computer could have an IP address of 10.1.1.1 and the second should be 10.1.1.2 and so on. You'll also be asked to provide a network mask. This is a bit beyond the scope of this article, but it affects what other computers your machine can talk to.

When in doubt on a small network at home, type 255.255.255.0 If you mess up these settings, right click your mouse on Network Neighborhood and select Properties. This will show you a dialogue box such as the one in **Fig 1**. Click where it says TCP/IP and hit Properties. You'll soon discover that any adjustment of these settings on most Windows operating systems requires a reboot.

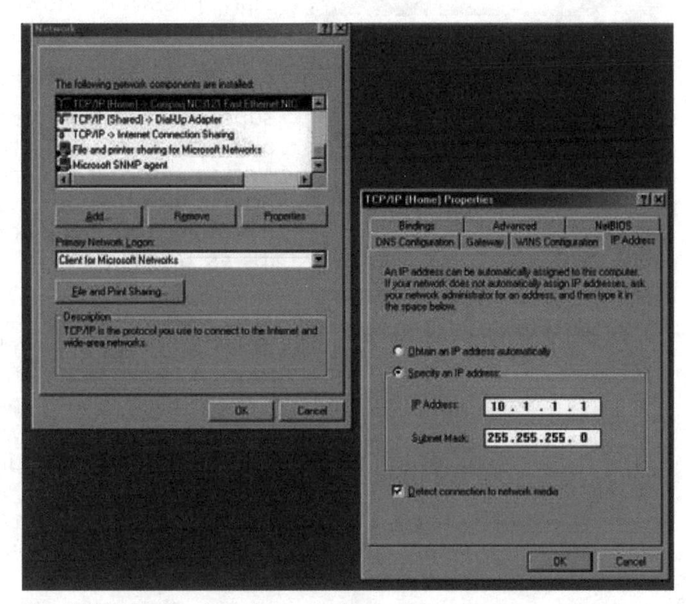

Fig 1: Dialogue box for setting network properties.

Once you've installed network cards and TCP/IP on each of your machines, and put the crossover cable between the machines, they should be ready to talk to each other. Make sure any security programs / firewalls are disabled for now (but do remember to re-enable them when your machines are talking and before you connect to the Internet!).

So the machines are talking to each other? That's excellent! With a couple of simple steps, you'll soon be ready to share files and printers on your two machines. On each machine, go back into the network settings (right mouse click from Network Neighborhood and select properties).

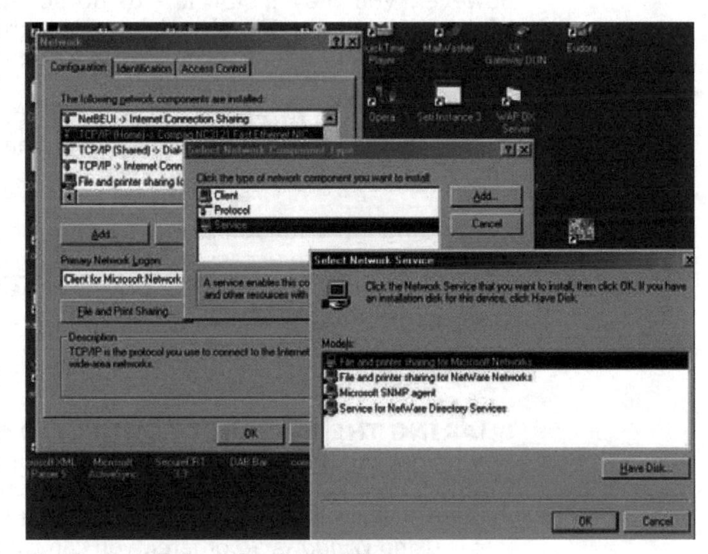

Fig 2: Specifying sharing options.

This time, hit Add, then select Client and from the list, choose Client for Microsoft Networks (**Fig 2**). Finally, hit Add, select Service from the list and choose File and Printer Sharing for Microsoft Networks.

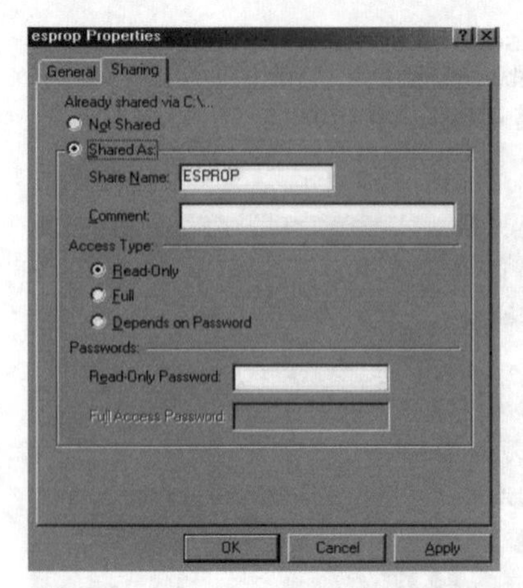


**Fig 3: Setting up a directory to share.**

Do this on both of your machines and then reboot. You should notice that when the machines start up you are prompted for a user name and password. Assuming that you are at home - accept the default and have a blank password. However, you should *never* do this on a real network or even if you are at home with children where there may be network resources such as the Internet that you do not want them to use without supervision. If you connect to the Internet, you should also use a Firewall program, such as *Zone Alarm* or *Norton Internet Security* to prevent hackers trying to access files on your machine. Do not think that this is a remote risk: if you don't have a firewall *it will happen*, it's as simple as that.

Then, on one of the machines (let's say it's called 'SHACK'), start up Windows Explorer. Select a directory that you want to be accessible from the other machine (this could be C:\). Right click it with the mouse and select Sharing. It will show a dialogue box such as **Fig 3**, where you enter the name of the share, in other words the name that your directory will be known as.

Then on the other machine, type Start/Run and then in the box that appears \\SHACK\ESPROP (in this case). All being well, you should see all the files in that directory.

You may find that where you are using computers running a variety of *Windows* versions that it is easier and advantageous to install a protocol called *NETBEUI*. This makes *Windows* networking very simple over a small network. *NETBEUI* is not immediately available under *Windows XP*, however, you may install it - to do so, follow the instructions on the Internet (see 'Web Search' below). Note that you should install TCP/IP if you want to share an Internet connection on your PC.

Next, you should be able to share your printer. On the machine to which it's connected, you can go into Control Panel and Printers and select the option to share the printer.

Depending on the operating system that you are using, you may automatically have access to the printer from the other machine, or you may need to install the printer driver. Either way, you will soon be able to print across the network.

Congratulations: your basic network is in place.

## SHARING THE INTERNET CONNECTION

Having got the basics in place, one of the next things to do is to share the Internet connection, so that it is accessible by all computers on your network. Assuming that the computer that the Internet connection comes into is using *Windows 98* or later, you can run 'Internet Connection Sharing'. Simply install this - it will guide you through a series of questions - nothing complex. At the end of this, you will be prompted to create a disc which you need to apply to the other workstation( s) on your network. I have found this unnecessary and the following method to be simpler.

On the PC(s) where you wish to access the Internet, which do not have a direct connection to the Internet and will access it through the PC that you've just set up, do the following: Right mouse click on Network Neighborhood, select TCP/IP and hit Properties. You will have, previously, typed in a TCP/IP address for this. Change this to 'Obtain a TCP/IP address automatically'. The PC which shares the Internet Connection acts as a Dynamic Host Configuration Protocol (DHCP) server and issues these addresses.

Reboot this PC, connect to the Internet on your other PC, and you should find that you can access the Internet from both workstations. Again, if you have a firewall, you may find that some adjustment is necessary as it may not automatically permit communications to the outside world. If you are having problems, make sure that your 'client' PC has obtained a TCP/IP address (use WINIPCFG /ALL as before) and if not, temporarily disable the firewall and try again. It usually works fairly easily.

## WIRELESS NETWORKING

You may want to extend your network beyond two PCs. Until recently, the usual way to do this was to install a network hub. The hub has a number of ports and you plug a standard network cable (not a crossover cable) into each port from the network card on your PC. Network hubs or even switches are very cheap and you should be able to get one for £30 or so. Better still, if you have a good relationship with a network manager of an IT department, see if he has any redundant hubs or switches. They are frequently upgraded, so small workgroup hubs or switches may often be lying in a box unused and thus available for the cost of a polite enquiry!

Instead of network hubs, over the last year or so wireless networking has become very affordable for the home user. Sometimes known as Wi-Fi (Wireless-Fidelity), there are several standards available for use. These standards are defined by the IEEE. Standard 802.11b operates in the 2.4GHz band and provides data transfer to a maximum of 11Mb/s. 802.11g provides data transfer up to 54Mb/s. Equipment conforming to the 802.11g standard is compatible with the earlier 802.11b specification.

To use wireless networking technology, you'll need an access point, which can be plugged into your network card or hub and configured and one or more PCs with a Wireless LAN card. The access point is the most expensive piece of gear we've discussed so far, but if you go for the lower 802.11b standard, can be obtained for around £50.

Configuration is very straightforward and generally achieved through a web interface. Remember to set some security on your Wireless network (you will be guided through how to do this) otherwise you may find others using your network!

You are able to select which channels the Wireless LAN uses and if you are an Oscar 40 user, you may wish to experiment with the choice of channel as the higher ones avoid interference with satellite operation.

Having configured your Wireless LAN, you will find that the network is suddenly so much more flexible than you've known before. Armed with your laptop and a wireless LAN card, you can look at the *DXCluster*, or web pages, from the comfort of the sofa or your garden!

**SOME USEFUL PROGRAMS**

You can look at the *DXCluster* easily on any Internet PC. Simply go to the command prompt and type 'telnet gb7djk.dxcluster.net 7300' and press enter. You will soon be logged on to the Cluster. The telnet program provided with Windows is a little unfriendly for amateur radio purposes, though adequate, so you might want to look at DX Telnet (see 'Web Search'). In addition to GB7DJK, there are many other Internet-connected *DXClusters* to which you can connect.

If you are a datamodes enthusiast, you might like to try remote controlling your datamodes programme. I found a very useful package (free) called *VNC* (Virtual Network Computing). You install this package on to a PC, and you can then connect to it across the network and see the screen it is displaying and type commands as if you were at the keyboard. It's great! Using this, I was able to make PSK and RTTY QSOs from the comfort of my sofa, when the PC and rig were in the shack!

**FINALLY**

Having a network at home is so useful - you can share printers, access your MP3 files from any machine on the network, check your e-mail and so on. Having Internet access on the shack PC is invaluable too, particularly for *DXCluster* access, leaving the VHF bands free to be used for VHF DXing.

I hope this article has stimulated some thought on how you could go about setting up a network at home (if you've not already done so) and the benefits and fun that will result if you do. Good luck and have fun!

**WEB SEARCH**

| | |
|---|---|
| Making a crossover cable | www.makeitsimple.com/how-to/dyi_crossover.htm |
| Installing NETBEUI | http://support.microsoft.com/default.aspx?scid=kb;en-us;301041 |
| DX Telnet | www.qsl.net/wd4ngb/telnet.htm |
| Virtual Network Computing (VNC) | www.realvnc.com |

# TRANSCEIVER / COMPUTER INTERFACING

**by Ian White, GM3SEK, from 'In Practice'**

***Q: How can I build an interface to control my transceiver from my PC? The interface from the transceiver manufacturer costs a fortune!***

***A:*** It isn't difficult. Almost all modern transceivers have an interface for computer control, and there is a wide variety of software available to communicate with the microprocessor inside your rig that controls the frequency synthesiser and many other functions. The PC software exchanges data with the transceiver using one of your PC's serial interface or COM ports [1]. The COM port uses a world-standard system called RS- 232 for exchanging serial data with the outside world. RS-232 is a bipolar signalling system: the logic zero or 'space' level is specified as +10V (actually anywhere in the range +5 to +15V) and the logic 1 or 'mark' level is -10V (actually -5 to -15V). The problem is that most transceiver serial ports provide TTL-level signals: +5V for logic 1 and 0V for logic zero. **Fig 1** shows why an interface is needed, to shift these voltage levels between the two systems. Controlling a transceiver by a PC is usually a two-way data exchange, so each direction needs its own level-shifter.

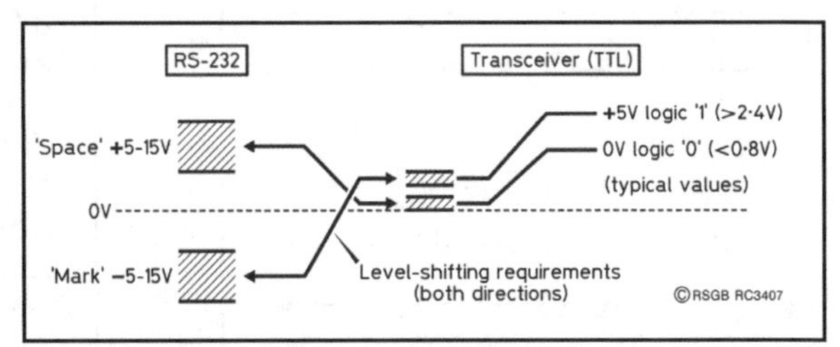


**Fig 1: Level-shifting requirements between RS-232 and TTL. As well as the polarity inversion in RS-232, the logic levels are also inverted.**

| Pin | Name | Function | Input/Output (for PC) | Connects to (other end) |
|---|---|---|---|---|
| 1 | | | | |
| 2 | **TXD** | **Transmitted Data** | **Output** | **RXD** |
| 3 | **RXD** | **Received Data** | **Input** | **TXD** |
| 4 | DSR* | Data Set Ready - handshake | Input | DTR* |
| 5 | **GND** | **Ground** | **(chassis)** | **GND** |
| 6 | DTR* | Data Terminal Ready - handshake | Output | DSR* |
| 7 | CTS* | Clear To Send - handshake | Input | RTS* |
| 8 | RTS* | Ready To Send - handshake | Output | CTS* |
| 9 | | | | |

**Table 1: RS-232 connections. Pin numbers are moulded into connectors, and are the same for both plugs and sockets. If handshake connections (marked *) are necessary, they can often be looped-back locally at the connector without using extra lines. In PC-transceiver interfacing, DTR and DSR are commonly used for other purposes: DC power, CW keying and / or PTT control.**

A few transceivers now provide direct plug-and-play RS-232 compatibility, using a similar 9-pin D-connector to the PC's COM port. All others need an interface, so let's examine the connector pins at the COM port to see what's needed. **Table 1** lists the signals we have to work with, at the PC COM port. Note the *Connects To* column: all the lines from the PC connect

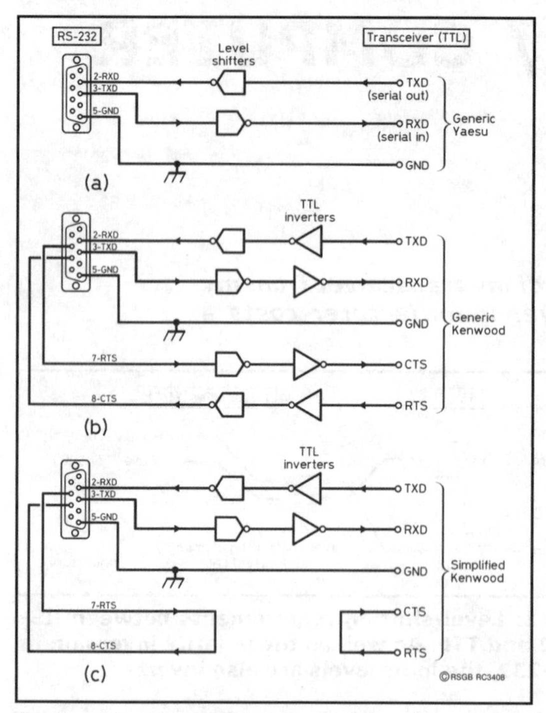


**Fig 2: Generic interfaces from RS-232 (a) 3-wire Yaesu, (b) 5-wire Kenwood with full RTS - CTS handshaking, (c) simplified 3-wire Kenwood, with handshaking emulated by CTS - RTS linking at each end.**

to a different-named line at the other end of the link, with the single obvious exception of Ground. The most important pair are TXD and RXD. The TXD output from the PC carries transmitted data from the PC to the transceiver, and is connected via the interface to the transceiver's RXD input (sometimes called 'Serial In'). The transceiver's TXD output (or 'Serial Out') sends data back to the PC's RXD port. The Big Three transceiver manufacturers each have different control protocols and different hardware interface requirements, and these also vary between different transceivers from the same manufacturer. The control protocols are usually handled for you by the software authors - and now, here's how to handle the hardware requirements.

All we require for two-way serial communication are just three wires: TXD, RXD and ground. That's what Yaesu rigs use (**Fig 2(a)**). In this kind of 'free streaming' communication, either end can send data whenever it wants, and this works fine for the kinds of simple, direct links we need for transceiver control. The Kenwood interface is more complex: it uses reversed-TTL logic polarities (indicated by the TTL inverters in **Fig 2(b**) and it also requires Request-To-Send (RTS) and Clear-To-Send (CTS) handshaking on two additional lines. This RS-232 jargon means that no data is supposed to be transmitted until the other end has been asked if it's ready to receive (RTS), and has sent back a confirmation (CTS). Mercifully, RS-232 communication doesn't always insist on such formalities. The hardware handshaking needs of Kenwood rigs can usually be satisfied by linking the RTS output directly back to the CTS input in the same connector (**Fig 2(c)**) - in effect, each device tells itself to go ahead.

Next let's look at some hardware for level shifting (don't worry, I haven't forgotten Icom).

Remember that the aim is to translate the TTL levels from the transceiver into RS-232 (or compatible) levels for the PC (Fig 1). The first practical interface I'll describe (**Fig 3**) uses a MAX232 IC to generate true RS-232 voltage levels. To fall securely inside the RS- 232 specification, this requires DC supply rails at about +10V and -10V. The MAX232 revolutionised TTL / RS-232 interfacing by generating both of these voltages on-chip from a standard +5V TTL supply, which makes the whole design very simple. The MAX232 contains two level-shifters in each direction, so it can provide a more complete RS-232 interface with handshaking if required. Fig 3 shows a MAX232 interface for Yaesu transceivers such as the FT-980 and

FT-990 that don't already have a PC-compatible COM port. It's really pretty simple: wire it up in a small shielded box close to the transceiver, apply +12V (which the 78L05 regulates down to +5V) and away you go with real RS-232 bipolar signalling [2]. More elaborate interfaces for Yaesu, Icom and Kenwood rigs using the MAX232 are described in recent editions of the *ARRL Handbook*.

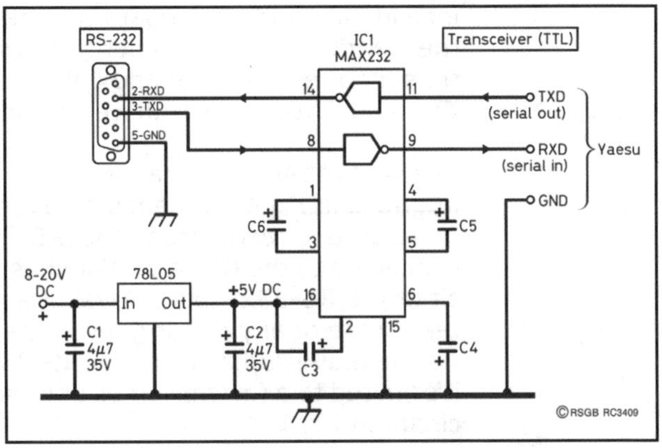

**Fig 3: Practical Yaesu interface using the MAX232. C3 - C6 are typically 10µF 35V tantalum electrolytic, but some varieties of MAX232 allow lower values (see data sheets).**

But interfaces can often be much simpler than that. We don't really need true bipolar signalling, because the RS-232 receiver ICs used in PC COM ports don't insist on the full-specification voltage levels. To the disgust of RS-232 purists, they treat any incoming voltage above about +1.8V as a valid 'space' level, and anything below about +1.6V as a valid 'mark'. This means you can fake an RS-232 interface by simply pulling the relevant line from a normal +5V TTL level down to ground. The noise margins are much reduced compared with true bipolar RS-232 with its 20V swing, but they are generally adequate for short screened runs between a PC and a transceiver. **Fig 4** shows an example for Yaesu rigs. R1 connects the PC's TXD port to the base of TR1, and a positive RS-232 'space' level will pull TR1's collector voltage down to ground. The negative RS-232 'mark' condition biases TR1 into cutoff (negative base voltage is limited by D1) so that R2 can pull the output line up to a valid TTL 'logic 1' level. TR2 works in a similar way: whenever the transceiver puts a TTL 'logic 1' on the line, TR2 pulls the RXD voltage at the PC COM port almost to zero, which the PC's RS-232 receiver IC interprets as a valid 'mark'.

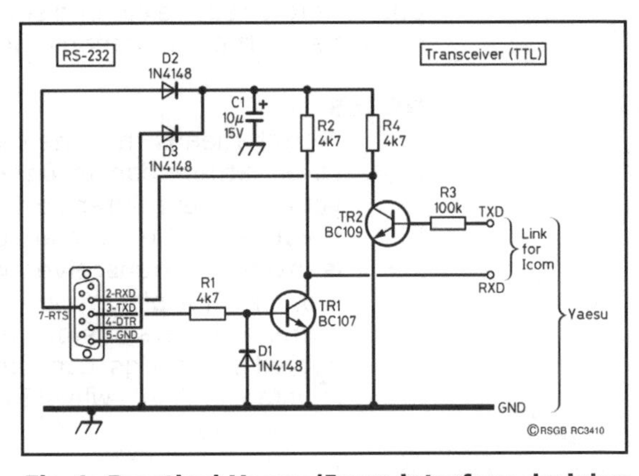


**Fig 4: Practical Yaesu/Icom interface deriving power from the COM port. Note the link for the Icom 'single-wire' data bus (Fig 5).**

Icom rigs are different because they use a simple 'one wire' interface (actually a single wire plus ground screen). The Icom CI-V control protocol is quite sophisticated: it not only handles signalling both ways along the single wire, but also allows the PC interface and several Icom rigs to *share* that one data line (**Fig 5**). The CI-V system sends

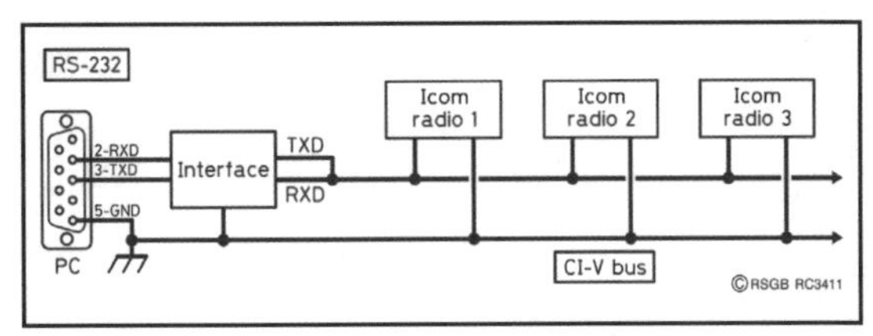


**Fig 5: Icom 'single-wire' (plus ground ) CI-V data bus allows multiple rig control; also used by Ten-Tec. The 'In Practice' website has many links to PC control interface circuits.**

individually-addressed data packets to each rig, and likewise recognises the identity of each rig when receiving data. The Icom hardware interface thus requires the input and output to be commoned - in Fig 4, simply link TXD and RXD at the transceiver side as shown [3].

You've probably noticed another simplification in the two-transistor interface in Fig 4 – it needs no external power supply. The positive rail is generated directly from the RS-232 port via D2 and D3, which are connected to the RTS and DTR lines. Either (or both) of these lines usually sit at the RS-232 'space' level of about +10V. C1 stores this voltage and takes care of any minor gaps in the supply. Stealing a small positive and / or negative supply from the RS-232 port is a well-known technique, but it's limited to a few milliamps only, so it won't work with the basic MAX232 circuit in Fig 3.

There.s also a completely different class of RS-232 / TTL interfaces using optoisolators. If you have multiple transceiver / PC interconnections including audio, these interfaces can be very useful because they also break the ground connection, which may help prevent hum loops. However, optoisolators require either a separate power supply or 'strong' TTL outputs from the PC that are capable of sourcing 10 – 20mA.

**NOTES**

1. 'Serial' means that the data bits travel in sequence down a single wire, rather than in parallel down several wires (as happens at your parallel printer port).
2. If you're a little more adventurous, the best place for this interface is inside the transceiver, where the manufacturer should have put it in the first place.
3. Some web versions of the Fig 4 interface show R3 as 4.7k&!, but not all Icom rigs can deliver a good TTL 'logic 1' into this low resistance. That's why R3 is 100kΩ, and TR2 is a high-beta transistor.

# INDEX

## RSGB Book Order Form

| Invoice Address | Delivery Address |
|---|---|
| | |

| Ordered By | | Order Number | |
|---|---|---|---|
| Date | | Contact Tel no. | |

| ISBN | Title | Price | Qty | Value |
|---|---|---|---|---|
| 1-872309-95-X | Advance - The Full Licence Manual | £ 11.99 | | |
| 1-872309-55-0 | Amateur Radio ~ the first 100 years | £ 49.99 | | |
| 1-872309-70-4 | Amateur Radio Explained | £ 9.99 | | |
| 1-872309-72-0 | Antenna File | £ 18.99 | | |
| 1-872309-89-5 | Antenna Topics | £ 18.99 | | |
| 1-872309-59-3 | Backyard Antennas | £ 18.99 | | |
| 1-872309-94-1 | Command | £ 16.99 | | |
| 1-872309-82-8 | Digital Modes for all Occasions | £ 16.99 | | |
| 1-872309-80-1 | Foundation Licence Now! | £ 4.99 | | |
| 1-872309-48-8 | Guide to EMC | £ 19.99 | | |
| 1-872309-58-5 | Guide to VHF/UHF Amateur Radio | £ 8.99 | | |
| 1-872309-75-5 | HF Amateur Radio | £ 15.99 | | |
| 1-872309-08-9 | HF Antenna Collection | £ 19.99 | | |
| 1-872309-15-1 | HF Antennas for all Locations | £ 19.99 | | |
| 1-872309-86-0 | Intermediate Licence Handbook | £ 6.99 | | |
| 1-872309-93-3 | International Antenna Collection | £ 12.99 | | |
| 1-905086-01-6 | International Antenna Collection 2 | £ 12.99 | | |
| 1-872309-83-6 | International Microwave Handbook | £ 24.99 | | |
| 1-872309-96-8 | IOTA 40th Anniversary Directory | £ 9.99 | | |
| 1-872309-99-2 | LF Today | £ 11.99 | | |
| 1-872309-65-8 | Low Frequency Experimenter's Handbook | £ 18.99 | | |
| 1-872309-73-9 | Low Power Scrapbook | £ 14.99 | | |
| 1-872309-90-9 | Microwave Projects | £ 14.99 | | |
| 1-905086-09-1 | Microwave Projects 2 | £ 14.99 | | |
| 1-872309-77-1 | Mobile Radio Handbook | £ 13.99 | | |
| 1-905086-00-8 | Operating Manual - 6th Edition | £ 19.99 | | |
| 1-872309-88-7 | Practical Projects | £ 12.99 | | |
| 1-905086-04-0 | Practical Wire Antennas 2 | £ 11.99 | | |
| 1-872309-85-2 | Prefix Guide | £ 8.99 | | |
| 1-872309-91-7 | QRP Basics | £ 14.99 | | |
| 1-905086-08-3 | Radio Communication Handbook *8th Edition* | £ 29.99 | | |
| 1-872309-97-6 | Radio Propagation - Principles & Practice | £ 14.99 | | |
| 1-905086-02-4 | RSGB Rig Guide | £ 3.99 | | |
| 1-905086-07-5 | RSGB Yearbook 2006 | £ 18.99 | | |
| 1-872309-71-2 | Technical Compendium | £ 17.99 | | |
| 1-872309-20-8 | Technical Topics Scrapbook 1985-89 | £ 9.99 | | |
| 1-872309-51-8 | Technical Topics Scrapbook 1990-94 | £ 13.99 | | |
| 1-872309-61-3 | Technical Topics Scrapbook 1995-99 | £ 14.99 | | |
| 1-905086-05-9 | Technical Topics Scrapbook 2000-2004 | £ 14.99 | | |
| 1-872309-76-3 | VHF/UHF Antennas | £ 13.99 | | |
| 1-872309-42-9 | VHF/UHF Handbook | £ 19.99 | | |
| 1-905086-03-2 | Whos Who in amateur Radio | £ 14.99 | | |
| 0-900612-09-6 | World at Their Fingertips | £ 9.99 | | |

**Post & Packing**

UK only - £1.75 for 1 item, £3.30 for 2 or more items
Rest of World - £3.00 for 1 item, £6.00 for 2 & £1.00 for each extra item
Air Mail - £6.00 for 1 item, £10.00 for 2 & £2.00 for each extra item
**NOTE:** All prices are subject to change without notice. E&OE

| | |
|---|---|
| Sub Total | |
| Discount % | |
| Total | |

RSGB, Lambda House, Cranborne Road, Potters Bar, Herts EN6 3JE UK
Tel: 0870 904 7373 Fax: 0870 904 7374 E-mail sales@rsgb.org.uk